Applied Condition Monitoring

Volume 13

Series editors

Mohamed Haddar, National School of Engineers of Sfax, Tunisia

Walter Bartelmus, Wrocław University of Technology, Poland

Fakher Chaari, National School of Engineers of Sfax, Tunisia
e-mail: fakher.chaari@gmail.com

Radoslaw Zimroz, Wrocław University of Technology, Poland

The book series Applied Condition Monitoring publishes the latest research and developments in the field of condition monitoring, with a special focus on industrial applications. It covers both theoretical and experimental approaches, as well as a range of monitoring conditioning techniques and new trends and challenges in the field. Topics of interest include, but are not limited to: vibration measurement and analysis; infrared thermography; oil analysis and tribology; acoustic emissions and ultrasonics; and motor current analysis. Books published in the series deal with root cause analysis, failure and degradation scenarios, proactive and predictive techniques, and many other aspects related to condition monitoring. Applications concern different industrial sectors: automotive engineering, power engineering, civil engineering, geoengineering, bioengineering, etc. The series publishes monographs, edited books, and selected conference proceedings, as well as textbooks for advanced students.

More information about this series at http://www.springer.com/series/13418

Tahar Fakhfakh · Chafik Karra
Slim Bouaziz · Fakher Chaari
Mohamed Haddar
Editors

Advances in Acoustics and Vibration II

Proceedings of the Second International Conference
on Acoustics and Vibration (ICAV2018),
March 19–21, 2018, Hammamet, Tunisia

 Springer

Editors
Tahar Fakhfakh
National School of Engineers of Sfax
Sfax, Tunisia

Fakher Chaari
National School of Engineers of Sfax
Sfax, Tunisia

Chafik Karra
Sfax Preparatory Engineering Institute
Sfax, Tunisia

Mohamed Haddar
National School of Engineers of Sfax
Sfax, Tunisia

Slim Bouaziz
Faculty of Sciences of Sfax
Sfax, Tunisia

ISSN 2363-698X   ISSN 2363-6998   (electronic)
Applied Condition Monitoring
ISBN 978-3-319-94615-3   ISBN 978-3-319-94616-0   (eBook)
https://doi.org/10.1007/978-3-319-94616-0

Library of Congress Control Number: 2016944485

Preface

The Second International Conference on Acoustics and Vibration (ICAV'2018) was held at Hammamet, Tunisia, from 19 to 21 March 2018. ICAV'2018 aims to collect the broadest range of high-level contributions, covering theoretical and practical case studies in all the fields of acoustics and vibration. After a successful first edition in 2016 published under ACM book series, the Tunisian Association of Industrial Acoustics and Vibration which is the organizer of this conference continues to promote communication and collaboration between international and local communities involved in the fields of acoustics and vibration. The book contains 33 chapters issued from the presentation done by eminent scientists which were rigourously peer reviewed. About 100 attendees discussed several topics such as:

- Dynamics and vibration of structures and machinery,
- Fault diagnosis and prognosis,
- Fluid–structure interaction and vibroacoustics,
- Nonlinear dynamics,
- Modelling and simulation,
- Computational vibroacoustics/numerical techniques,
- Material behaviour in dynamics.

According to the several topics discussed during the conference, the book is divided into four parts.

Part 1—Dynamics and fault detection of machinery
Part 2—Multiphysics system dynamics
Part 3—Structure dynamics and fluid–structure interaction
Part 4—Material behaviour in dynamic systems

We would like to thank the organizing committee, scientific committee and all participants coming from Tunisia, Algeria, France, Saudi Arabia, Spain and Portugal. Thanks to Springer for continuous support of ICAV2018.

March 2018   Tahar Fakhfakh
                                                                       Chafik Karra
                                                                       Slim Bouaziz
                                                                     Fakher Chaari
                                                                  Mohamed Haddar

Contents

Mutiphysics Systems Dynamics

Structures Dynamics and Fluid-Structure Interaction

Dynamics and Fault Detection
of Machinery

L-Kurtosis and Improved Complete Ensemble EMD in Early Fault Detection Under Variable Load and Speed

Hafida Mahgoun[1(✉)], Fakher Chaari[2], Ahmed Felkaoui[1],
and Mohamed Haddar[2]

[1] Applied Precision Mechanics Laboratory,
Institute of Optics and Precision Mechanics, Sétif 1 University, Setif, Algeria
mahafida006@yahoo.fr
[2] Mechanical Engineering Department, National School of Engineers of Sfax,
Sfax, Tunisia

Abstract. In this work, we propose to follow the progression of different gearbox defects under the effect of variable load and speed. The non-stationary vibration signals are obtained by using a physical model of a spur gear transmission. In order to detect the presence of the fault characterized by transient signals which are usually masked by other vibration signals and noise. We can use the Improved Complete Ensemble empirical mode decomposition with adaptive noise (ICEEMDAN) to decompose the non-stationary vibration signals into many components that represent mechanical behaviour of the machine, transient component and noise. The ICEEMDAN method is based on the estimation of the local mean and the white noise is not used directly. this method eliminates the mode mixing introduced by EMD and reduces the amount of noise contained in the modes given by using EEMD and gives better results than EEMD. To analyze IMFs given by ICEEMDAN method we can use statistical methods like kurtosis which is very used to detect impulsion in the signal. In this work, we also use a statistical method, the L-Kurtosis, as an indicator to compare the IMFs given by ICEEMDAN, the results given by this indicator are compared to the results given by the Kurtosis.

Keywords: Rotating machines · EEMD · CEMDAN · Gear · Fault detection
L-moments · L-Kurtosis

1 Introduction

Rotating machines are very used in industry (McFadden 1986). This mechanical equipment operates under complicated conditions like the variation of load and speed and is therefore subject of faults which lead generally to breakdowns.

Vibration signal analysis is largely used in fault diagnosis of rotating machines (McFadden 1986), the vibration signals picked up from these machines are non-stationary and non-linear.

In order to detect the presence of the fault characterized by transient signals which are usually masked by other vibration signals and noise (Capdessus and Sidahmed

© Springer Nature Switzerland AG 2019
T. Fakhfakh et al. (Eds.): ICAV 2018, ACM 13, pp. 3–15, 2019.
https://doi.org/10.1007/978-3-319-94616-0_1

1992; Wang and Mcfadden 1997), we should choose the proper signal processing method.

Until now, many methods were applied to detect the fault at an early stage, among these methods traditional ones including statistical analysis based on the signal its self such as (root mean square, crest factor kurtosis, and so on (Sharma and Parey 2016)) and the frequency domain analyses based essentially on the Fourier transform. Therefore, the Fourier analysis gives good results if the vibration signal is stationary and linear and it is inapt to analyze the non-stationary signal, which may lead to false information about the mechanical faults. To solve this problem new methods have been introduced. The time-frequency analysis methods such as Wigner Ville decomposition (WVD) (Forrester 1989), short Fourier Transform (STFT) (Staszewski 1997) and wavelet transform (WT) (Wang and Mcfadden 1997) seem to be the suitable tool to identify the signal frequency and to provide information about the time variation of the frequency. Therefore, the STFT uses the same window to analyze the whole length of the signal, which means that we have the same resolution for the high and low frequencies, if we desire to improve the resolution by changing the width of the window we lose the time resolution or the frequency resolution because the resolution frequency and the time localization is imposed to the Heisenberg uncertainty principle. Thus, the STFT is appropriate only to analyze signals with slow variation (Cohen 1989) and it is inefficient for the case of non-stationary signals.

The WT transform was widely applied because it's a multiresolution analysis, is very used to detect the transient features to extract impulses and for denoising (Mallat 1998). Nevertheless, the wavelet analysis is also a linear transform and it uses functions named wavelet as window function like the STFT. The window changes its width by using a dilatation parameter. Then, at the high frequency, we have high time resolution and a low frequency resolution. While at low frequencies, we have low time resolution and high-frequency resolution. Then, we can't have a good resolution for all time-scale map due to the Heisenberg uncertainty principle (Mallat 1998).

In addition, this method gives a time-scale representation which is difficult to interpret as a time-frequency representation; we must have a relation between the scale and the frequency to understand the obtained results and to identify the fault frequencies. Another limitation of the WT is how to select the mother wavelet used in the analyses of the signal, since different wavelets have different time-frequency structures, also, how to calculate the range scale used in the WT is another deficiency of the transform. Many researchers demonstrated that the use of the WT introduces border distortion and energy leakage (Yang et al. 2011). In mechanical application, W. Yang et al. confirm that results given by using this method are highly dependent on the rotational speed and pre-knowledge of the machine (Yang et al. 2011).

The WVD method is also a time-frequency representation and doesn't involve any window function, and has high time-frequency resolution, however, it suffers from the cross terms interferences and the aliasing problem. To overcome these problems the pseudo-Wigner Ville was proposed but the correction of the interference lead to loss of the resolution.

To overcome the deficiencies of these methods empirical mode decomposition (EMD) was proposed by Huang et al. (1998) for nonlinear and non-stationary signals and was applied in fault diagnosis of rotating machinery (Yang et al. 2011; Liu et al. 2005;

Yu et al. 2006). It does not use a priori determined basis functions and can iteratively decompose a complex signal into a finite number of zero mean oscillations named intrinsic mode functions (IMFs). Each resulting elementary component (IMF) can represent the local characteristic of the signal (Huang et al. 1998). However, one of the problems of EMD is mode mixing as a result of intermittency. Mode mixing (Huang et al. 2003; Rilling and Flandrin 2008) occurs when different frequencies that should appear separately in different IMFs are presented in one IMF. This problem gives a vague physical significance of the IMF. EMD is unable to separate different frequencies in separate IMFs. Also, the IMFs are not orthogonal to each other, which produce end effects. To solve the problem of mode mixing the ensemble empirical mode decomposition EEMD method was proposed by Wu and Huang (2009) by adding several realizations of Gaussian white noise to the signal, and then using the EMD to decompose the noisy signal, multiple IMFs can be obtained and the added noise is canceled by averaging the IMFs. The addition of white Gaussian noise solves the mode mixing problem, however, it creates some other problems, the reconstructed signal includes residual noise and different realizations of signal plus noise may produce a different number of modes and take a large number of sifting iterations to achieve the decomposition. To reduce the computational cost and to provide a better spectral separation of the modes, the improved complete ensemble EMD with adaptive noise (ICEEMDAN) was proposed. The ICEEMDAN method was tested by analyzing biomedical and seismic signals and has given good result compared to EEMD method (Colominas et al. 2014).

To analyze IMFs given by ICEEMDAN method we can use statistical methods like kurtosis which is very used to detect impulsion in the signal.

In this paper, we propose, first, to use the ICEEMDAN method to decompose the non-stationary vibration signal. Then, we calculate the L-kurtosis of each IMF as a new indicator and we compare the results given by this indicator to those given by the kurtosis.

The structure of the paper is as follows: Sect. 2 introduces the basic of EMD, EEMD and ICEEMDAN. In Sect. 3, we give the notion of L moment. Section 4 is dedicated to results. In Sect. 5, a conclusion of this paper is given.

2   EMD, EEMD and ICEEMDAN

2.1   EMD Algorithm

The EMD consists to decompose iteratively a complex signal into a finite number of intrinsic mode functions (IMFs) which verify the two following conditions:

(a) The number of extrema and the number of zeros of an IMF must be equal or differ at most by one,
(b) An IMF must be symmetric with respect to local zero mean.

For a given a signal $x(t)$ the EMD algorithm used in this study is summarized as follows (Huang et al. 1998):

(1) Identify the local maxima and minima of the signal x(t)
(2) Generate the upper $x_{up}(t)$ and lower $x_{low}(t)$ envelopes of $x(t)$ by the cubic spline interpolation of the all local maxima and the all local minima.

6   H. Mahgoun et al.

(3) Average the upper and lower envelopes of $x(t)$ to obtain the local mean function:

$$m(t) = \frac{x_{up}(t) + x_{low}(t)}{2} \qquad (1)$$

(4) Calculate the difference

$$d(t) = x(t) - m(t) \qquad (2)$$

(5) If d(t) verifies the above two conditions, then it is an IMF and replaces $x(t)$ with the residual

$$r(t) = x(t) - d(t) \qquad (3)$$

(6) Otherwise, replace $x(t)$ with $d(t)$.

Repeat steps (1)–(5) until the residual satisfies the criterion of a monotonic function. At the end of this algorithm, the signal can be expressed as:

$$x(t) = \sum_{n=1}^{N} IMF_n(t) + r_N(t) \qquad (4)$$

Where $IMF_n(t)$ are IMFs, N is the number of IMFs extracted and $r_N(t)$ is the final residue.

2.2   EEMD Algorithm

To alleviate the mode mixing effect of EMD, the EEMD was used. The EEMD decomposition algorithm of the original signal $x(t)$ used in this work is summarized in the following steps (Wu and Huang 2009):

(1) Add a white noise $n(t)$ with given amplitude β_k to the original signal $x(t)$ to generate a new signal:

$$x_k(t) = x(t) + \beta_k n(t) \qquad (5)$$

(2) Use the EMD to decompose the generated signals $x_k(t)$ into N IMFs, $IMF_{nk}(t), n = 1, \ldots, N$, where $IMF_{nk}(t)$ is the nth IMF of the kth trial.

Repeat steps (1) and (2) K times with different white noise series each time to obtain an ensemble of IMFs: $IMF_{nk}(t), k = 1, \ldots, K$.
Determine the ensemble mean of the K trials for each IMF as the final result:

$$IMF_n(t) = \lim_{k \to \infty} \frac{1}{K} \sum_{k=1}^{K} IMF_{nk}(t), \quad n = 1, \ldots, N \qquad (6)$$

The relationship among the amplitude of the added white noise and the number of ensemble trials is given by (Wu and Huang 2009):

$$\delta_k = \frac{\beta_k}{\sqrt{K}} \tag{7}$$

where K is the number of ensemble trials, β_k is the amplitude of the added noise and δ_k is the variance of the corresponding IMF(s).

2.3 CEEMDAN Algorithm

To overcome the difficulties of EEMD, (Colominas et al. 2014) propose the ICEEM-DAN method, which is based on the estimation of the local mean; the white noise is not used directly but uses two operators $M(.)$ the operator which produces the local mean and the operator $E_k(.)$; where:

$$E_1(x) = x - M(x) \tag{8}$$

Let w^i is a realization of a white Gaussian noise with zero mean and unit variance. the algorithm of ICEEMDAN is as follows:

(1) Calculate by EMD the local means of l realizations

$$x^i = x + \beta_0 E_1(w^i) \tag{9}$$

to get the first residue $r_1 = M(x^i)$. where

$$\beta_0 = \varepsilon_0 std(x)/std(E_1(w^i)) \tag{10}$$

(2) For k = 1 we calculate the first IMF, $IMF_1 = x - r_1$.
(3) Estimate the second residue as the average of local means of the realization $r_2 = M(r_1 + \beta_1 E_2(w^i))$,   where $\beta_1 = \varepsilon_1 std(r_1)$ and   the   second   IMF, $IMF_2 = r_1 - r_2$.
(4) For I = 3,...N calculate the Ith residue(r_I)

$$r_I = M(r_{I-1} + \beta_{I-1} E_I(w^i))$$

(5) Then the Ith IMF $IMF_I = r_{I-1} - r_I$
(6) iterate the steps 4 to 6 until the obtained residue cannot be further decomposed because it has not three local extrema. Then the signal can be expressed as:

$$x = \sum_{I=1}^{N} IMF_1 + r_N \tag{11}$$

This algorithm is summarized in Fig. 1.

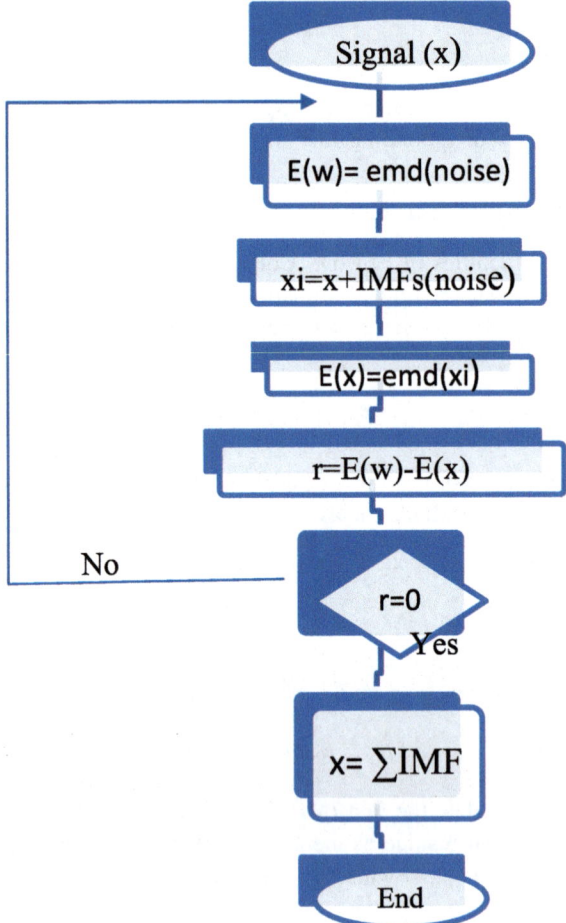

Fig. 1. CEEMDAN AlgorithmL-kurtosis

3 L Moment

L-moments are linear combinations of order statistics and the L-kurtosis is the fourth-order L-moment, it is similar to kurtosis and the L-kurtosis value 0.1226 replaces the kurtosis value 3 (Liu et al. 2018), the L-kurtosis is defined as:

$$\frac{L_4}{L_2},\tag{12}$$

$$\text{where: } L_4 = 20\beta_3 - 30\beta_2 + 12\beta_1 - \beta_0 \tag{13}$$

and where:

$$\beta_0 = n^{-1} \sum\nolimits_{i=1} x_i \tag{14}$$

$$\beta_1 = n^{-1} \sum\nolimits_{i=2} x_i[(i-1)/(n-1)] \tag{15}$$

$$\beta_2 = n^{-1} \sum\nolimits_{i=3} x_i[(i-1)(i-2)]/[(n-1)(n-2)] \tag{16}$$

$$\beta_3 = n^{-1} \sum\nolimits_{i=4} x_i[(i-1)(i-2)(i-3)]/[(n-1)(n-2)(n-3)] \tag{17}$$

4 Application

In this section, we propose to analyze different signals that correspond to the different variations of load (constant load, a variation of load fluctuation between 10% and 50%) and we have also different sizes of fault (healthy gear, a variation of fault severity between 1% and 20%). These signals are the results of a dynamic modelling of the gears transmission (Bartelmus and Zimroz 2009; Chaari et al. 2013). The characteristics of the spur gear system driven by a squirrel cage motor are given in Table 1. The load on the machine is fluctuating in a saw-tooth shape with a frequency $f_L = 5\,\text{Hz}$ rising from 32 to 100 Nm as presented in the Fig. 2a.

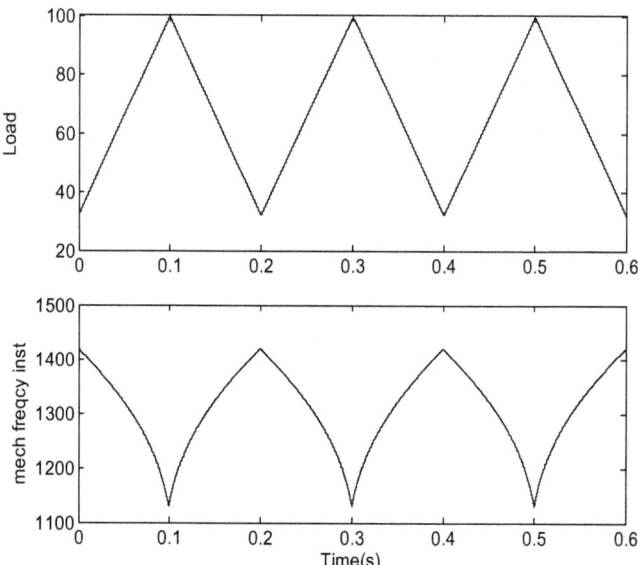

Fig. 2. (a) Evolution of the applied load, (b) Evolution of the mech frequency.

Table 1. Parameters of the spur gear transmission

	Pinion	Wheel
Teeth numbers	14	45
Mass (kg)	0.6	1.5
Mass moment of inertia (kg.m^2)	2.7×10^{-4}	0.0027
Full load torque (N.m)	10	−20
Rotation speed (rpm)	3630	1130
Modulus (mm)	3	
Primitive diameter (mm)	60	120
Base circle diameter (mm)	56.38	112.76
Material	42CrMo4	
Density (kg/m^3)	$\rho = 7860$	
Poisson ratio	0.33	
Pressure angle	$\alpha = 20°$	
Teeth width (mm)	23	
Contact ratio	$c = 1.6$	
Bearing stiffness (N/m)	$kx1 = ky1 = kx2 = ky2 = 10^8$	
Torsional stiffness (N rd/m)	$k\theta1 = k\theta2 = 10^5$	
Mean gearmesh stiffness (N/m)	$Kg = 2 \times 10^8$	

The variation of the rotational speed leads to a variation of the gear mesh frequency (Fig. 2). The mean value of the motor rotational speed is nr = 1320 rpm which corresponds to a mean gear mesh frequency of fgm = 308 Hz. The sampling frequency is 30800 Hz for all signals; the frequency default is 22 Hz.

In this work, we propose to study acceleration signals for different loads, The acceleration signals for healthy gear for a different variation of load (constant load, a variation of load fluctuation between 10%, 25% and 50%) are shown in Fig. 3 and for the faulty gear we have different case the first one is early stage for 5% of severity for different variation of load (constant load, variation of load fluctuation between 10%, 25% and 50%), the second one is an intermediate stage for 25% of severity for different variation of load (constant load, variation of load fluctuation between 10%, 25% and 50%) and the last one is an advanced stage for 50% of severity for different variation of load (constant load, variation of load fluctuation between 10%, 25% and 50%) as shown in Fig. 4.

The spectrum of the acceleration signal an advanced stage for 50% of severity and for load fluctuation of 50% is presented in Fig. 4. It is well known from Wang and Mcfadden (1997), Sharma and Parey (2016), Forrester (1989) that the frequency content of a faulty gear in stationary conditions is dominated mainly by the mesh frequency and its harmonics. From the presented spectrum zoomed (Fig. 4), it is observed the high activity of sidebands around mesh frequency and its harmonics. Sidebands can be a serious indicator of the presence of some kind of local damage on teeth or in bearings in stationary conditions if the frequency between sidebands is equal to fr, from the spectrum we can see that this frequency is different from the rotational

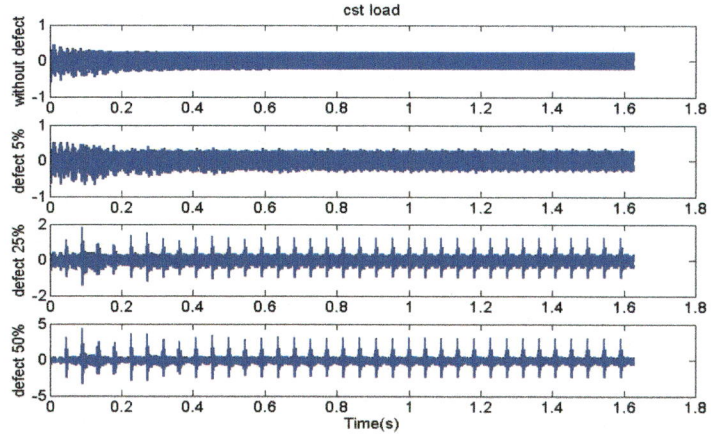

Fig. 3. The acceleration signals for a constant load.

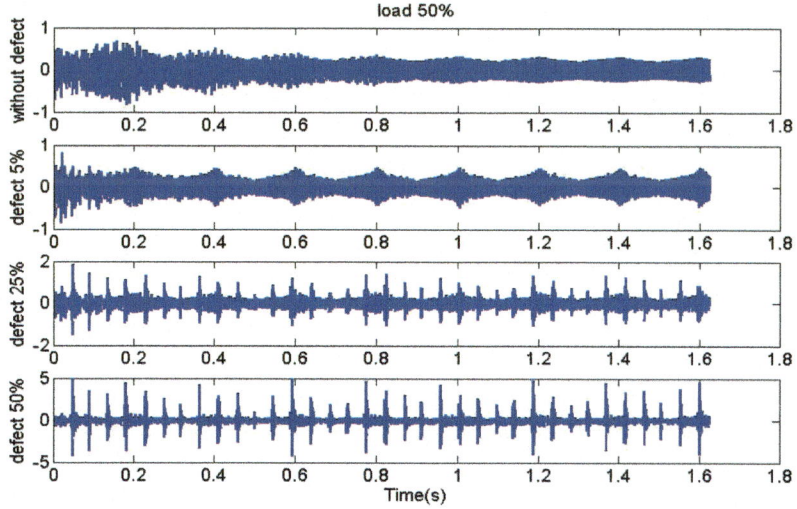

Fig. 4. The acceleration signals for 50% fluctuation of the load.

frequency fr. So it can be concluded that for non-stationary operating conditions separated time and frequency analysis for the studied case may induce in error during condition monitoring process. In order to overcome this difficulty, we propose to use ICEEMDAN to analyze such signals. The figure presents only the decomposition of the signal for 50% of severity and for load fluctuation of 25%, we can see clearly to compare between the IMFs given by ICEEMDAN for different signals, we have calculated L-kurtosis and kurtosis for different IMFs for three different cases. Figure 5 shows the variation of Kurtosis and L-kurtosis for a constant load and for four types of gear:

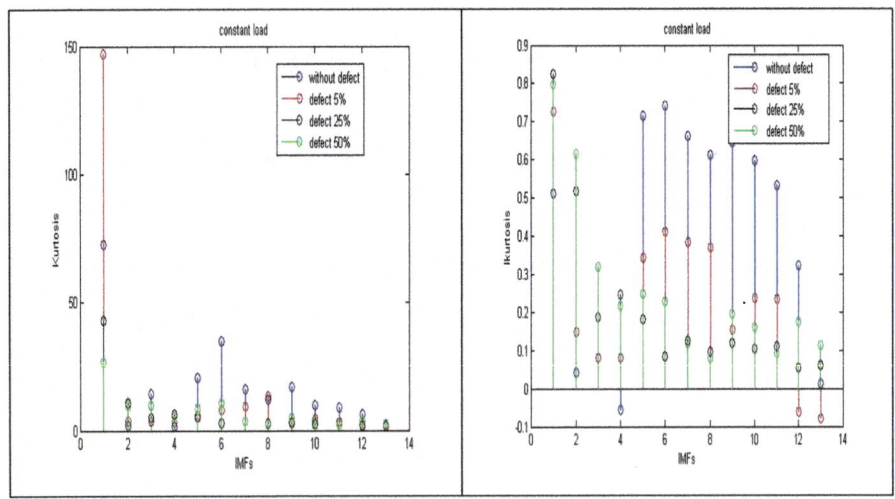

Fig. 5. Variation of Kurtosis and L-kurtosis for a constant load

(1) gear without defect (0%),
(2) gear with a defect that the severity is 5%
(3) gear with a defect that the severity is 25%
(4) gear with a defect that the severity is 50%

Figure 6 shows the variation of Kurtosis and L-kurtosis for a for a load fluctuation of 10% and for four types of gear.

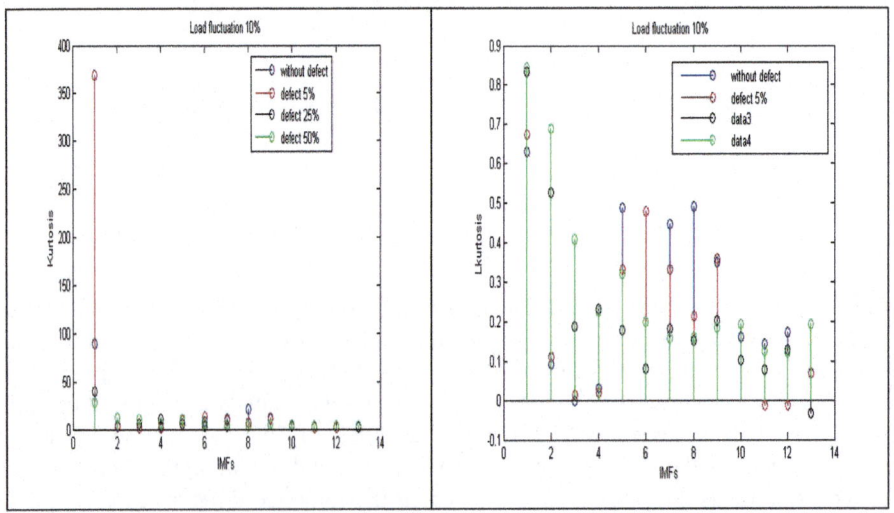

Fig. 6. Variation of Kurtosis and L-kurtosis for a for a load fluctuation of 10%.

(1) *gear without defect (0%),*
(2) *gear with a defect that the severity is 5%*
(3) *gear with a defect that the severity is 25%*
(4) *gear with a defect that the severity is 50%*

Figure 7 shows the variation of Kurtosis and L-kurtosis for a for a load fluctuation of 25% and for four types of gear.

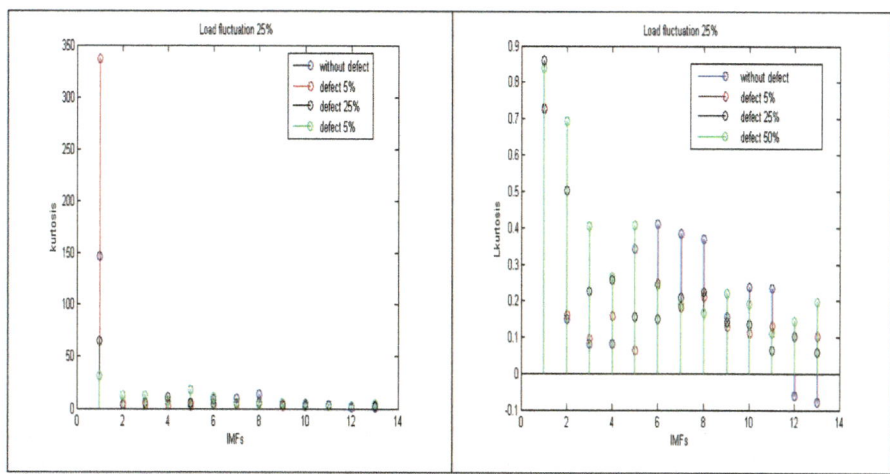

Fig. 7. Variation of Kurtosis and L-kurtosis for a for a load fluctuation of 25%.

(1) *gear without defect (0%),*
(2) *gear with a defect that the severity is 5%*
(3) *gear with a defect that the severity is 25%*
(4) *gear with a defect that the severity is 50%*

Figure 8 shows the variation of Kurtosis and L-kurtosis for a for a load fluctuation of 50% and for four types of gear.

(1) *gear without defect (0%),*
(2) *gear with a defect that the severity is 5%*
(3) *gear with a defect that the severity is 25%*
(4) *gear with a defect that the severity is 50%*

We can from these figures that the information given by the L-Kurtosis without using any zoom, is more clear than the information given by the Kurtosis, and the variation of the value of L-Kurtosis is logical for all IMFs, then we can use this indicator in classification and in the early fault detection.

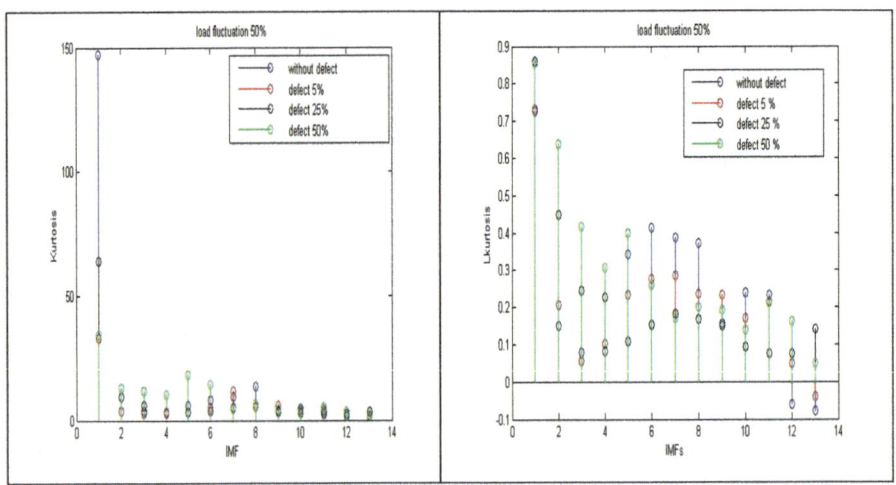

Fig. 8. Variation of Kurtosis and L-kurtosis for a load fluctuation of 50%.

5   Conclusion

In this study, we have used ICEEMDAN method to analyze non-stationary signals that give information about the variable conditions like (variable speed and load). The ICEEMDAN method achieves good frequency and modes of separation and gives better results than EEMD which use a noise that can mask the impulsions due to impacts that are generated by the presence of a defect. To detect the fault masked by the variation of the load, we have used the ICEEMDAN which separate the different modes that correspond to the variation of load and the effect of fault. We have also used the L-kurtosis to compare between the different IMFs and we have compared the result given by L-Kurtosis to those given by the kurtosis and we have seen that the figures given by the L-kurtosis are more clear than those given by kurtosis, and we have observed that the kurtosis cannot give any indication information of the presence of the defect, but if we use the L-kurtosis, we have observed that the L-kurtosis can be used as an indicator of the fault in variable conditions.

References

McFadden, P.D.: Detecting fatigue cracks in gears by amplitude and phase demodulation of the meshing vibration. Trans. ASME, J. Vib. Acoust. Stress Reliab. Des. **108**, 165–170 (1986)

Capdessus, C., Sidahmed, M.: Analyse des vibrations d'un engrenage cepstre, corrélation, spectre. Traitement du signal **8**(5), 365–371 (1992)

Wang, W.J., Mcfadden, P.D.: Application of orthogonal wavelet to early gear damage detection. Mech. Syst. Signal Process. **9**(5), 497–507 (1997)

Sharma, V., Parey, A.: A review of gear fault diagnosis using various condition indicators. Procedia Eng. **144**, 253–263 (2016)

Forrester B.D.: Use of Wigner Ville distribution in helicopter transmission fault detection. In: Proceedings of thethe Australian, Symposium on Signal Processing and Applications, ASSPA89, Adelaide, Australia, 17–19 April 1989, pp. 77–82 (1989)

Staszewski, W.J.: Local tooth fault detection in gearboxes using a moving window procedure. Mech. Syst. Signal Process. 11(3), 331–350 (1997)

Cohen, L.: Time-frequency distributions a review. Proc. IEEE 77(7), 941–981 (1989)

Mallat, S.G.: A Wavelet Tour of Signal Processing. Academic, San Diego (1998)

Yang, W., Court, R., Tavner, P.J., Crabtree, C.J.: Bivariate empirical mode decomposition and its contribution to wind turbine condition monitoring. J. Sound Vib. 330(15), 3766–3782 (2011)

Huang, N.E., Shen, Z., Long, S.R.: The empirical mode decomposition and the Hilbert spectrum for nonlinear and non-stationary time series analysis. Proc. R. Soc. Lond. Ser. 454, 903–995 (1998)

Liu, B., Riemenschneider, S., Xub, Y.: Gearbox fault diagnosis using empirical mode decomposition and Hilbert spectrum. Mech. Syst. Signal Process. 17(9), 1–17 (2005)

Yu, Y., Junsheng, C.: A roller bearing fault diagnosis method based on EMD energy entropy and ANN. J. Sound Vib. 294(1–2), 269–277 (2006)

Huang, N.E., Wu, M.L., Long, S.R.: A confidence limit for the empirical mode decomposition and Hilbert spectral analysis. Proc. R. Soc. Lond. 459, 2317–2345 (2003)

Rilling, G., Flandrin, P.: One or two frequencies? The empirical mode decomposition answers. IEEE Trans. Signal Process. 56(1), 85–95 (2008)

Wu, Z., Huang, N.E.: Ensemble empirical mode decomposition: a noise-assisted data analysis method. Adv. Adapt. Data Anal. 1(1), 1–41 (2009)

Colominas, M.A., Schlotthauer, G., Torres, M.E.: Improved complete ensemble EMD: a suitable tool for biomedical signal processing. Biomed. Signal Process. Control 14, 19–29 (2014)

Liu, S., Hou, S., Yang, W.: L-Kurtosis and its application for fault detection of rolling element bearings. Measurement 116, pp. 523–532 (2018). https://www.sciencedirect.com/science/article/pii/S0263224117307595

Chaari, F., Abbes, M.S., Rueda, F.V., del Rincon, A.F., Haddar, M.: Analysis of planetary gear transmission in non-stationary operations. Front. Mech. Eng. 8(1), 88–94 (2013)

Bartelmus, W., Zimroz, R.: Vibration condition monitoring of planetary gearbox under varying external load. Mech. Syst. Signal Process. 23(1), 246–257 (2009)

Dynamic Behavior of Back to Back Planetary Gear in Presence of Pitting Defects

Ayoub Mbarek[1,2]([⊠]), Alfonso Fernandez Del Rincon[2],
Ahmed Hammami[1], Miguel Iglesias[2], Fakher Chaari[1],
Fernando Viadero Rueda[2], and Mohamed Haddar[1]

[1] Laboratory of Mechanics, Modeling and Production (LA2MP),
National School of Engineers of Sfax, BP 1173, 3038 Sfax, Tunisia
ayoubmbarekenit@gmail.com,
ahmed.hammami2109@gmail.com,
fakher.chaari@gmail.com, mohamed.haddar@enis.rnu.tn
[2] Department of Structural and Mechanical Engineering,
Faculty of Industrial and Telecommunications Engineering,
University of Cantabria, Avda de los Castros s/n, 39005 Santander, Spain
{alfonso.fernandez, miguel.iglesias,
fernando.viadero}@unican.es

Abstract. The diagnosis of rotating machinery such as planetary gearbox running under operating condition is a complex task. Depending on their nature, the defects can be classified into teeth, geometrical or bearing defects. This paper studies the dynamic behavior of a back-to-back spur planetary gear running in stationary condition in the presence of teeth defects. This damage is the pitting and it is located in one tooth of the sun gear. An experimental test is performed by the measurement of the instantaneous accelerations on a test gear. The tests are carried out under a fixed load and speed. The presence of the tooth pitting defect activates repeated-like transient in ring's vibration displacement because of phase change and amplitude reduction of the mesh stiffness when the damage tooth comes into contact. The Fast Fourier Transform (FFT) is used to transform the time signal into the frequency domain for signature analysis. The acceleration spectra show the influence of damage both in frequency and amplitude. The frequency of pitting which is the frequency of rotation of the test sun and the frequency of rotation of the motor appears along of frequency bandwidth. Moreover, the impulses of the defect tooth with the planets become important in amplitude.

Keywords: Pitting defect · Planetary gear · Stationary condition
Back-to-back configuration

1 Introduction

Planetary gears transmissions are running in different conditions. They can be affected by several damage and defects like the pitting defects. This kind of defects is caused by the teeth contact condition and it can introduce a variation on mesh stiffness function. The pitting defects was studied by many research such as Chaari et al. (2008). The pits

© Springer Nature Switzerland AG 2019
T. Fakhfakh et al. (Eds.): ICAV 2018, ACM 13, pp. 16–22, 2019.
https://doi.org/10.1007/978-3-319-94616-0_2

were modeled as rectangular shapes, this defect was analytically studied, and it was modeled in mesh stiffness function. Del Rincon et al. (2012) studied also the effect of pitting on mesh stiffness function for an enhanced model gear.

The pit was modeled in elliptical shape in three different locations. They evaluated the influence of pitting on gear mesh stiffness using finite element method. Recently, the influence of pitting on mesh stiffness function was studied by Liang et al. (2016) who proposed an analytical method on mesh stiffness calculation based on potential energy method external spur gear and they validated the model through by finite element method. In their study, the pits are modeled in circular shapes and they studied in three severity level. Furthermore, the influence of this local damage on dynamic response is fewer studied in literature. Choy et al. (1996) studied also the influence of pitting on the vibration of a gear transmission system. They made a modification on the shape of gear mesh stiffness to simulate the pitting fault. Wigner Ville time distribution are used to examine gear vibration response. Chaari et al. (2006) studied the impact of pitting on the dynamic response planetary gear. They modified the mesh stiffness function's shape in order to simulate the defects.

Abouel-seoud et al. (2012) presented a single rectangular tooth pit effect on the mesh stiffness of a wind turbine gearbox and analyzed vibration signal fault signatures in the time and frequency domain.

Only numerical results were introduced by the aforementioned works. However, in this paper an experimental investigation of the effects of pitting on dynamic behavior of back to back planetary gear is studied. In fact, Pitting defect is located only one tooth of the test sun.

2   Experimental Test Bench

Figure 1 shows the experimental test bench where the tests are achieved. It consists of two identical planetary gears driven by an electrical motor. A planetary gear set called test gear where the measurements are carried out. It characterized by a fixed ring. The gear set near the motor is the reaction gear. Its ring is free. The sun's gears and carrier's gears relate to each other by two rigid shafts. This test bench was used by Mbarek et al. (2017a, b) and it is well described in Mbarek et al. (2017a, b).

The system was loaded by adding mass on a rigid arm fixed on the free ring.

The experimental tests were carried out without defects and in presence of pitting defects which is located on sun's test planetary gear. Figure 2 shows a photo of pitting location.

Sensors are mounted on the test ring in order to measure the instantaneous accelerations. The signals issued from the sensors will be processed with the software "LMS Test.Lab signature acquisition" to obtain the instantaneous accelerations. The test bench layout and the instruments are displayed in Fig. 3.

Fig. 1. Experimental test bench

Fig. 2. Pitting located on the test sun

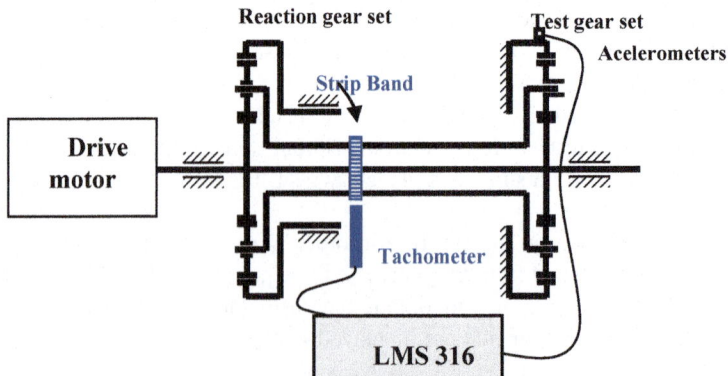

Fig. 3. Test bench instruments

3 Results

The pitting defects impact on the principal function in planetary gear which is the mesh stiffness function. Because of the pitting is located in the test sun, the sun-planets mesh stiffness functions are modified as consequences the time response and the spectra of acceleration will be varied.

The experimental tests were carried out with fixed speed (1498.5 rpm) and fixed load (300 N m).

The measured signal on the test ring, presented in Fig. 4, repeats many times. Therefore, the amplitude modulation is defined by the force due to rotation of carrier.

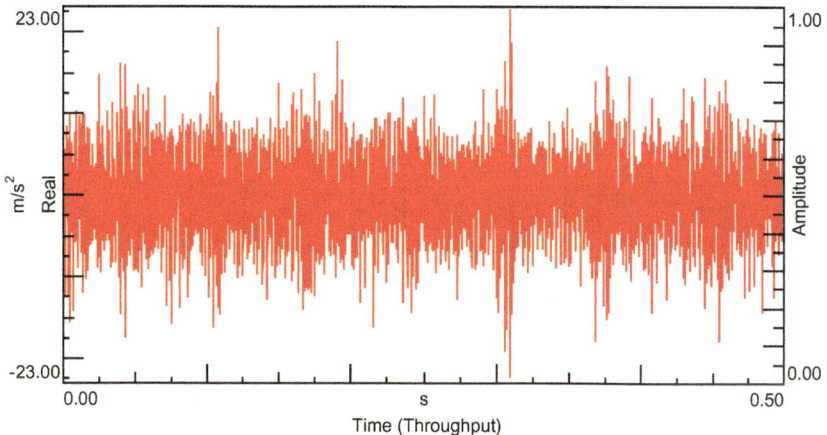

Fig. 4. Time response measured on the fixed ring

The presence of a tooth pitting defect activates repeated-like transient in ring's vibration displacement because of phase change and amplitude reduction of the mesh stiffness when the failure tooth comes into contact.

The meshing is considered as the main excitation source of the system (Hidaka et al. 1979; Lin and Parker 1999). The spectrum shows that the test ring is excited by the gear mesh frequency and its harmonics, which are defined by:

$$f_m = \frac{Z_s Z_r}{Z_s + Z_r} \frac{N}{60}$$

Where N is the electrical motor rotational speed, Zs and Zr are the tooth numbers of the sun and the ring gear respectively. In our study, as mentioned previously, the motor was running at 1498.5 rpm. So, the structure was excited by the gear mesh frequency $f_m = 320$ Hz.

The defects frequency be like the test sun rotational frequency as well as the motor rotational frequency. However, the pitting frequency $f_d = f_s = N/60 = 25$ Hz, appears along the frequency bandwidth which explain the presence of defects. Figure 5 shows the spectra of acceleration measured on the test ring.

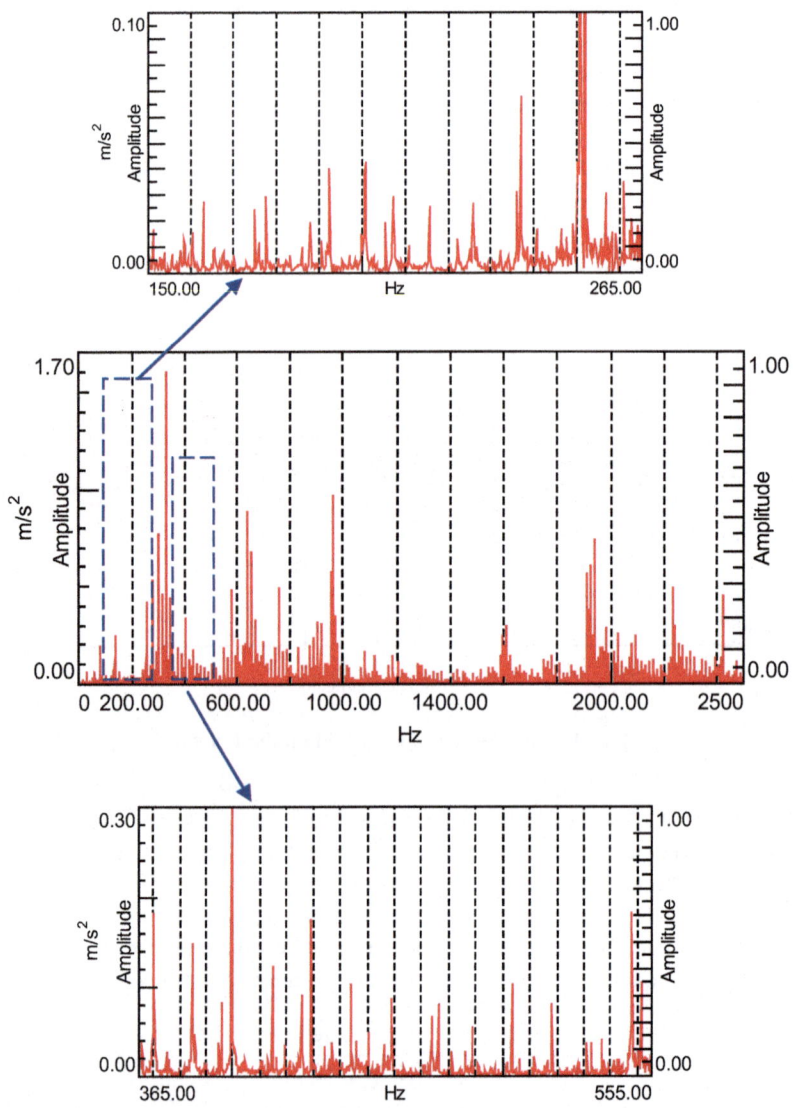

Fig. 5. Spectra of acceleration on the fixed ring

Figure 5 shows a zoom around the two first meshing frequency. It can be noticed the influence of pitting on dynamic response of gear through the spectrum. The frequency of pitting is 25 Hz which is the frequency of rotation of the test sun as well the frequency of rotation of the motor. The impulses of the defect tooth with the planets become important in amplitude. Sidebands are more active around the gear mesh frequency and its harmonics.

4 Conclusion

This paper investigated the influence of pitting defects on the dynamic response of back to back planetary gear. The measurements were carried out in stationary conditions where the fix input load and speed are fixed.

The sensors were mounted on the fix ring in order to measure the instantaneous accelerations. The FFT was utilized to transform the averaged time signal into the frequency domain for signature analysis. The spectra were also used to examine the planetary gear system. The damage frequency which is the sun test rotational frequency as well as the rotational frequency of the motor appears along of frequency bandwidth. Moreover, the impulses of the defect tooth with the planets become important in amplitude. Sidebands are more active around the gear mesh frequency and its harmonics.

Future investigation will be focused on the effects of pitting defects in non-stationary condition (run-up and run-down regimes).

Acknowledgements. This paper was financially supported by the Tunisian-Spanish Joint Project No. A1/037038/11. The authors would like also to acknowledge project "Development of methodologies for the simulation and improvement of the dynamic behavior of planetary transmissions DPI2013-44860" funded by the Spanish Ministry of Science and Technology.

Acknowledgment to the University of Cantabria cooperation project for doctoral training of University of Sfax's students.

References

Chaari, F., Baccar, W., Abbes, M.S., et al.: Effect of spalling or tooth breakage on gearmesh stiffness and dynamic response of a one-stage spur gear transmission. Eur. J. Mech. A Solids **27**(4), 691–705 (2008)

del Rincon, A.F., Viadero, F., Iglesias, M., et al.: Effect of cracks and pitting defects on gear meshing. Proc. Inst. Mech. Eng. Part C J. Mech. Eng. Sci. **226**(11), 2805–2815 (2012)

Liang, X., Zhang, H., Liu, L., et al.: The influence of tooth pitting on the mesh stiffness of a pair of external spur gears. Mech. Mach. Theory **106**, 1–15 (2016)

Choy, F.K., Polyshchuk, V., Zakrajsek, J.J., et al.: Analysis of the effects of surface pitting and wear on the vibration of a gear transmission system. Tribol. Int. **29**(1), 77–83 (1996)

Chaari, F., Fakhfakh, T., Haddar, M.: Dynamic analysis of a planetary gear failure caused by tooth pitting and cracking. J. Fail. Anal. Prev. **6**(2), 73–78 (2006)

Abouel-Seoud, S.A., Dyab, E.S., Elmorsy, M.S.: Influence of tooth pitting and cracking on gear meshing stiffness and dynamic response of wind turbine gearbox. Int. J. Sci. Adv. Technol. **2**(3), 151–165 (2012)

Mbarek, A., Hammami, A., del Rincon, A.F., et al.: Effect of gravity of carrier on the dynamic behavior of planetary gears. In: International Conference Design and Modeling of Mechanical Systems, pp. 975–983. Springer, Cham (2017a)

Mbarek, A., Hammami, A., del Rincon, A.F., et al.: Effect of load and meshing stiffness variation on modal properties of planetary gear. Appl. Acoust. (2017b)

Hidaka, T., Terauchi, Y., Dohi, K.: On the relation between the run-out errors and the motion of the center of sun gear in a stoeckicht planetary gear. Bull. JSME **22**(167), 748–754 (1979)

Lin, J., Parker, R.G.: Analytical characterization of the unique properties of planetary gear free vibration. J. Vib. Acoust. **121**(3), 316–321 (1999)

Losses in Thrust Ball Bearings Lubricated with Axle Gear Oils

Maroua Hammami[1,2]([✉]), Mohamed Slim Abbes[1], Ramiro Martins[3],
Jorge H. O. Seabra[2], and Mohamed Haddar[1]

[1] Laboratory of Mechanical, Modelling and Manufacturing,
National Engineers School of Sfax (ENIS), BP 1173, 3038 Sfax, Tunisia
hammamimaroi@gmail.com, {slim.abbes,Mohamed.Haddar}@enis.rnu.tn
[2] Faculdade de Engenharia da Universidade do Porto (FEUP),
Rua Dr. Roberto Frias S/n, 4200-465 Porto, Portugal
jseabra@fe.up.pt
[3] Institute of Science and Innovation in Mechanical and Industrial Engineering
(INEGI), Rua Dr. Roberto Frias 400, 4200-465 Porto, Portugal
rmartins@inegi.up.pt

Abstract. This work provide more new knowledge about rolling bearings lubricated with axle gear oils. Extensive tests were performed and a considerable amount of experimental results of power loss in rolling bearings, difficult to find in literature, were obtained. Five fully formulated axle gear oils with different base oil, viscosity and different formulations and additive packages were selected. Their chemical and physical properties were measured. Thrust ball bearings lubricated with axle gear oils were tested using a modified Four-Ball Machine where the four-ball arrangement was replaced by a rolling bearing assembly. These tests were performed under a constant temperature of 70 °C and under the following operating conditions: speed between 75 and 1200 rpm and two axial loads (4000 N and 7000 N). The rolling bearing friction torque was measured and the effect of speed, temperature and axial load have been evaluated. Based on experimental results, a rolling bearing torque loss model was calibrated using SKF model for ball contacts. The model allows a better understanding of the behaviour of the rolling bearing geometries and of the influence of oil formulation on rolling bearing friction torque. The rolling bearing power loss model will be relevant for the global axle differential power loss model predictions.

Keywords: Axle gear oils · Chemical and physical properties
Thrust ball bearings · Friction torque · SKF model

1 Introduction

The main function of rolling bearings in axles is to support the pinion and the differential gear under high load carrying capacity and high stiffness. However, the rolling bearings are a major contributor to axle system power loss (Matsuyama et al. 2004).

© Springer Nature Switzerland AG 2019
T. Fakhfakh et al. (Eds.): ICAV 2018, ACM 13, pp. 23–36, 2019.
https://doi.org/10.1007/978-3-319-94616-0_3

To achieve high efficiency in axle differentials, the reduction of internal friction torque in rolling bearings is of major concern. Thus, the importance of understanding internal friction in rolling bearings becomes relevant. The energy saving and the bearings performance optimization are required (Cousseau et al. 2010).

Recently, the automotive manufacturers and the rolling bearings manufacturers are trying to improve rolling bearing designs in order to reduce the power loss generated, reduce the energy consumption, reduce the operating temperatures and improve the lubrication conditions. At the same time, they claim the lubricant manufacturers to provide new products that increase rolling bearing life, while reducing the energy dissipated (Matsuyama et al. 2004; Hosokawa et al. 2009).

Several authors have studied the rolling bearings friction torque. Spindler and Von Petery (2003) developed a new bearing design, where the tapered roller bearings used previously have been replaced by angular contact ball bearing, that meets the requirements for high rigidity, long life and no preload loss during operation. Matsuyama et al. (2004) developed a super-low friction torque tapered roller bearing (80% torque reduction from that of a standard bearing). Hoshokawa et al. (2009) developed a new bearing concept which is the double row angular contact ball bearing -so-called Tandem Ball Bearings for rear axle drives. This bearing concept not only increases the service life but also make significant contribution to lower fuel consumption.

This work provide more new knowledge about rolling bearings lubricated with axle gear oils. Extensive tests were performed and a considerable amount of experimental results of power loss in thrust ball bearings, difficult to find in literature, were obtained. The SKF rolling bearing friction torque model will be calibrated with the experimental results. The model allows a better understanding of the behaviour of the rolling bearing geometry and of the influence of oil formulation. The rolling bearing power loss model will be relevant for the global axle differential power loss model predictions.

2   Lubricant Properties

Five multigrade oils, suitable for axle lubrication, were selected. All the lubricants are polyalphaolephin base oils (PAO) except for the 80W90-A product which has mineral base oil (MIN). Three among them (75W90-A, 80W90-A and 75W140-A), are reference oils (A) and labelled as "Fuel Efficient" while the other two products (75W85-B and 75W90-B) are candidate oils (B). The reference lubricants 75W90-A and 80W90-A meet the requirements of API GL-4 and/or GL-5 and/or MT-1 standards and reference oil 75W140-A meets the requirements of API GL-5 standard. The candidate lubricants, 75W90-B and 75W85-B have not yet been assessed in what concerns the API (American Petroleum Institute) standards. Axle gear oils physical properties as well as their chemical composition were displayed in Table 1. A detailed description on the physical and chemical properties can be found in previous works respectively (Hammami et al. 2017a; Hammami et al. 2017b).

Table 1. Physical and chemical properties of axle gear oils used.

Parameter	Unit	75W85-B candidate	75W90-A reference	75W90-B candidate	80W90-A reference	75W140-A reference
Base oil	[-]	PAO	PAO	PAO	Mineral	PAO
API/standard	[-]	-	GL-4/GL-5/MT-1	-	GL-4/GL-5/MT-1	GL-5
Chemical composition						
Boron (B)	[ppm]	0	-	81	-	-
Calcium (Ca)	[ppm]	1795	18	2891	97	33
Magnesium (Mg)	[ppm]	6	1087	17	936	1093
Phosphorus (P)	[ppm]	783	1622	958	1436	1686
Sulfur (S)	[ppm]	2954	23262	3271	26947	22784
Zinc (Zn)	[ppm]	899	7	1120	23	12
Physical properties						
Density @ 15 °C	[g/cm^3]	0.853	0.87	0.861	0.886	0.885
Thermal expansion coefficient ($\alpha_t \times 10^{-4}$)	[/]	−8.1	−7.3	−7.6	−7.7	−6.8
Viscosity @ 40 °C	[cSt]	68.95	112.35	114.42	123.3	200.7
Viscosity @ 70 °C	[cSt]	23.86	36.7	38.14	34.86	61.86
Viscosity @ 100 °C	[cSt]	11.44	16.37	17.18	14.38	26.21
a_A	[/]			0.7		
n_A	[/]	7.6655	7.5833	7.407	8.5027	7.1537
m_A	[/]	2.9663	2.9133	2.842	3.2783	2.7211
Thermoviscosity @ 40 °C ($\beta \times 10^3$)	[K^{-1}]	40.2	44.3	43.3	50.7	46.3
Thermoviscosity @ 70 °C ($\beta \times 10^3$)	[K^{-1}]	28.5	31.3	30.9	34.8	33.2
Thermoviscosity @ 100 °C ($\beta \times 10^3$)	[K^{-1}]	21.1	23.1	22.9	25	24.7
s @ 0,2 GPa	[/]	0.7382	0.7382	0.7382	0.9904	0.7382
t @ 0,2 GPa	[/]	0.1335	0.1335	0.1335	0.139	0.1335
Piezoviscosity @ 40°C ($\alpha \times 10^{-8}$)	[Pa^{-1}]	1.291	1.387	1.39	1.934	1.498
Piezoviscosity @ 70°C ($\alpha \times 10^{-8}$)	[Pa^{-1}]	1.128	1.194	1.2	1.623	1.28
Piezoviscosity @ 100°C ($\alpha \times 10^{-8}$)	[Pa^{-1}]	1.022	1.072	1.079	1.435	1.142
VI	[/]	162	147	163	118	169

3   Rolling Bearing Assembly, Operating Conditions and Test Procedure

Rolling bearing tests were performed on a modified Four-Ball machine where the four-ball arrangement was replaced by a rolling bearing assembly as shown in Fig. 1. This assembly allows to test several rolling bearings and to obtain reliable friction torque measurements at different operating temperature. A detailed description of this assembly can be found in (Cousseau et al. 2010). In operation, the internal friction torque is transmitted to the torque cell (11) through the bearing housing (1). The friction torque was measured with a piezoelectric torque cell KISTLER9339A and five thermocouples (IV) were used to monitor the temperature at strategic locations (see Fig. 1).

In order to maintain a constant temperature during the test, the rolling bearing assembly is exposed to a continuous air flow, provided by two 38 mm diameter fans running at 2000 rpm, cooling the chamber surrounding the bearing house. Also, a heater controlled with PID system was used to increase and maintain a constant operating temperature at a desired value (70 °C for the study case).

The rolling bearing is lubricated by an oil volume of 14 ml. This volume was selected for the purpose that the centre of the ball is reached by the oil level, such as indicated by the manufacturer. For each oil tested, a new rolling bearing sample was used in order to reduce the influence of the surface finish and to avoid possible chemical interactions between oils tested. The surface roughness of different new rolling bearings was measured using an absolute probe in a HommelWereke T4000 device. The results obtained present similar finishing on the rolling bearing raceways.

The rotational speeds from 75 up to 1200 rpm were selected for the friction torque tests considering the limits of test machine and also the rotational speeds usually used in axles covering all the lubrication regimes (boundary, mixed and full film lubrication). For example, under no load conditions, the pinion speeds range from 300 to 3000 rpm which is 4 to 5 times higher than the axle speed and under loaded conditions, the speed of the pinion is adapted from 'the standard 29-point test schedule' proposed by (Xu et al. 2012; Matsuyama et al. 2004).

The operating temperature of 70 °C was selected according to the axle lubricant temperatures measured during an EPA (Environmental Protection Agency) driving cycle including both city and highway cycles (Hammami et al. 2017a).

Usually the preload range in axle differentials is between 4000 and 7000 N, assuring a sufficiently high bearing rigidity (Spindler and Von Petery 2003, Hosokawa et al. 2009).

For axial load tested (4000 and 7000 N), the rolling element (ball) and the raceway contact present the characteristics presented in Table 2.

The experimental tests were performed using a thrust ball bearing (TBB, ref. 51107 SKF) geometry. The dimensions and characteristics of the selected geometry are reported in Table 3.

The friction torque was measured for different rotational speeds, in the range 75 up to 1200 rpm and the operating temperature of 70 °C was selected. All thrust ball bearings were submitted to two axial loads (4000 N and 7000 N).

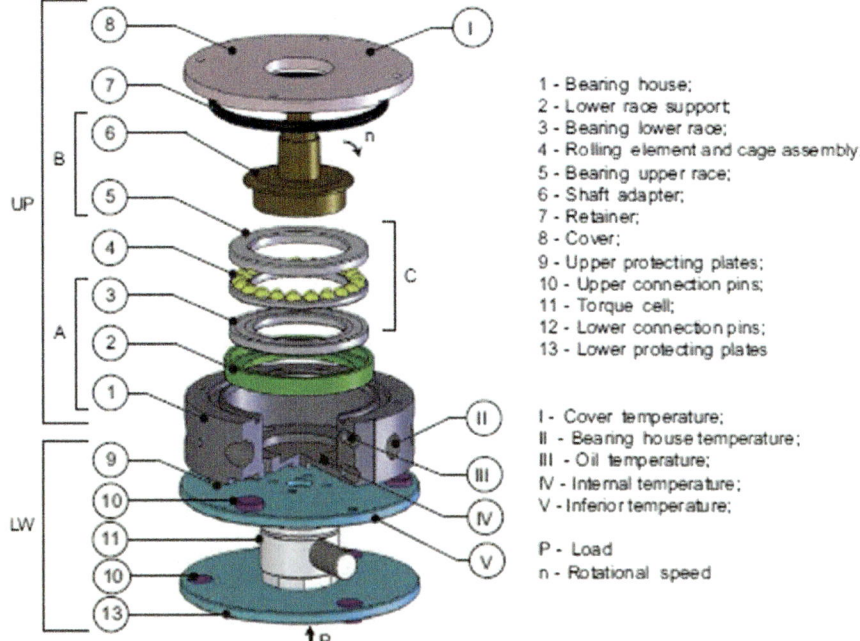

1 - Bearing house;
2 - Lower race support;
3 - Bearing lower race;
4 - Rolling element and cage assembly;
5 - Bearing upper race;
6 - Shaft adapter;
7 - Retainer;
8 - Cover;
9 - Upper protecting plates;
10 - Upper connection pins;
11 - Torque cell;
12 - Lower connection pins;
13 - Lower protecting plates

I - Cover temperature;
II - Bearing house temperature;
III - Oil temperature;
IV - Internal temperature;
V - Inferior temperature;

P - Load
n - Rotational speed

Fig. 1. Schematic view of the rolling bearing assembly.

The test started at the selected speed and ran until it reached a constant operating temperature (70 °C). The measurements were taken after a 30 min time operation at the desirable speed, load and operating temperature. Four friction torque measurements were performed: three values are stored and the most dispersed one was disregarded. Due to the drift effect, which affects the measurements of the piezoelectric sensors after long periods of operation, the friction torque measurements should be made in a short period of time (less than 120 s) and at constant temperature (70 °C).

4   Friction Torque Model

In order to understand the experimental results as well as the friction torque behavior of the TBB lubricated with axle gear oils, a friction torque model is required. For that SKF proposed a detailed model (SKF 2013) which divides the total friction torque in its true physical forms. It takes into account four different torque losses as shown in the following equation:

$$M_t = M'_{rr} + M_{sl} + M_{seal} + M_{drag} \tag{1}$$

Since, the TBB (51107) used for experimental tests do not have seals, the M_{seal} component was disregarded in the calculation. The drag losses are very

Table 2. Ball-raceway contact parameters.

Contact element	Unit	Raceway	Ball
R_{Xi}	[m]	∞	3×10^{-3}
R_{Yi}	[m]	-3.38×10^{-3}	3×10^{-3}
Axial load	[N]	4000	7000
R_X	[m]	6×10^{-3}	
R_Y	[m]	53.4×10^{-3}	
σ_c	[μm]	0.18	
a	[μm]	102.8	123
A_c	[mm^2]	0.14	0.198
p_0	[GPa]	2.06	2.47
δ	[μm]	5.22	7.48

Table 3. Characteristics of thrust ball bearing (TBB).

			TBB
Principal dimensions	d	mm	35
	H	mm	12
	D	mm	52
Basic load ratings	Dynamic C	kN	19.9
	Static C_0	kN	51
Speed ratings	Reference speed	rpm	5600
	Limiting speed	rpm	7500

small because the operating speeds and the mean diameter ($d_m = 43.5$ mm) of the TBB are also small, consequently, the drag torque loss term was also disregarded.

Therefore, the total friction torque of the cylindrical thrust roller bearing had only two contributions: the rolling and sliding torques, respectively, M'_{rr} and M_{sl}, as presented in the following equation:

$$M_t = M'_{rr} + M_{sl} \tag{2}$$

Assuming that the friction torque obtained from experimental measurements was equal to the total torque loss predicted by the SKF model ($M_t = M_t{}^{exp}$) and the rolling torque was accurately calculated, the sliding torque can be directly determined (see Eq. (3)).

$$M_{sl} = M_t - M'_{rr} = M_t^{exp} - M'_{rr} \tag{3}$$

All the following Eqs. (4)–(11) are useful to calculate the rolling and sliding torques.

$$M'_{rr} = \phi_{ish} \cdot \phi_{rs} \cdot [G_{rr} \cdot (n \cdot v)^{0.6}] \tag{4}$$

$$\phi_{ish} = \frac{1}{1 + 1.84 \cdot 10^{-9} \cdot (n \cdot d_m)^{1.28} \cdot \nu^{0.64}} \tag{5}$$

$$\phi_{rs} = \frac{1}{e^{[K_{rs} \cdot \nu \cdot n \cdot (d+D) \cdot \sqrt{\frac{K_z}{2 \cdot (D-d)}}]}} \tag{6}$$

$$G_{rr} = R1 \cdot d_m^{1.83} \cdot Fa^{0.31} \tag{7}$$

$$M_{sl} = G_{sl} \cdot \mu_{sl} \tag{8}$$

$$G_{sl} = S1 \cdot d_m^{0.05} Fa^{4/3} \tag{9}$$

$$\mu_{sl} = \phi_{bl} \cdot \mu_{bl} + (1 - \phi_{bl}) \cdot \mu_{EHL} \tag{10}$$

$$\phi_{bl} = \frac{1}{e^{2.6 \cdot 10^{-8} \cdot (n \cdot \nu)^{1.4} \cdot d_m}} \tag{11}$$

For thrust ball bearing, the constants $R1$, $S1$, K_{rs} and K_z are given in Table 4.

Table 4. SKF rolling bearing constants.

	Thrust ball bearings (51107)
$R1$	1.03×10^{-6}
$S1$	1.6×10^{-2}
K_z	3.8
K_{rs}	3×10^{-8}

The rolling torque (Eq. (4)) was mainly dependent on the bearing type, rotational speed, oil viscosity and two factors which are the inlet shear heating ϕ_{ish} and the kinematic replenishment reduction factor ϕ_{rs}.

The sliding torque (Eq. (8)) was highly influenced by the bearing type, the coefficient of friction and the lubrication regime. The lubrication regime on the model is quantified by the ϕ_{bl} quantity where the coefficient of friction in full film μ_{EHL} and boundary μ_{bl} lubrication were presented.

This model is applied on the experimental results in order to determine the experimental sliding coefficient of friction (μ_{sl}^{exp}) (see Eq. (12)).

$$\mu_{sl}^{exp} = \frac{M_{sl}}{G_{sl}} = \frac{M_t^{exp} - M_{rr}'}{G_{sl}} \tag{12}$$

Also the model can predict the sliding coefficient of friction using Eq. (10), where the coefficient of friction is dependent on two coefficients: μ_{bl} that is related to the additive package in the lubricant and μ_{EHL} that is related to the base oil and the bearing type.

Using the provided values of μ_{bl} and μ_{EHL} from the SKF model, the coefficient of friction obtained is slightly different from the experimental value. To minimize the difference between μ_{sl}^{exp} and μ_{sl}, the two coefficients μ_{bl} and μ_{EHL} should be calculated for each operating temperature and for the selected rolling bearing type. They are clearly dependent on the load and the speed range used in the rolling bearing tests, as will be shown later on.

5   Results and Discussion

The SKF friction torque model was used to predict the values of the rolling M_{rr} and sliding M_{sl} friction torques and the EHL (μ_{EHL}) and the boundary (μ_{bl}) coefficients of friction for all testing conditions considered in the thrust ball bearing tests under an axial load of 4000 N and 7000 N, respectively 20% and 35% of dynamic load capacity of the thrust ball bearing.

5.1   Axial Load 7000 N

The TBB lubricated with axle gear oils tests were carried out under constant temperature of 70 °C and an axial load of 7000 N. Figure 2(a) shows that, in general, the measured total friction torque of the TBB decreases when the operating speed increases from 75 rpm to 1200 rpm, except in the case of oil 75W140-A, for which the friction torque increased as the speed increases. For low rotational speeds, the two candidate (B) oils generated higher values of the total friction torque than the reference (A) oils (see Fig. 2(a)) due do the significant differences between them in terms of the additive packages present in their formulations.

Figure 2(b) shows that when the speed increased the specific film thickness inside the TBB increased from 0.22/0.46 up to 1.30/2.75, depending on the tested oil, meaning that the lubrication regime evolved from boundary to mixed film lubrication. The values presented for the specific film thickness were calculated using Hamrock and Dowson (Hamrock and Dowson 1981) equation for point contact (see Appendix B in previous work (Hammami et al. 2017a)).

Figure 2(c) shows the rolling torque inside TBB. As expected, at constant temperature (70 °C) the rolling torque increases when the speed increases ($M'_{rr} \propto (n \cdot v)^{0.6}$). This Figure also shows that the 75W140-A oil, with the highest operating viscosity, generated the highest rolling torques, while 75W85-B oil, with the lowest viscosity, has the lowest rolling torques. Other oil formulations produced very similar rolling torques, because they have very close viscosity values at 70 °C.

The sliding torque M_{sl}, presented in Fig. 2(d), is obtained by subtracting the rolling friction torque to the experimental friction torque (see Eq. 3). The sliding torque values are higher than those calculated for the rolling torque and they show the same trend of the total friction torque curves. This Figure shows that the sliding torque decreases with the increase of the rotational speeds. Such behaviour was anticipated since the coefficient of friction μ_{sl} decreases when speed increases, at constant temperature, as presented in Fig. 2(e).

The experimental sliding coefficient of friction μ_{sl}^{exp}, calculated using Eq. (12), is presented in Fig. 2(e) where the sliding coefficient of friction follows the trend of the sliding torque.

Figure 2(f) presents the sliding coefficient of friction for candidate oils (B). Their experimental results with the error bar for each value are shown with markers and the model simulations are shown by the continuous lines, in function of the modified Hersey parameter $Sp = Ur.\eta.\alpha^{0.5}.Fn^{-0.5}$. The approximation of the sliding coefficient of friction μ_{sl}, predicted by the model, is quite good,

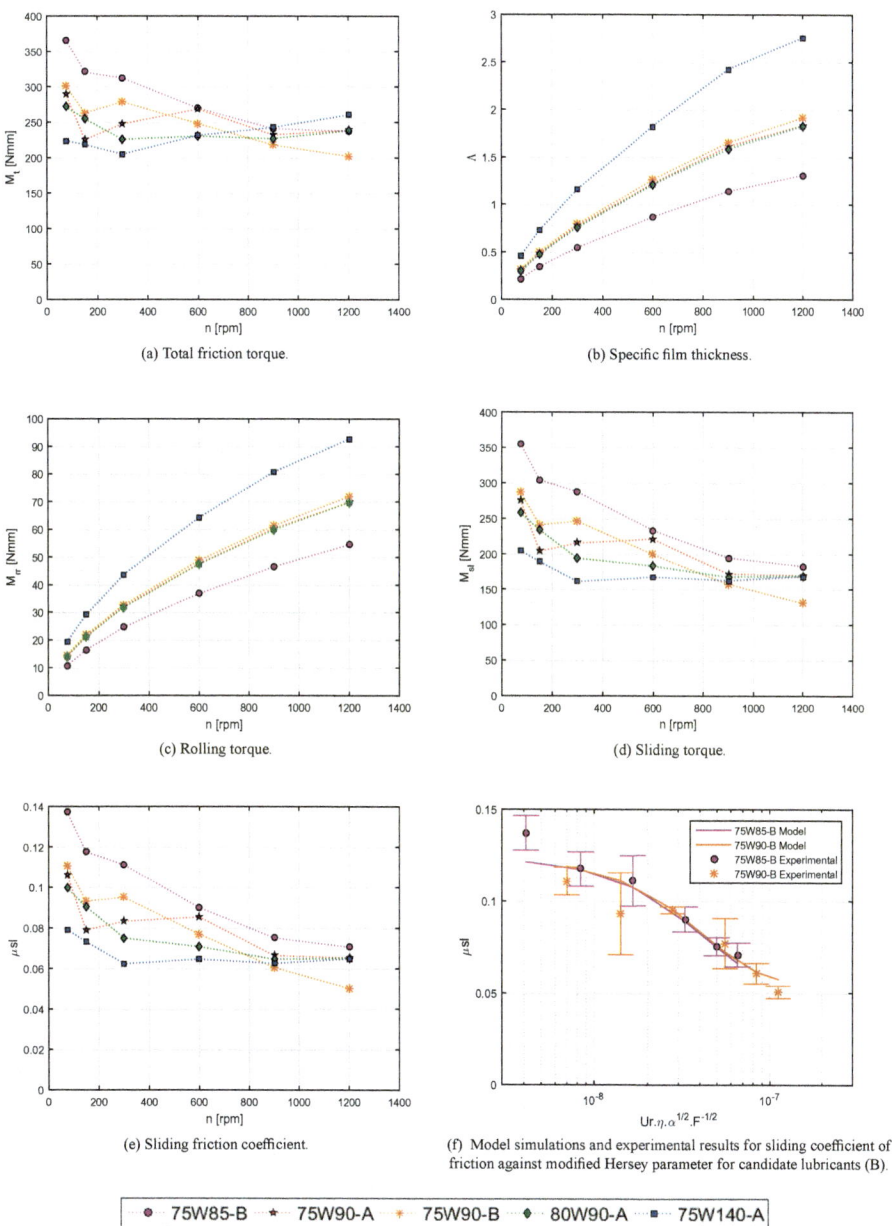

(a) Total friction torque.

(b) Specific film thickness.

(c) Rolling torque.

(d) Sliding torque.

(e) Sliding friction coefficient.

(f) Model simulations and experimental results for sliding coefficient of friction against modified Hersey parameter for candidate lubricants (B).

75W85-B   75W90-A   75W90-B   80W90-A   75W140-A

Fig. 2. Results of TBB 51107 lubricated with axle gear oils at constant temperature of 70 °C under an axial load of 7000 N.

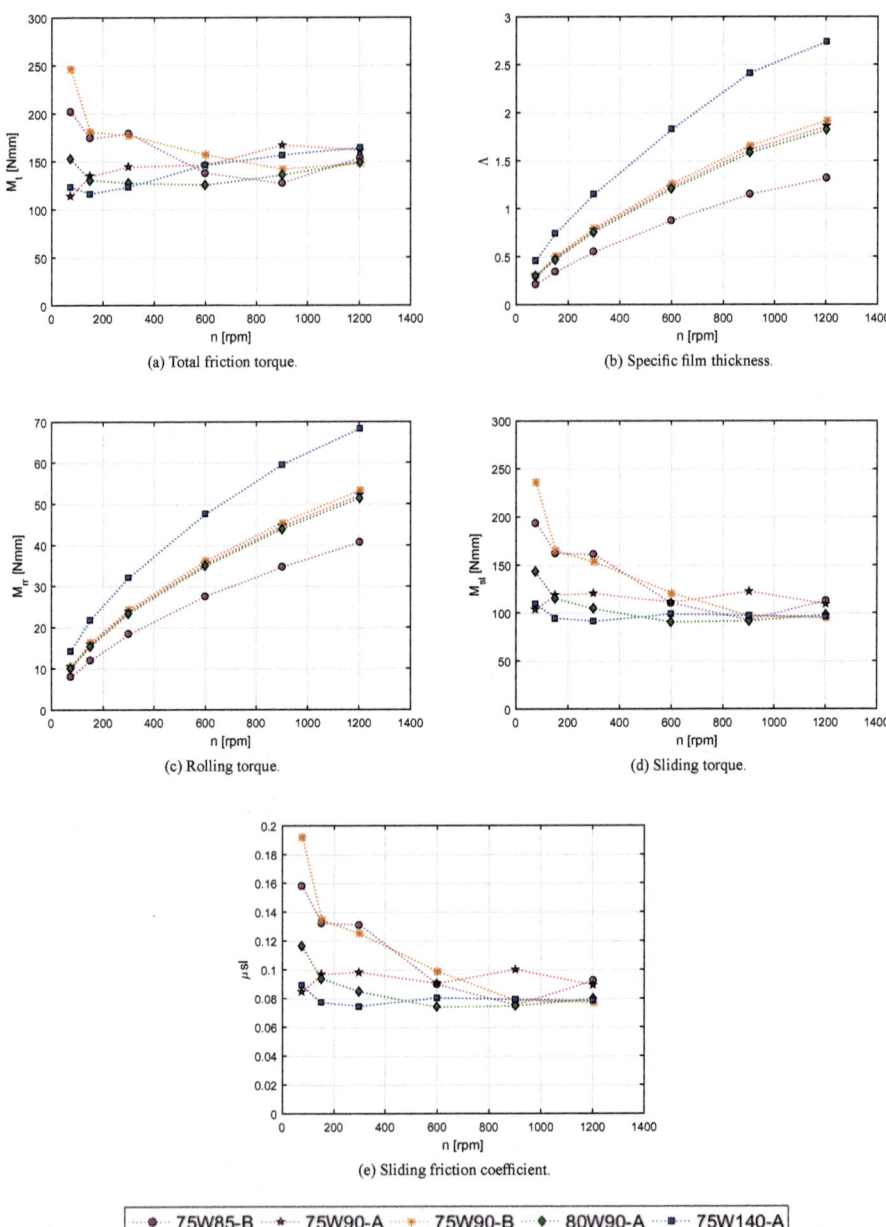

Fig. 3. Results of TBB 51107 lubricated with axle gear oils at constant temperature of 70 °C under an axial load of 4000 N.

whatever the axle gear oil formulation and rolling bearing type. It is also clear that there is a better approximation at high speeds and high viscosities (larger values of Sp). At low speed, under boundary film lubrication conditions, the scatter of the μ_{sl} values is larger.

5.2   Axial Load 4000 N

Figure 3(b) shows that when the operating speed increases from 75 rpm to 1200 rpm the specific lubricant film thickness inside the TBB increases from 0.220.46 to 1.322.74, depending on the oil tested, meaning that the lubrication regime evolved from boundary to mixed film lubrication. All axle gear oils exhibited a similar trend, but the 75W140-A oil produced the highest Λ, because of its high viscosity at 70 °C, and the 75W85-B oil generated the lowest Λ, because of its low viscosity and piezo-viscosity.

Figure 3(c) shows the rolling torque estimated for the TBB in all operating conditions. Since, the tests were performed at constant temperature (70 °C), when the speed increases the rolling torque also increases. It is observed that the oils 75W90-B, 75W90-A and 80W90-A generated very similar rolling torques, because they have similar viscosities at 70 °C. The 75W140-A oil produces very high rolling torques because it has the highest viscosity at 70 °C, while the 75W85-B oil, on the other hand, has the lowest viscosity at 70 °C.

The experimental friction torque (see Fig. 3(a)) as well as the sliding torque (see Fig. 3(d)) presented similar behaviour, but at different rates. The sliding torque (see Fig. 3(d)), shows that the sliding torque decreases with the speed for all operating conditions except for the 75W90-A oil show almost constant sliding torque.

Table 5. Values of the coefficients μ_{bl} and μ_{EHL} for TBB 51107, under 4 kN and 7 kN and at 70 °C.

Valid for $3262.5 < n \cdot d_m < 52200$			
		TBB	
Lubricant	Parameter	4000 N	7000 N
75W85-B and 75W90-B	μ_{bl}	0.149	0.124
	μ_{EHL}	0.076	0.056
75W90-A and 80W90-A	μ_{bl}	0.095	0.097
	μ_{EHL}	0.086	0.064
75W140-A	μ_{bl}	0.076	0.078
	μ_{EHL}	0.079	0.063

5.3   7000 N Vs 4000 N

The rolling torque is influenced by the load that means that when the load increased from 4000 N to 7000 N the rolling torque increases about 1.2 times for each speed. The load increment results in higher total friction torque within the rolling bearing that increases with the speed.

At lower load (4000 N) the amount of the sliding torque in the total friction torque is very important for all speed range and higher than for 7000 N. This behaviour is observed since the sliding coefficient of friction is much higher for low load than for higher.

A good approximation of the sliding coefficient of friction μ_{sl} was obtained through the reference values of μ_{EHL} and μ_{bl} for each axle gear oil formulation and thrust ball bearing type as shown in Table 5.

6   Conclusions

The results achieved with the torque loss tests for TBB 51107, showed that:

- The total friction torque M_t decreases when the speed increases, for the two different axial loads at low rotational speed below 600 rpm, although its values are much higher for an higher axial load. The candidate lubricants (B) generally generated higher total friction torque than reference oils (A) which is related to their different additive package in their formulations.
- For the friction torque components, the rolling torque increased with the increase of speed and the sliding torque decreased with the increase of speed, no matter what the axial load applied.
- The calculation of the sliding coefficient of friction covering all lubrication regimes can be considered as representative of the axle gear oils tested. It is clear that the candidate (B) formulations have higher values of the coefficient of boundary friction than the standard (A) formulations ($\mu_{bl}^B > \mu_{bl}^A$), thus generating higher torque loss values for the candidate oils (B) at low speed. It is also clear that the candidate (B) formulations have lower values of the full-film coefficient of friction than the standard (A) formulations ($\mu_{EHL}^B < \mu_{EHL}^A$), thus generating lower torque loss values at high speed, mainly because they have lower kinematic viscosities.
- The influence of axial is clear in term of sliding coefficient of friction. When the axial load increases from 4000 N to 7000 N, both boundary and full film coefficient of friction generally decrease for all axle gear oils tested.

Notation and Units

a	radius of the contact circle [m]	p_0	maximum Hertz pressure [Pa]
a_A	D341 viscosity parameter [-]	$R1$	geometry constant for rolling bearing friction torque [-]
A_c	area of contact [mm^2]	R_X	equivalent radius of curvature in x direction [m]

C basic dynamic load rating capacity [N]

C_0 basic static load rating [N]

d bearing bore diameter [mm]

D bearing outside diameter [mm]

d_m rolling bearing mean diameter [mm]

F_a axial load [N]

F_n normal load [N]

G_{rr} factor depending on the bearing type, bearing mean diameter and applied load [N mm]

G_{sl} factor depending on the bearing type, bearing mean diameter and applied load [N mm]

H width of the bearing [mm]

K_{rs} starvation constant for oil bath lubrication [-]

K_z bearing type related geometry constant [-]

m_A D341 viscosity parameter [-]

M_{drag} friction torque of drag losses [Nmm]

M'_{rr} rolling friction torque [Nmm]

M_{seal} friction torque of seals [Nmm]

M_{sl} sliding friction torque [Nmm]

M_t internal bearing friction torque [Nmm]

M_t^{exp} total bearing friction torque measured experimentally [Nmm]

n rotational speed [rpm]

n_A D341 viscosity parameter [-]

R_{Xi} curvature radius of element i in x direction [m]

R_Y equivalent radius of curvature in y direction [m]

R_{Yi} curvature radius of element i in y direction [m]

$S1$ geometry constant for sliding friction torque [-]

s pressure-viscosity parameter [-]

Sp modified Stribeck parameter [-]

t pressure-viscosity parameter [-]

Ur rolling speed [m/s]

VI Viscosity Index [-]

α pressure viscosity coefficient [Pa^{-1}]

α_t thermal expansion coefficient [-]

β thermoviscosity coefficient [K^{-1}]

η dynamic viscosity [Pa s]

ϕ_{bl} sliding friction torque weighting factor [-]

ϕ_{ish} inlet hear heating reduction factor [-]

ϕ_{rs} kinematic replenishement/starvation reduction factor [-]

μ_{bl} coefficient of friction in boundary film lubrication [-]

μ_{EHL} coefficient of friction in full film lubrication [-]

μ_{sl} sliding coefficient of friction [-]

ν kinematic viscosity [cSt]

δ penetration [μm]

Acknowledgements. Authors gratefully acknowledge the funding of Project NORTE-01–0145-FEDER-000022 - SciTech - Science and Technology for Competitive and Sustainable Industries, cofinanced by Programa Operacional Regional do Norte (NORTE2020), through Fundo Europeu de Desenvolvimento Regional (FEDER).

References

Cousseau, T., Graca, B., Campos, A., Seabra, J.: Experimental measuring procedure for the friction torque in rolling bearings. Lubr. Sci. **22**(4), 133–147 (2010)

Hammami, M., Martins, R., Abbes, M.S., Haddar, M., Seabra, J.: Axle gear oils: tribological characterization under full film lubrication. Tribol. Int. **106**, 109–122 (2017a)

Hammami, M., Rodrigues, N., Fernandes, C., Martins, R., Seabra, J., Abbes, M.S., Haddar, M.: Axle gear oils: friction, wear and tribofilm generation under boundary lubrication regime. Tribol. Int. **114**, 88–108 (2017b). https://doi.org/10.1016/j.triboint.2017.04.018

Hamrock, B.J., Dowson, D.: Ball Bearing Lubrication, p. 386. Wiley, Hoboken (1981)

Hosokawa, R.S., de Oliveira, R.A.A., Franco, D.M., de Aguiar Vendrasco, A.: Double row regular contact ball bearing for axle drives. SAE Technical Paper (2009)

Matsuyama, H., Dodoro, H., Ogino, K., Ohshima, H., Toda, K.: Development of super-low friction torque tapered roller bearing for improved fuel efficiency. SAE Technical Paper (2004)

SKF.: SKF Rolling Bearings General Catalogue (2013)

Spindler, D., Von Petery, G.: Angular contact ball bearings for a rear axle differential. SAE Technical Paper (2003)

Xu, H., Singh, A., Kahraman, A., Hurley, J., Shon, S.: Effects of bearing preload, oil volume, and operating temperature on axle power losses. J. Mech. Des. **134**(5), 054501 (2012)

Dynamic Behavior and Stability of a Flexible Rotor

Abdelouahab Rezaiguia[1(✉)], Oussama Zerti[1(✉)],
Salah Guenfoud[1(✉)], and Debra F. Laefer[2(✉)]

[1] Applied Mechanics of New Materials Laboratory and Department
of Mechanical Engineering, University of 8 Mai 1945 Guelma, Guelma, Algeria
a.rezaiguia@cmail.com,
rezaiguia.abdelouahab@univ-guelma.dz
[2] Center for Urban Science and Progress and Department of Civil Engineering,
Tandon School of Engineering, New York University, New York, USA

Abstract. Among the problems encountered in rotordynamics, the phenomena of instability generally due to the hydrodynamic bearings and interns damping, the gyroscopic effect due to the discs and shafts, the excitations due to the unbalance as well as the nonlinear phenomena related on the bearings which support the rotor and the elements carried by the rotor.

The objective of this paper is to investigate the effects of damping and rigidity of hydrodynamic bearing on the stability of a Lalanne-Ferraris rotor. This study was conducted with respect to the stability criterion seen in a natural frequencies' equation. The process begins with the establishment of the characteristics of rotor elements. This is to assess the expressions of the kinetic and deformation energies, as well as the corresponding virtual work for rotor components: disk, shaft, unbalance and hydrodynamic bearing. The Rayleigh-Ritz method and Lagrange's equations were used to determine the equations of motion. A computational program in FORTRAN is elaborated to solve the characteristic equation in free vibration. The roots are pairs of complex conjugate quantities. In forced vibration, the resolution of the linear algebraic system is conducted by the Gauss-Jordan direct method. Also a computational program in FORTRAN is elaborated. A numerical example of Lalanne and Ferraris rotor is computed. According to the results presented herein, the presence of damping for certain values of rigidities was found to be a possible source for instability of the rotor at defined speeds.

Keywords: Flexible rotor · Stability · Damping · Rigidity
Hydrodynamic bearing

1 Introduction

The study of the stability of rotating machines particularly concerns rotors supported by hydrodynamic bearings. These bearings are influenced by internal phenomena. When the rotational speed increases, the vibration amplitude often goes through a maximum value at a specific speed called the critical speed. This amplitude is usually excited by an imbalance. In addition, rotating machines often develop instabilities that are related to the internal composition of the rotors. There are several criteria for rotor stability

© Springer Nature Switzerland AG 2019
T. Fakhfakh et al. (Eds.): ICAV 2018, ACM 13, pp. 37–50, 2019.
https://doi.org/10.1007/978-3-319-94616-0_4

analysis such as the Nyquist, Routh-Hurwitz and Liapunov stability criterion. Liapunov criterion is a general theory valid for any system described by linear or nonlinear differential equations. The definition of stability introduced by Liapunov is based on a state space representation of the motion of the system. The term unstable can have several meanings, and different definitions of stability exist (Bigret 1980; Muszynska 2005). One of the most common was introduced by Liapunov (Genta 2005). A clear definition is given by Lalanne and Ferraris 1998. They said that in free vibration and under initial conditions, the rotor is unstable when the amplitude of its motion grows indefinitely with time. The research presented herein will adopt the above definition by Lalanne and Ferraris (1998).

The rotating damping of rotors is one of the destabilizing factors of rotating machines. It comes from the internal damping of the shafts, the interfaces between the discs and the shafts, and the coupling joints made of rubber elements (Bigret 1997; Erich 1992). These phenomena give rise to tangential forces called circulatory forces in the direction of precession and in the opposite direction to the external damping forces. When these forces, proportional to the displacement and the rotating speed, become greater than the external damping forces, instability develops.

As the rotating speed can increase above the first critical speed, self-excited vibrations has become a serious problem. In certain circumstances and on very particular operations, it has been observed on the dynamic behavior of rotating machines the phenomenon of "oïl whip". Newkirk (1924) and Kimball (1924), published two papers entitled "Shaft Whipping" and "Internal friction theory of shaft whirling", respectively in the review "General Electric". These two papers showed that the internal friction of materials could cause unstable movements. These phenomena (there named "oil whip" and "oil whirl"), in which vibration-damping friction usually causes a self-excited vibration has attracted the attention of many researchers. Newkirk and Taylor 1925 published "Shaft whipping to oil action in journal bearings". They studied an unstable vibration called "oil whip", due to a film of oil in the bearings. Hummel 1926 and Newkirk 1930 both confirmed through experimental studies the destabilizing effect caused by the oil film bearings. They also observed that below a certain eccentricity of operation; that is to say when the bearing is weakly loaded; the tree is animated by a precession called "whipping" with a frequency close to half the rotating speed (called "half frequency whirl"). While for larger eccentricities (i.e. large loads), the system becomes stable again. Muszynska 2005, experimentally demonstrated the two phenomena "oil whip" and "oil whirl" on a rotor on rigid supports including a cylindrical bearing lubricated with oil.

In this work, we study the stability of a Lalanne-Ferraris rotor model by the stability criterion seen by the natural frequencies equation. After formulating the general expressions of the characteristics of the various elements of the rotor studied, the development of the equations of the motion is necessary to predict their dynamic behavior and their stability. The Rayleigh-Ritz method is used as a method of resolution to provide a model for highlighting and treating basic phenomena. The application of Lagrange's equations leads to obtaining equations of motion.

2   Characteristics of Rotor Elements

The physical model studied herein is that of Lalanne and Ferraris 1998. It consists of a flexible shaft of length, l, with a constant circular cross-section, S, a rigid balanced disk of center, C, and mass, M_D, situated at $y = l_1$, a hydrodynamic bearing located at $y = l_2$ and two rigid bearings at the ends (Fig. 1). The rotor rotates at a constant speed Ω. One degree of freedom is used for each motion in the directions X_0 and Z_0 that are, respectively, $u(y,t)$ and $w(y,t)$.

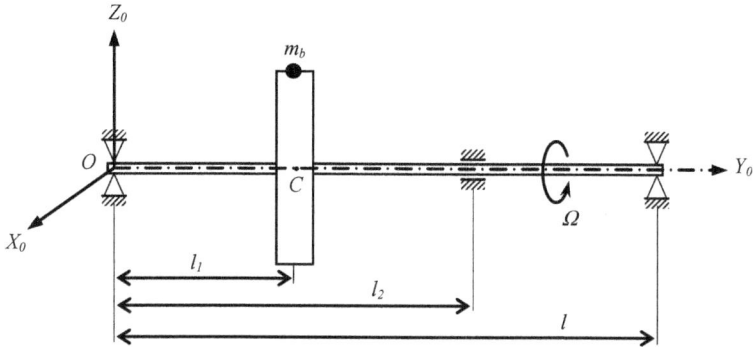

Fig. 1. Physical model of the rotor (after Lalanne-Ferraris 1998)

The different frames used to study the kinetic and dynamic behavior of the rotor are presented in Fig. 2. R_0 (O, X_0, Y_0, Z_0) is an inertial frame fixed to the fixed part of the rotating machine. The term $R(C, X, Y, Z)$ is the final frame fixed to the rotating disk, while the (C, Y) axis is perpendicular to the disk coinciding with the rotating disk axis at the deformed state of the shaft. The (x, y, z) coordinate system is related to the (x_0, y_0, z_0) coordinate system through a set of three angles, ψ, θ and ϕ. The two first angles are characterized the gyroscopic effect of both the disk and the shaft. To achieve the instantaneous orientation of the disk, one first rotates it by an amount ψ around the (C, Z_0) axis, then by an amount θ around the new axis (C, X_1), and finally by an amount ϕ around the free rotating axis of the rotor (C, Y) (Fig. 3).

The instantaneous rotation vector of the disk $\vec{\omega}\,(R/R_0)$ is as per Lalanne and Feraris (1998) shown in (1):

$$\vec{\omega}\,(R/R_0) = \begin{bmatrix} \omega_x \\ \omega_y \\ \omega_z \end{bmatrix}_R = \begin{bmatrix} -\dot{\psi}\cos\theta\sin\phi + \dot{\theta}\cos\phi \\ \dot{\phi} + \dot{\psi}\sin\theta \\ \dot{\psi}\cos\theta\cos\phi + \dot{\theta}\sin\phi \end{bmatrix}_R \qquad (1)$$

where (C, X), (C, Y), (C, Z) are the principal inertia directions. The inertia tensor of the disk on point C is as per (2):

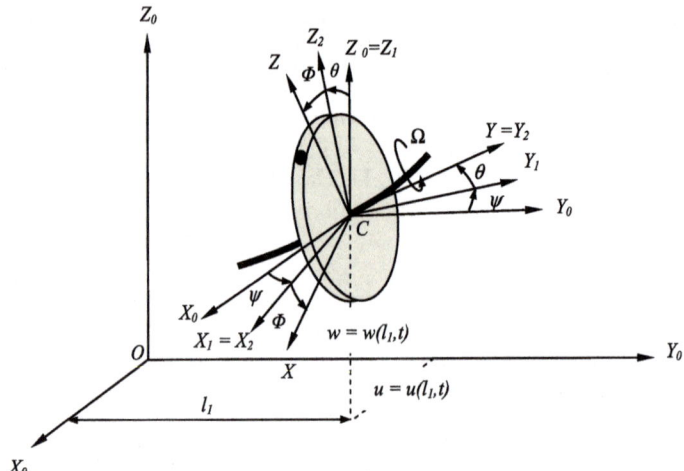

Fig. 2. Different frames for a disk on rotating flexible shaft

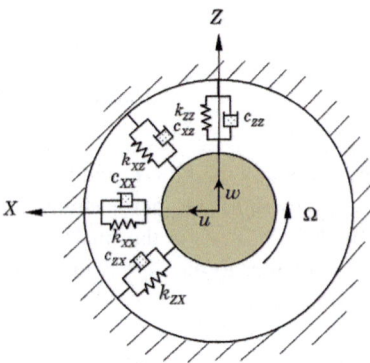

Fig. 3. Bearing stiffness and damping

$$[\mathrm{I}_C(D)] = \begin{bmatrix} I_{Dx} & 0 & 0 \\ 0 & I_{Dy} & 0 \\ 0 & 0 & I_{Dz} \end{bmatrix}_{(\vec{x},\vec{y},\vec{z})} \tag{2}$$

The kinetic energy of the disk is the sum of the translation and rotation energies of the center C of the disk as per Lalanne and Feraris (1998) as shown in (3):

$$T_D = \frac{1}{2} M_D \left(\vec{V}(C/R_0) \right)^2 + \frac{1}{2} \{\omega\}_R^T [\mathrm{I}_C(D)] \{\omega\}_R \tag{3}$$

Assuming that the disk is symmetric, the angles ψ and θ are small, and the angular velocity and displacement along Y_0 are constant, the kinetic energy expression (3) can be simplified as (4):

$$T_D = \frac{1}{2}M_D(\dot{u}^2 + \dot{w}^2) + \frac{1}{2}I_{Dx}(\dot{\theta}^2 + \dot{\psi}^2) + \frac{1}{2}I_{Dy}(\Omega^2 + 2\Omega\dot{\psi}\theta) \qquad (4)$$

where the last term, $I_{Dy}\Omega\dot{\psi}\theta$, represents the gyroscopic (Coriolis) effect.

The flexible shaft is characterized by the kinetic and strain energies. It is represented by Timoshenko's rotating beam. The general formulation of the kinetic energy of the shaft comes from an extension of the disk expression (4) (as per Lalanne and Feraris 1998). The energy of a beam section of an infinitesimal length dy is the same energy of a disc of the same dimensions. Integrating (4) along the shaft length generates (4):

$$T_a = \frac{\rho S}{2}\int_0^l (\dot{u}^2 + \dot{w}^2)dy + \frac{\rho I}{2}\int_0^l (\dot{\psi}^2 + \dot{\theta}^2)dy + \rho I l\Omega^2 + 2\rho I\Omega\int_0^l \dot{\psi}\,\theta\, dy \qquad (5)$$

where ρ is the mass per unit volume and I is the area moment of inertia of the cross-section beam.

The first term of (5) is the classical expression for the kinetic energy of a beam bending in two planes, the second term is the secondary effect of a rotary inertia (Timoshenko's rotating beam), the third term constant is the rotation energy of the shaft, and the last term represents the gyroscopic effect.

The strain energy of a symmetric rotating beam is shown as per Lalanne and Ferraris (1998) as (6):

$$U_a = \frac{EI}{2}\int_0^l \left[\left(\frac{\partial^2 u}{\partial y^2}\right)^2 + \left(\frac{\partial^2 w}{\partial y^2}\right)^2\right] dy \qquad (6)$$

The imbalance is defined by a mass m_b situated at a distance d from geometric center C of the disk, and its kinetic energy must be calculated. By supposing that the mass m_b remains in a plane perpendicular to the (O, Y_0) axis, its coordinates along this axis are constant, and the mass m_b is much smaller than the rotor mass, then the expression for the kinetic energy of imbalance can be written as (7)

$$T_b = m_b\,\Omega\, d\,(\dot{u}\cos\Omega t - \dot{w}\sin\Omega t) \qquad (7)$$

The stiffness and viscous damping of the hydrodynamic bearing are assumed to be known. The virtual work of the forces acting on the shaft can be written as (8):

$$\begin{aligned}\delta W &= -(k_{XX}u + k_{XZ}w + c_{XX}\dot{u} + c_{XZ}\dot{w})\delta u - (k_{ZZ}w + k_{ZX}u + c_{ZZ}\dot{w} + c_{ZX}\dot{u})\delta w \\ &= F_u\delta u + F_w\delta w\end{aligned} \qquad (8)$$

where F_u and F_w are components of the generalized force along the X_0 and Z_0 directions, respectively.

3 Numerical Application

3.1 Numerical Data

The disk is of steel with a density of $\rho = 7800$ kg/m^3, a Young's modulus $E = 2 \times 10^{11}$ N/m^2, an inner radius $R_1 = 0.01$ m, an outer radius $R_2 = 0.15$ m, and a thickness $h = 0.03$ m situated at $l_1 = l/3$. The shaft is also of steel of length $l = 0.4$ m and a cross-sectional radius $R_1 = 0.01$ m. The imbalanced mass is $m_b = 10^{-4}$ kg, situated at a distance $d = 0.15$ m from geometric center, C, of the disk. The hydrodynamic bearing is situated at $l_2 = 2\,l/3$, characterized by: $k_{XX} = 2 \times 10^5$ N/m, $k_{ZZ} = 5 \times 10^5$ N/m, $c_{XX} = \beta\,k_{XX}$, $c_{ZZ} = \beta\,k_{ZZ}$, $k_{XZ} = k_{ZX} = c_{XZ} = c_{ZX} = 0$, for which β is a damping factor (as per Lalanne and Feraris 1998).

3.2 Equations of Motion

For the equations of motion, the numerical Rayleigh-Ritz method is used. According to the bending beam theory, the first mode shape is the dominant response. So the displacements u and w projected in a modal basis can be uncoupled as per (10)

$$u(y,t) = \phi(y)\,q_1(t); \quad w(y,t) = \phi(y)\,q_2(t) \tag{10}$$

where $\phi(y) = sin(\pi y/l)$, is the exact first bending mode shape of the shaft simply supported at both ends, and q_1 and q_2 are generalized coordinates.

The angles ψ and θ are small and as such are approximated by (11)

$$\theta(y,t) = \frac{\partial w(y,t)}{\partial y}; \quad \psi(y,t) = -\frac{\partial u(y,t)}{\partial y} \tag{11}$$

Taking into account all numerical data presented in Sect. 3.1 and all previous expressions, after integration the total kinetic energy of the system and strain energy of the shaft becomes (12, 13)

$$T = 7.14(\dot{q}_1^2 + \dot{q}_2^2) - 2.87\Omega\dot{q}_1 q_2 + 1.3 \times 10^{-5}\Omega(\dot{q}_1 \cos \Omega t - \dot{q}_2 \sin \Omega t) \tag{12}$$

$$U_a = 5.97 \times 10^5 (q_1^2 + q_2^2) \tag{13}$$

and final generalized forces (14)

$$\begin{aligned} F_{q_1} &= -0.15 \times 10^6 q_1 - \beta \times 1.5 \times 10^5 \dot{q}_1; \\ F_{q_2} &= -0.37 \times 10^6 q_2 - \beta \times 3.75 \times 10^5 \dot{q}_2 \end{aligned} \tag{14}$$

The application of Lagrange's Eq. (15)

$$\frac{d}{dt}\left(\frac{\partial T}{\partial \dot{q}_i}\right) - \frac{\partial T}{\partial q_i} + \frac{\partial U_a}{\partial q_i} = F_{q_i}; \; i = 1, 2 \tag{15}$$

gives the equations of motion of geometric center, C, of the disk (16.1 and 16.2):

$$m\ddot{q}_1 - a\Omega \dot{q}_2 + c_1 \dot{q}_1 + k_1 q_1 = m_b d\phi(l_1)\,\Omega^2 \sin \Omega t \tag{16.1}$$

$$m\ddot{q}_2 + a\Omega \dot{q}_1 + c_2 \dot{q}_2 + k_2 q_2 = m_b \phi(l_1)\,d\Omega^2 \cos \Omega t \tag{16.2}$$

with $m = 14.29$; $a = 2.87$; $c_1 = \beta \times 1.5 \times 10^5$; $c_2 = \beta \times 3.75 \times 10^5$;
$k_1 = 1.34 \times 10^6$; $k_2 = 1.57 \times 10^{6}$; and $m_b\, d\, \phi\,(l_1) = 1.3 \times 10^{-5}$.

3.3   Numerical Results

3.3.1   Campbell Diagram
To draw the Campbell diagram, which represents the evolution of natural frequencies as a function of the rotating speed, the rotor is studied in free motion. Only solutions of (16.1 and 16.2) without second member are considered:

$$m\ddot{q}_1 - a\Omega \dot{q}_2 + c_1 \dot{q}_1 + k_1 q_1 = 0 \tag{17.1}$$

$$m\ddot{q}_2 + a\Omega \dot{q}_1 + c_2 \dot{q}_2 + k_2 q_2 = 0 \tag{17.2}$$

The solutions of (17.1 and 17.2) have to be sought in the following form:

$$q_1(t) = Q_1 e^{rt}; \quad q_2(t) = Q_2 e^{rt} \tag{18}$$

Substituting (18) into (17.1 and 17.2) gives (19)

$$\begin{bmatrix} mr^2 + c_1 r + k_1 & -a\Omega r \\ a\Omega r & mr^2 + c_2 r + k_2 \end{bmatrix} \begin{Bmatrix} Q_1 \\ Q_2 \end{Bmatrix} = \begin{Bmatrix} 0 \\ 0 \end{Bmatrix} \tag{19}$$

For non-trivial solutions of (19), the value of the determinant of the matrix must be zero from which the forth order characteristic equation can be generated (20):

$$m^2 r^4 + m(c_1 + c_2)r^3 + \left(k_1 m + k_2 m + c_1 c_2 + a^2\Omega^2\right)r^2 + (k_2 c_1 + k_1 c_2)r + k_1 k_2 = 0 \tag{20}$$

In general, the values of the damping coefficients c_1 and c_2 are such that the roots of (20) are pairs of complex conjugate quantities, which can be written as (21):

$$r_i(\Omega) = \sigma_i(\Omega) + j\omega_i(\Omega) \tag{21}$$

where $r_i(\Omega)$ is the ith complex frequency, $\omega_i(\Omega)$ is the ith natural frequency, and $\sigma_i(\Omega)$ is the ith decay rate (the rate at which the amplitude decreases over time) that changes sign: a negative value of σ_i characterizes a movement that decreases over time (stable movement), while a positive value characterizes an exponential growth of motion over time (unstable movement).

3.3.2   Unbalanced Response

In the case of the presence of an excitation force due to an imbalance, the study of the particular solution is done by considering the system (16.1 and 16.2) with a second member. As the system is damped, the response is generally not synchronous with the strength of the imbalance. Solutions are sought in the form of (22):

$$
\begin{aligned}
q_1(t) &= A_1 \sin \Omega t + B_1 \cos \Omega t \\
q_2(t) &= A_2 \sin \Omega t + B_2 \cos \Omega t
\end{aligned}
\tag{22}
$$

These expressions (22) are reported in the equations of motion (16.1 and 16.2), and each equation leads to two equations representing the equality of the $\sin \Omega t$ and $\cos \Omega t$. This gives a set of linear algebraic equations that can be written in the following matrix form (23):

$$
\begin{bmatrix}
k_1 - m\Omega^2 & -c_1\Omega & 0 & a\Omega^2 \\
c_1\Omega & k_1 - m\Omega^2 & -a\Omega^2 & 0 \\
0 & -a\Omega^2 & k_2 - m\Omega^2 & -c_2\Omega \\
a\Omega^2 & 0 & c_2\Omega & k_2 - m\Omega^2
\end{bmatrix}
\begin{Bmatrix}
A_1 \\ B_1 \\ A_2 \\ B_2
\end{Bmatrix}
=
\begin{Bmatrix}
m^* d\Omega^2 \\ 0 \\ 0 \\ m^* d\Omega^2
\end{Bmatrix}
\tag{23}
$$

For a value of Ω, the resolution of (23) gives $Q_1(\Omega)$ and $Q_2(\Omega)$ from:

$$
Q_1(\Omega) = \sqrt{A_1^2 + B_1^2}; \quad Q_2(\Omega) = \sqrt{A_2^2 + B_2^2}
\tag{24}
$$

The resolution of (20) is done using a program in FORTRAN. The resolution of the linear algebraic system (23) is conducted by the Gauss-Jordan direct method. To see the influence of damping, Figs. 4 and 5 show the evolution of the natural frequencies of the rotor as a function of the rotating speed $N = 30\,\Omega/\pi$, as well as the magnitude of point C of the unbalanced response for the four damping factor values assigned to β (Lalanne and Feraris 1998). In this figure, when $\beta = 0.0002$, two peaks corresponding to the critical speeds 2759 rpm and 3431 rpm are apparent. These are the same critical speeds obtained by the Campbell diagram (points A and B). For other answers ($\beta = 0.015$, $\beta = 0.02$, $\beta = 0.026$), the amplitudes of the peaks corresponding to the critical speeds reduced with the increased value of the damping factor β until they disappear altogether. For the very important damping factor of $\beta = 0.026$, Campbell's diagram is very different to the others. Specifically, at rest only one frequency appears, and the other natural frequency does not appear before the speed of rotation reaches the value of $N = 3867$ rpm.

Figure 6 illustrates the variation of the direction of precession of the geometric center of the disk as a function of the rotating speed. Precession can be defined as a change of direction of the rotating axis. There are two types of precession: direct and inverse. For direct precession, the geometrical rotor center rotates in the same direction of rotation and vice versa for the inverse precession. The figures shows how the

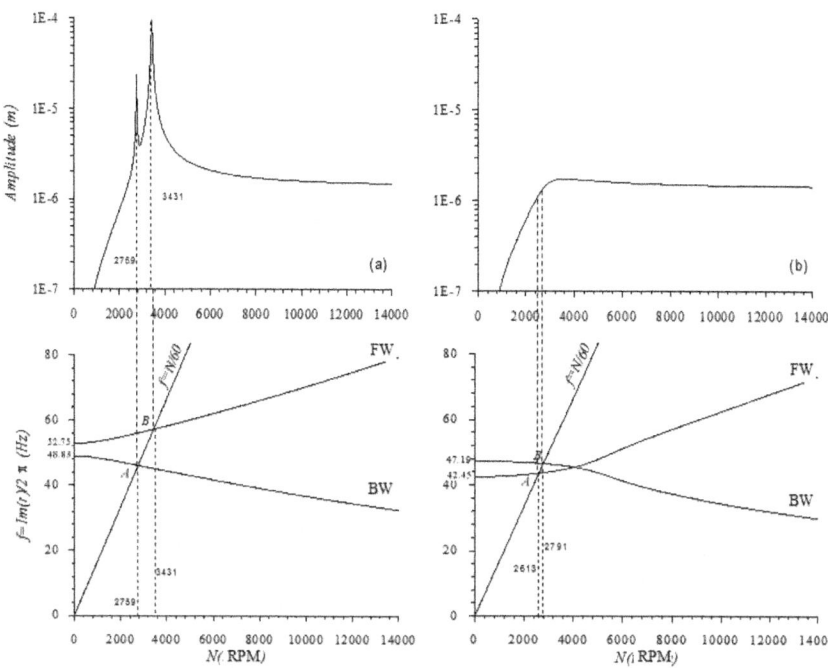

Fig. 4. Unbalanced response and Campbell's diagram, (a) $\beta = 0.0002$, (b) $\beta = 0.015$

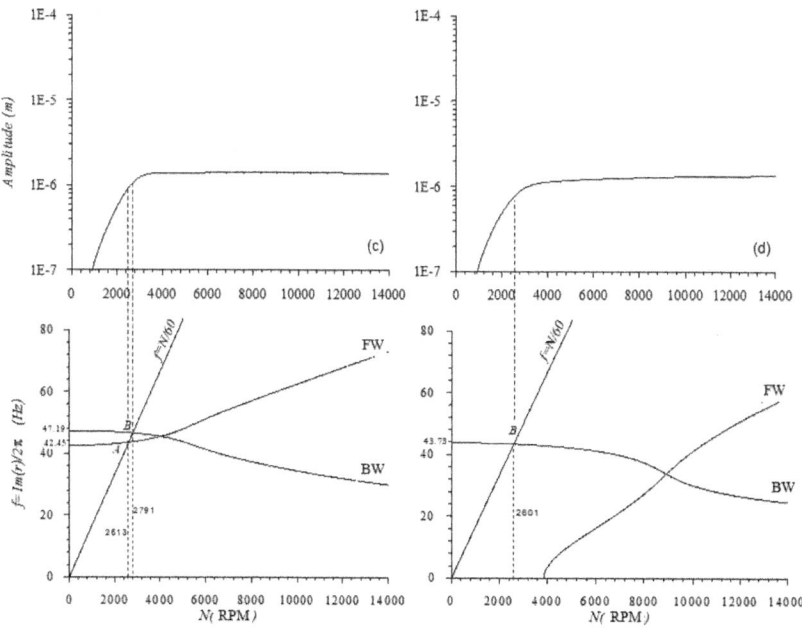

Fig. 5. Unbalanced response and Campbell's diagram, (c) $\beta = 0.02$, (d) $\beta = 0.026$

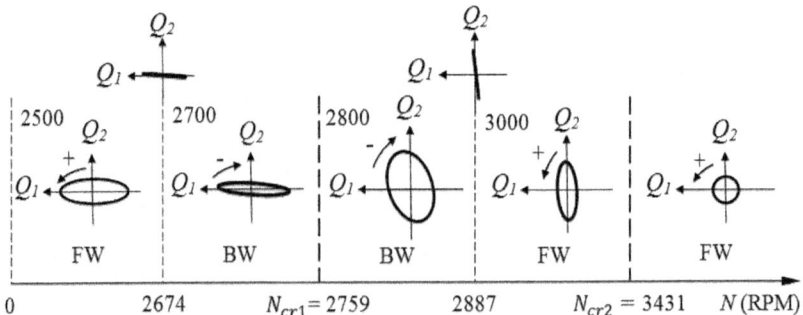

Fig. 6. Variation of the direction of the precession as a function of the speed of rotation

changes in precession direction do not occur at critical speeds. When the rotor rotates at very high speeds, the trajectory of the center of the disk becomes circular.

Figure 7 shows the variation of the decay rate, σ, according to the rotating speed for different values of the damping factor. A distinction is made between the variation of the decay rate and the increase in the damping factor and that all the values of the decay rate for all values of β are negative. These negative values of σ characterize a movement that decreases over time. So, the rotor is stable.

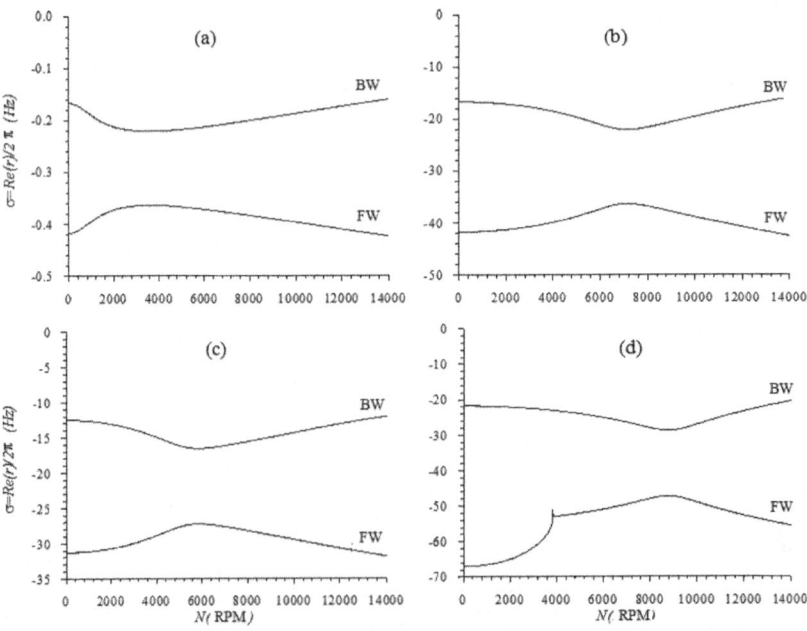

Fig. 7. Decay rate, (a): $\beta = 0.0002$, (b): $\beta = 0.015$, (c): $\beta = 0.02$, (d): $\beta = 0.026$

3.3.3 Influence of Stiffness on Rotor Stability

If the stiffness and damping characteristics are as follows (Lalanne and Feraris 1998): $k_{XX} = 2 \times 10^5$ N/m; $k_{ZZ} = 5 \times 10^5$ N/m; $k_{XZ} = -k_{ZX} = 4 \times 10^4$ Ns/m; $c_{XX} = c_{ZZ} = 10^2$ Ns/m; $c_{XZ} = c_{ZX} = 0$. Then the equations of free motion (17.1 and 17.2) become as per (25):

$$
\begin{aligned}
m\ddot{q}_1 - a\Omega\dot{q}_2 + c\dot{q}_1 + k_1 q_1 + k_{12} q_2 &= 0 \\
m\ddot{q}_2 + a\Omega\dot{q}_1 + c\dot{q}_2 + k_2 q_2 + k_{21} q_1 &= 0
\end{aligned}
\tag{25}
$$

with

$$
\begin{aligned}
k_1 &= k + k_{XX}\phi^2(l_2) = 1.34 \times 10^6 \text{ N/m} \\
k_2 &= k + k_{ZZ}\phi^2(l_2) = 1.57 \times 10^6 \text{ N/m} \\
k_{12} &= k_{XZ}\phi^2(l_2) = 3 \times 10^4 \text{ N/m} \\
k_{21} &= k_{ZX}\phi^2(l_2) = -3 \times 10^4 \text{ N/m} \\
c &= c_{XX}\phi^2(l_2) = c_{ZZ}\phi^2(l_2) = 75 \text{ Ns/m}
\end{aligned}
\tag{26}
$$

The solutions of free motion Eq. (25) have the same form as (22); substituting (22) into (25) results in (27):

$$
\begin{bmatrix} mr^2 + cr + k_1 & -a\Omega r + k_{12} \\ a\Omega r + k_{21} & mr^2 + cr + k_2 \end{bmatrix} \begin{Bmatrix} Q_1 \\ Q_2 \end{Bmatrix} = \begin{Bmatrix} 0 \\ 0 \end{Bmatrix}
\tag{27}
$$

Hence the characteristic polynomial of (28):

$$
r^4 + \frac{2c}{m}r^3 + \left(\frac{k_1}{m} + \frac{k_2}{m} + \frac{c^2}{m^2} + \frac{a^2\Omega^2}{m^2} \right) r^2 + \left(\frac{ck_1}{m^2} + \frac{ck_2}{m^2} + \frac{a}{m^2}(k_{21} - k_{12})\Omega \right) r + \frac{k_1 k_2 - k_{12} k_{21}}{m^2} = 0
\tag{28}
$$

To see the influence of bearing rigidities on the stability of the rotor, Fig. 8 shows the Campbell diagram and the decay rate as a function of the rotation speed. Note that this system becomes unstable for $N > 1395$ rpm. This speed (said rate of instability) corresponds to the change of sign of the decay rate from a negative value to a positive value. The rotor is unstable after this speed.

3.3.4 Response Due to Harmonic Force Fixed in Space

In rotation, the rotor can be excited by a harmonic force fixed in space with excitation frequency ω. This force is assumed to act only on the rotor along the direction X_0 at $y = l_3 = 2\, l/3$. These components are as per (29):

$$
F_{q_1} = F_0\phi(l_3)\sin\omega t = F\sin\omega t; \quad F_{q_2} = 0
\tag{29}
$$

So, the equations to solve are shown in (30):

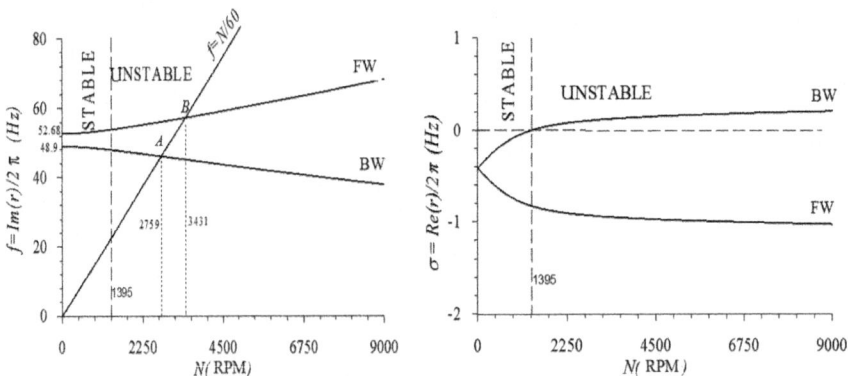

Fig. 8. Influence of hydrodynamic bearing rigidities on rotor stability

$$m\ddot{q}_1 - a\,\Omega\,\dot{q}_2 + c_1\,\dot{q}_1 + k_1 q_1 = F \sin \omega t$$
$$m\ddot{q}_2 + a\,\Omega\,\dot{q}_1 + c_2\,\dot{q}_2 + k_2 q_2 = 0 \tag{30}$$

The solutions of (30) can be sought in the form of (31):

$$q_1(t) = A_1 \sin \omega t + B_1 \cos \omega t$$
$$q_2(t) = A_2 \sin \omega t + B_2 \cos \omega t \tag{31}$$

By substituting (31) into (30), each equation leads to two equations representing the equality of the factors of $\sin \omega t$ and $\cos \omega t$. This gives a set of linear algebraic equations that can be written in the matrix form (32):

$$\begin{bmatrix} k_1 - m\omega^2 & -c_1\omega & 0 & a\omega\Omega \\ c_1\omega & k_1 - m\omega^2 & -a\omega\Omega & 0 \\ 0 & -a\omega\Omega & k_2 - m\omega^2 & -c_2\omega \\ a\omega\Omega & 0 & c_2\omega & k_2 - m\omega^2 \end{bmatrix} \begin{Bmatrix} A_1 \\ B_1 \\ A_2 \\ B_2 \end{Bmatrix} = \begin{Bmatrix} F \\ 0 \\ 0 \\ 0 \end{Bmatrix} \tag{32}$$

For every value of ω, the resolution of (32) gives $A_1(\omega,\Omega)$, $B_1(\omega,\Omega)$, $A_2(\omega,\Omega)$, $B_2(\omega,\Omega)$ and so $Q_1(\omega,\Omega)$ and $Q_2(\omega,\Omega)$ from (33):

$$Q_1(\omega, \Omega) = \sqrt{A_1^2 + B_1^2}; \quad Q_2(\omega, \Omega) = \sqrt{A_2^2 + B_2^2} \tag{33}$$

Figure 9 shows a cascade representation of the amplitude of vibration due to a fixed harmonic force in the amplitude space $F_0 = 100$ N for a damping factor $\beta = 0.0002$. Notably, this representation shows the evolution of the vibration amplitude as a function of the excitation frequency and the rotating speed at the same time. For a given speed at each time, a natural frequency coincides with an excitation frequency; a peak is obtained at the points of intersection. At the start, two natural frequencies are distinctive, because of the dissymmetry of the rotor and are the same as that obtained by the Campbell diagram (Fig. 4a). Thus Campbell's diagram can be observed through this representation.

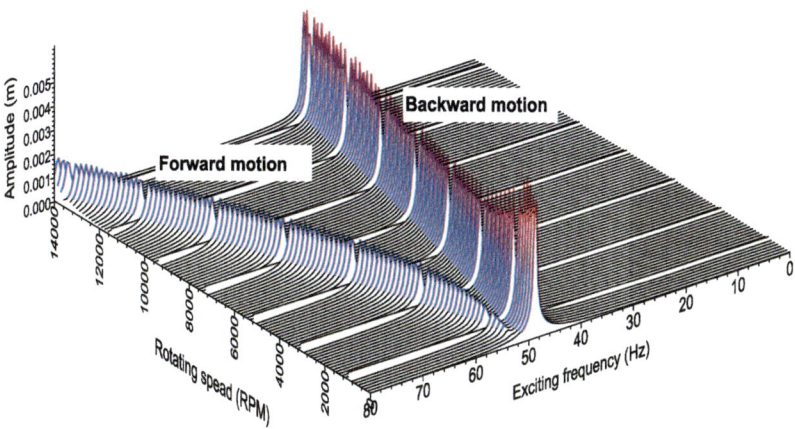

Fig. 9. Cascade representation of the amplitude of vibration due to a harmonic force fixed in space, $\beta = 0.0002$, $F_0 = 100$ N

4 Conclusions

The objective of this work was to study the stability of a simple mono-rotor model with a hydrodynamic bearing. The different expressions that characterize the rotor elements are determined. The equations of motion are obtained using the Rayleigh-Ritz method and Lagrange's equations. Taking into account the gyroscopic effect due to the disc or the shaft, as well as the characteristics of the hydrodynamic bearing, makes the natural frequencies vary according to the rotating speed. The free behavior of the rotor is summarized by Campbell's diagram. The latter is one of the basic tools for determining the critical speeds of the system. In this studied model, asymmetry is introduced by the stiffness and damping of the hydrodynamic bearing. In the presence of damping, the roots of the characteristic polynomial are pairs of complex conjugate quantities. Increasing the damping factor influences the evolution of the natural frequencies and the variation of the decay rate, as a function of the rotating speed. Stability analysis is one of the most powerful tools for studying and improving the dynamic behavior of rotating machinery. The criterion used is very reliable for the study of the stability and the determination of the rate at which instability manifests itself.

References

Bigret, R.: Vibrations des machines tournantes et des Structures. Technique et documentation (1980)

Bigret, R.: Stabilité des machines tournantes et des systèmes, Cetim (1997)

Erich, F.E.: Handbook of Rotordynamics. McGraw-Hill, New York (1992)

Genta, G.: Dynamic of Rotating Systems. Mechanical Engineering Series. Springer, Berlin (2005)

Hummel, B.L: Kritische Drehzahlen als folge der Nachgiebigkeit der Schmiermittels im lager, VDI-Forschift, p. 287 (1926)

Kimball, A.L.: Internal friction theory of shaft whirling. Gen. Electric Rev. **27**(4), 244–251 (1924)

Lalanne, M., Ferraris, G.: Rotordynamics Prediction in Engineering. Wiley, New York (1998)

Muszynska, A.: Rotordynamics, 2nd edn. Taylor & Francis Ltd., Boca Raton (2005)

Newkirk, B.L.: Shaft whipping. Gen. Electr. Rev **27**, 169–178 (1924)

Newkirk, B.L., Taylor, H.D.: Shaft whirling due to oil action in journal bearings. Gen. Electr. Rev. **28**(7), 559–568 (1925)

Newkirk, B.L.: Whirling balance shafts. In: Proceeding of 3rd ICAM, Stockholm, pp. 105–110 (1930)

Operational Modal Analysis for a Half Vehicle Model

Dorra Ben Hassen[1]([✉]), Mariem Miladi[1], Mohamed Slim Abbes[1],
S. Caglar Baslamisli[2], Fakher Chaari[1], and Mohamed Haddar[1]

[1] Mechanics, Modeling and Production Laboratory, National Engineering
School of Sfax (ENIS), BP 1173, 3038 Sfax, Tunisia
`dorra.benhassen@yahoo.fr`, `mariam.mi@hotmail.fr`,
`ms.abbes@gmail.com`, `fakher.chaari@gmail.com`,
`Mohamed.haddar@enis.rnu.tn`
[2] Department of Mechanical Engineering, Hacettepe University,
06800 Beytepe, Ankara, Turkey
`caglar.baslamisli@gmail.com`

Abstract. The objective of this paper is to use the Independent Component Analysis technique (ICA) in the Operational Modal Analysis (OMA) in order to determine the modal parameters of a half car model with four degrees of freedom. The ICA method is a major technique of the Blind Source Separation (It considers the studied system as a black box and knowing only its responses it can estimate its modal parameters) which is based on the inverse problem. In our case, this technique can be used to reconstruct the modal responses of the half car model knowing only its vibratory responses. In this paper, these vibratory responses are numerically computed using the Newmark approach and they constitute the observed signals for the ICA algorithm. So that based only on the knowledge of these responses, the ICA estimates the modal characteristics (modal responses, eigenfrequencies) of the studied half car model. Finally, the modal responses of the studied system obtained by the classical modal analysis are compared with those estimated by the ICA technique using some performance criteria which are the Modal Assurance Criterion (MAC number) and the relative error. The obtained results show a good agreement between the theoretical and estimated modal characteristics.

Keywords: Operational modal analysis · Independent component analysis
Half car · Modal parameters

1 Introduction

The knowledge of the modal parameters of the system is very important to avoid its excitation with its eigenfrequencies so its resonance. Thus the Operational Modal Analysis is used. The Classical modal recombination method necessitates the knowledge of the excitation forces applied on the structures to compute its dynamic responses. But in complex structures, there is no information about theses excitation forces so it is difficult to know them exactly such as the case of machines vibrations or the wind acting on the buildings. In this paper the Operational Modal Analysis

© Springer Nature Switzerland AG 2019
T. Fakhfakh et al. (Eds.): ICAV 2018, ACM 13, pp. 51–60, 2019.
https://doi.org/10.1007/978-3-319-94616-0_5

(OMA) based on the Independent Component Analysis (ICA) is used. This technique has the advantage that is able to identify the modal parameters of a system knowing only on its vibratory responses. The ICA was firstly used in a biological field (Hérault and Ans 1984; Ans et al. 1983) then it was extended to be used in the mechanical field such as (Akrout et al. 2012) who used the ICA in the OMA for a laminated double-glazing system in order to identify the modal parameters of this structure. Also (Abbès et al. 2011) showed that the ICA can be used in the OMA for a double panel system. Recently (Ben Hassen et al. 2017a) used the ICA algorithm to determine the modal parameters of a suspension system.

This paper is an extension to the work of (Ben Hassen et al. 2017a, b). Here a more complex system with four-degree of freedom is considered (Ben Hassen et al. 2017a, b). The aim of the use of the ICA is to determine the modal parameters of the half vehicle model. The ability of this technique to identify the modal parameters is studied by computing some performance criteria. In fact the identification of the natural frequencies of such system is very important to avoid the excitation of this system in the frequency band between 4–8 Hz which is harmful to the human body according to the ISO 2631.

The remainder of this paper was organized as follows: In a second section a brief presentation of the concept of the use of the ICA in the OMA is done, then, a description of the studied system is carried out. Finally, a comparison between the estimated modal parameters by the ICA and those identified by the classical modal analysis is performed. The validation of the results is done by computing some performance criteria. And a good agreement between the results is obtained.

2 Operational Modal Analysis

In this paper the ICA is used in the Operational Modal Analysis to determine the modal parameters of the studied system based on its dynamic responses.

2.1 Concept of the ICA

The ICA is a statistical method of Blind Source Separation (BSS). It has for objective the decomposition of a random signal X into a linear combination of statistically independent signals. This vector is defined as follows (Jutten and Hérault 1991)

$$X(t) = [A]\{S\} \tag{1}$$

where: A is the mixing matrix and S is the vector of the source signals.

2.2 Application of the ICA in the OMA

Using the classical modal analysis the vibratory responses $x(t)$ can be written as:

$$x = \psi y \tag{2}$$

This vector is considered as the only known input of the ICA algorithm. By appliying the ICA algorithm we can find the mixing matrix A which is close to the matrix ψ and the estimated signals S which are close to the modal responses. From these responses we can compute the eignfrequencies of the studied system.

3 Studied System

3.1 Presentation of the System

The proposed system in this study is a half car model. It is presented by the following figure (Fig. 1).

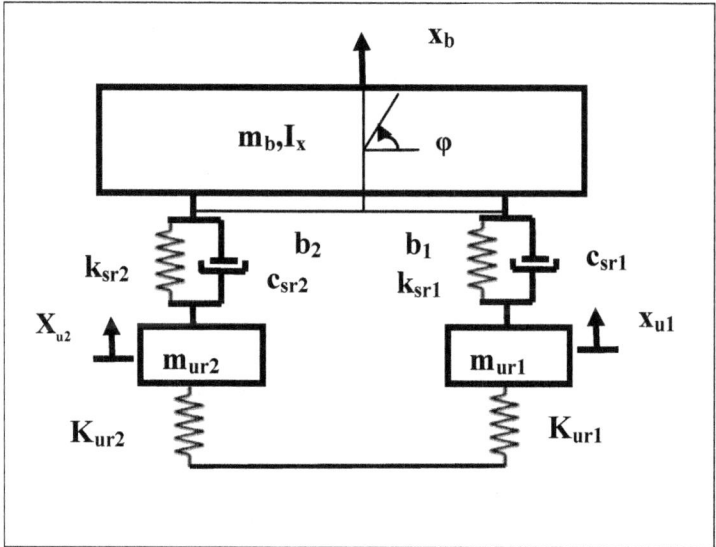

Fig. 1. Half car vehicle model

This model incorporates the rolling motion of the vehicle. It includes the displacement of the vehicle body x_b, the roll angle ϕ; the x_{u1} and x_{u2} wheel movements and the excitation of the h_1 and h_2 road profile. The motion equations of this system are presented in the following section (Ben Hassen et al. 2017a, b).

3.2   Equations of Motion

The equations of motion of the system can be written using Newton's law:

$$m_b\ddot{x}_b + c_{sr1}(\dot{x}_b - \dot{x}_{u1} + b_1\dot{\varphi}) + c_{sr2}(\dot{x}_b - \dot{x}_{u2} - b_2\dot{\varphi}) + k_{sr1}(x_b - x_{u1} + b_1\varphi) \\ + k_{sr2}(x_b - x_{u2} - b_2\varphi) = 0. \tag{3}$$

$$I_x\ddot{\varphi} + b_1 c_{sr1}(\dot{x}_b - \dot{x}_{u1} + b_1\dot{\varphi}) - b_2 c_{sr2}(\dot{x}_b - \dot{x}_{u2} - b_2\dot{\varphi}) + b_1 k_{sr1}(x_b - x_{u1} + b_1\varphi) \\ - b_2 k_{sr2}(x_b - x_{u2} - b_2\varphi) = 0 \tag{4}$$

$$m_{ur1}\ddot{x}_{u1} - c_{sr1}(\dot{x}_b - \dot{x}_{u1} + b_1\dot{\varphi}) + k_{ur1}x_{u1} - k_{sr1}(x_b - x_{u1} + b_1\varphi) = 0 \tag{5}$$

$$m_{ur2}\ddot{x}_{u2} - c_{sr2}(\dot{x}_b - \dot{x}_{u2} - b_2\dot{\varphi}) + k_{ur2}x_{u2} - k_{sr2}(x_b - x_{u2} - b_2\varphi) = 0 \tag{6}$$

These motion equations can be written in a matrix form as:

$$M\ddot{X} + C\dot{X} + KX = 0 \tag{7}$$

Where:

$$X = \begin{bmatrix} x_b \\ \varphi \\ x_{u1} \\ x_{u2} \end{bmatrix} \tag{8}$$

$$M = \begin{bmatrix} m_b & 0 & 0 & 0 \\ 0 & I_x & 0 & 0 \\ 0 & 0 & m_{ur1} & 0 \\ 0 & 0 & 0 & m_{ur2} \end{bmatrix} \tag{9}$$

$$k = \begin{bmatrix} k_{sr1} + k_{sr2} & -b_2 k_{sr2} + b_1 k_{sr1} & -k_{sr1} & -k_{sr2} \\ -b_2 k_{sr2} + a_1 k_{sr1} & b_1^2 k_{sr1} + b_2^2 k_{sr2} & -b_1 k_{sr1} & b_2 k_{sr2} \\ -k_{sr1} & -b_1 k_{sr1} & k_{sr1} + k_{ur1} & 0 \\ -k_{sr2} & -b_2 k_{sr2} & 0 & k_{sr2} + k_{ur2} \end{bmatrix} \tag{10}$$

$$C = \begin{bmatrix} c_{sr1} + c_{sr2} & -b_2 c_{sr2} + b_1 c_{sr1} & -c_{sr1} & -c_{sr2} \\ -b_2 c_{sr2} + b_1 c_{sr1} & b_1^2 c_{sr1} + b_2^2 c_{sr2} & -b_1 c_{sr1} & b_2 c_{sr2} \\ -c_{sr1} & -b_1 c_{sr1} & c_{sr1} & 0 \\ -c_{sr2} & -b_2 c_{sr2} & 0 & c_{sr2} \end{bmatrix} \tag{11}$$

The following table contains the values of the parameters of the half car model (Table 1).

Table 1. Parameters of the half car model

Parameters	Value	Unit
m_b	500	[kg]
m_{ur1}/m_{ur2}	53	[kg]
k_{sr1}/k_{sr2}	$11, 5.10^3$	[N/m]
k_{ur1}/k_{ur2}	20.10^4	[N/m]
c_{sr1}/c_{sr2}	1000	[N/ms]
I_x	410	[kg/m²]
b_1	0.7	[m]
b_2	0.75	[m]

Table 2. Comparison between the eignfrequencies

Mode	Theoretical frequency (Hz)	Estimated theory (Hz)	Relative error (%)
1	0.79	0.8	1.26
2	0.98	1	2.04
3	9.82	9.6	2.2
4	9.87	9.8	0.7

4 Classical Modal Analysis

Using the classical modal analysis, the modal matrix is calculated as:

$$[\phi] = \begin{bmatrix} 0.0031 & 0.0446 & 0.0007 & 0.0008 \\ -0.0493 & 0.0034 & -0.0006 & 0.0007 \\ -0.0017 & 0.0026 & 0.0019 & -0.1373 \\ 0.0019 & 0.0020 & -0.1373 & -0.0019 \end{bmatrix} \tag{12}$$

So the theoretical frequencies can be obtained:

$$[f] = \begin{bmatrix} 0.79 \\ 0.98 \\ 9.82 \\ 9.87 \end{bmatrix} (Hz) \tag{13}$$

The modal responses $\{Y\} = \begin{Bmatrix} y_1 \\ y_2 \\ y_3 \\ y_4 \end{Bmatrix}$ that represent the separate signals to be

determined from the ICA and their frequency spectrum are represented by the following Fig. 2.

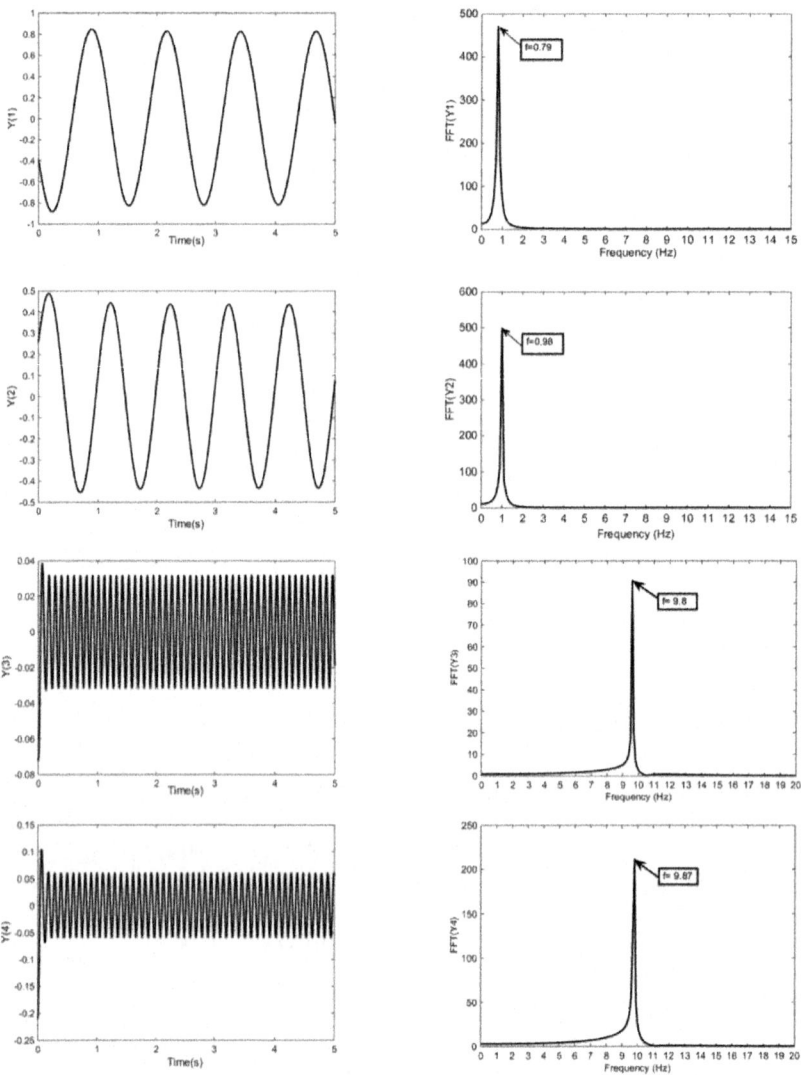

Fig. 2. Modal responses and their spectrum

For each modal response, the frequency spectrum shows the natural frequency of the correspondent mode i (i = 1, 2, 3, 4).

5 Results of the Application of the ICA in the OMA

Starting from the physical responses measured at the different degrees of freedom i (i = 1, 2, 3, 4), the ICA program is used to determine the estimated source signals that contain the estimated eigenfrequency information and the mixing matrix which has the estimated modal matrix.

The obtained dynamic responses (observed signals) and their spectrum are shown in the following Fig. 3.

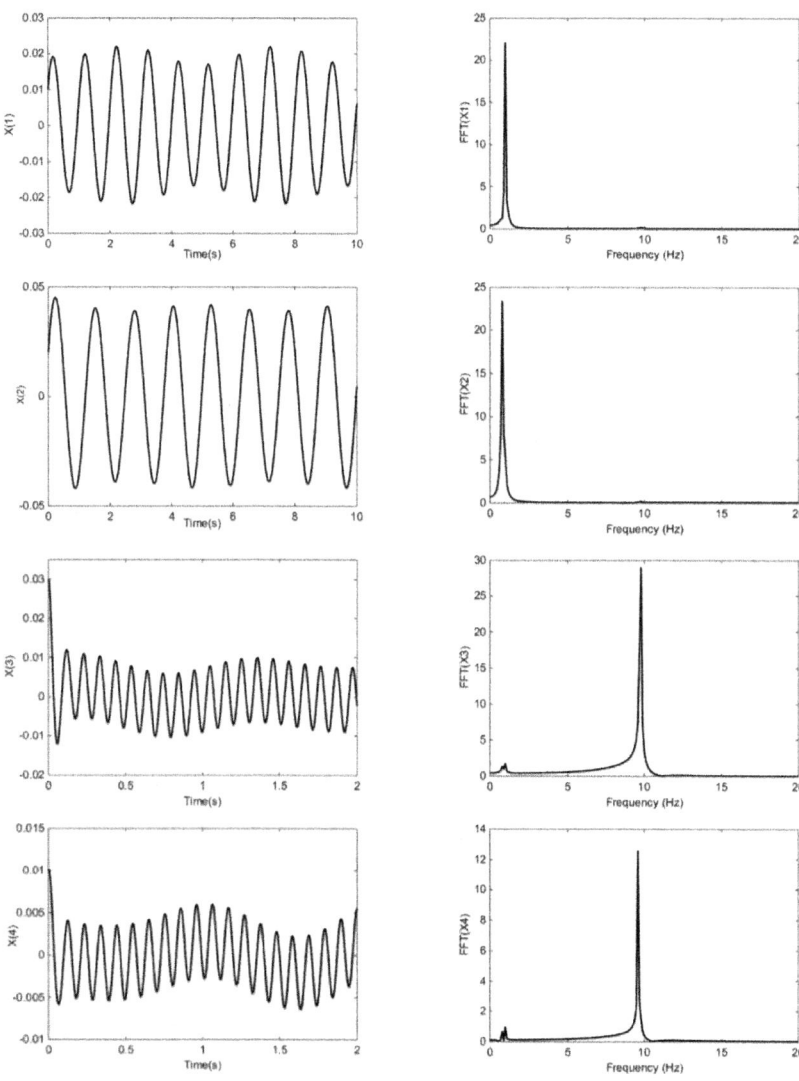

Fig. 3. Vibratory responses and their spectrum

It can be seen that the frequency spectrum spectral of the observed signals X presents the different natural frequencies of the discrete system studied. In order to find the separate signals (the modal responses), we apply the ICA starting from the vibratory responses of our system. The results are exposed below (Fig. 4).

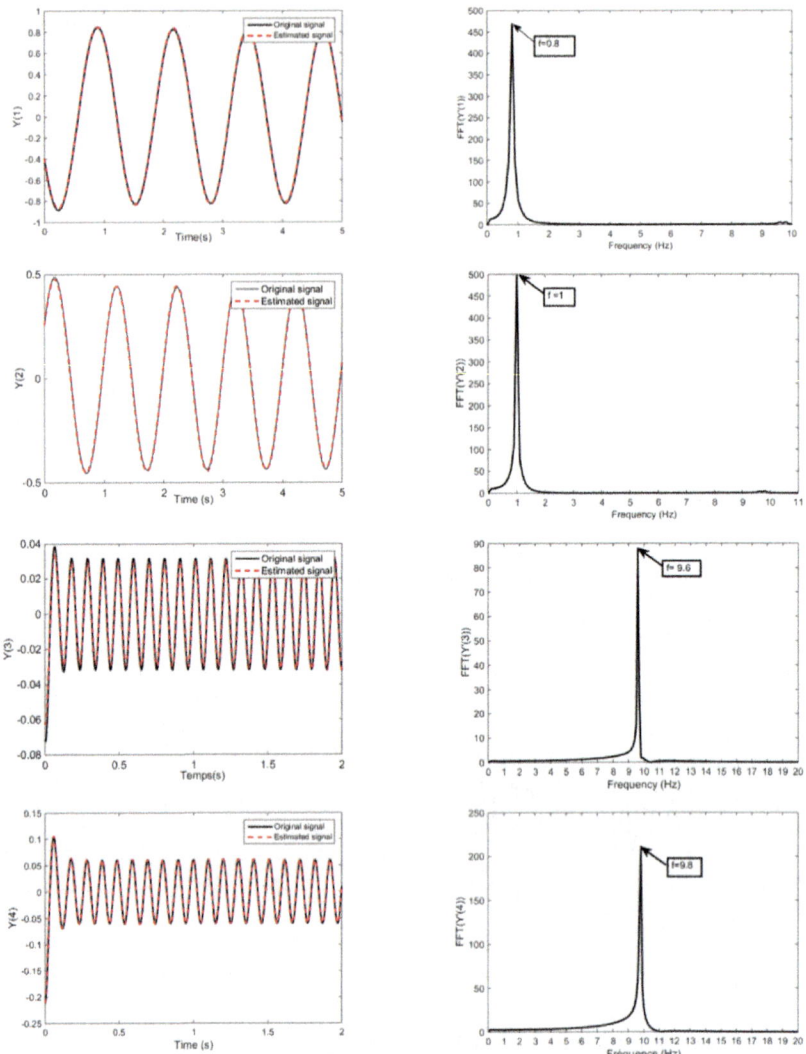

Fig. 4. Modal responses by the ICA and their frequency spectrum

The following diagram shows the Mac number of the different estimated modal deformities (Fig. 5):

It can be seen that the Mac value is close to 1 which prove that the ICA is able to reconstruct the modal responses of the half car model.

Also from the frequency spectrum, it can be noticed that the ICA is able to identify the eigenfrequency of each corresponding mode. The following Table 2 shows a comparison between the theoretical frequencies and the estimated ones using the relative error.

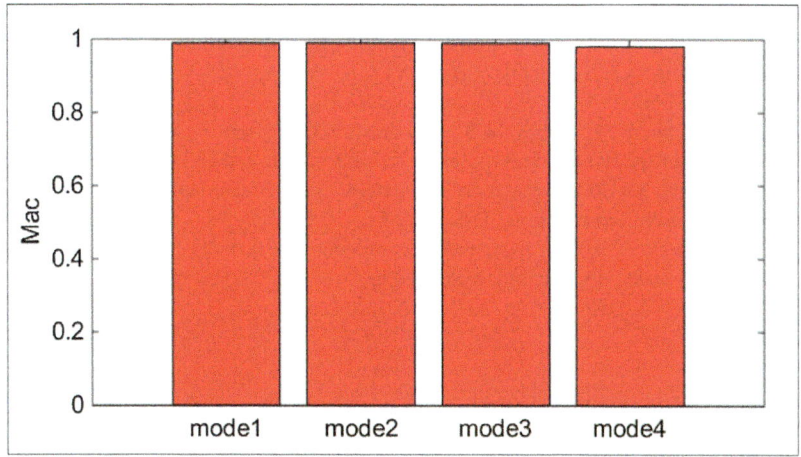

Fig. 5. Mac number

6 Conclusion

In this study the ICA is exploited and used in the OMA in order to determine, the modal properties (modal response, natural frequencies) of a half car model. By using some performance criteria like the Mac value and the relative error between the theoretical modal responses obtained by the classical modal analysis and those estimated by the ICA, we notice that the ICA give satisfactory results. Thus by using only the measured vibratory responses of the studied system, the Independent Component Analysis (ICA) can determine its modal characteristics.

Acknowledgements. The authors are grateful to Vicent Zarzoso and Pierre Comon for making ICA algorithm (Zarzoso 2010).

References

Hérault, J., Ans, B.: Réseau de neurones à synapses modifiables: Décodage de messages sensoriels composites par apprentissage non supervisé et permanent. Comptes rendus des séances de l'Académie des sciences. Série 3, Sciences de la vie **299**(13), 525–528 (1984)

Ans, B., Gilhodes, J.C., Hérault, J.: Simulation de réseaux neuronaux (SIRENE). II: Hypothése de décodage du message de mouvement porté par les afférences fusoriales IA et II par un mécanisme de plasticité synaptique. Comptes rendus des séances de l'Académie des sciences. Série 3, Sciences de la vie **297**(8), 419–422 (1983)

Akrout, A., Chaabene, M.M., Hammami, L., Haddar, M.: Edge stiffness effects on thin-film laminated double glazing system dynamical behavior by the operational modal analysis. J. Mech. Mater. Struct. (2012). https://doi.org/10.2140/jomms.2012.7.837

Abbès, M.S., Akrout, M.A., Fakhfekh, T., Haddar, M.: Vibratory behavior of double panel system by the operational modal analysis. Int. J. Model. Simul. Sci. Comput. **2**(4), 459–479 (2011)

Ben Hassen, D., Miladi, M., Abbes, M.S., Baslamisli, S.C., Chaari, F., Haddar, M. Application of the operational modal analysis using the independent component analysis for a quarter car vehicle model. In: Advances in Acoustics and Vibration, pp. 125–133. Springer International Publishing (2017a)

Ben Hassen, D., Miladi, M., Abbes, M.S., Baslamisli, S.C., Chaari, F., Haddar, M.: Road profile estimation using the dynamic responses of the full vehicle model. Appl. Acoust. (2017b)

Jutten, C., Hérault, J.: Blind separation of sources, part I: an adaptive algorithm based on neuromimetic architecture. Signal Process. **24**, 1–10 (1991)

Zarzoso, V., Comon, P.: Robust independent component analysis by iterative maximization of the kurtosis contrast with algebraic optimal step size. IEEE Trans. Neural Netw. **21**(2), 248–261 (2010)

Detecting Sound Hard Cracks in Isotropic Inhomogeneities

Lorenzo Audibert[1,2]([⊠]), Lucas Chesnel[2]([⊠]), Houssem Haddar[2]([⊠]),
and Kevish Napal[2]([⊠])

[1] Department PRISME, EDF R&D, 6 quai Watier, 78401 Chatou CEDEX, France
lorenzo.audibert@edf.fr
[2] INRIA/Centre de mathématiques appliquées, École Polytechnique Université
Paris-Saclay, Route de Saclay, 91128 Palaiseau, France
{lucas.chesnel,houssem.haddar}@inria.fr,
kevish.napal@cmap.polytechnique.fr

Abstract. We consider the problem of detecting the presence of sound-hard cracks in a non homogeneous reference medium from the measurement of multi-static far field data. First, we provide a factorization of the far field operator in order to implement the Generalized Linear Sampling Method (GLSM). The justification of the analysis is also based on the study of a special interior transmission problem. This technique allows us to recover the support of the inhomogeneity of the medium but fails to locate cracks. In a second step, we consider a medium with a multiply connected inhomogeneity assuming that we know the far field data at one given frequency both before and after the appearance of cracks. Using the Differential Linear Sampling Method (DLSM), we explain how to identify the component(s) of the inhomogeneity where cracks have emerged. The theoretical justification of the procedure relies on the comparison of the solutions of the corresponding interior transmission problems without and with cracks. Finally we illustrate the GLSM and the DLSM providing numerical results in 2D. In particular, we show that our method is reliable for different scenarios simulating the appearance of cracks between two measurements campaigns.

Keywords: Inverse scattering · Sampling methods
Sound-hard cracks

1 Introduction

This work is a contribution to sampling methods in inverse scattering theory when the issue is to determine the shape of an unknown inclusion from fixed frequency multi-static data. More precisely we extend the Generalized Linear Sampling Method (GLSM) and the Differential Linear Sampling Method (DLSM) [2,3] to inhomogeneous media containing sound-hard cracks. GLSM provides an exact characterization of the target shape from the far field operator, and

© Springer Nature Switzerland AG 2019
T. Fakhfakh et al. (Eds.): ICAV 2018, ACM 13, pp. 61–73, 2019.
https://doi.org/10.1007/978-3-319-94616-0_6

its implementation mainly requires two complementary factorizations of the far field operator, one used in the Linear Sampling Method (LSM) and another used in the Factorization Method (FM). From the measurements for both the damaged background and the initial background, it is possible to detect the defect thanks to the DLSM. This method consists in combining a result of comparison of two interior transmission problems associated with each background and the results of the GLSM.

The purpose of this article is to establish a similar factorization for a medium containing sound-hard cracks and to provide the theoretical results needed in the justification of the DLSM, the method we use to identify emergence of defects in an unknown background. For references of works dealing with qualitative methods to detect cracks, we mention, among others, [4–6,11].

2   The Forward Scattering Problem

We consider an isotropic medium embedded in $\mathbb{R}^d$, $d = 2$ or 3, containing sound-hard cracks. Following [6], a crack Γ is defined as a portion of a smooth nonintersecting curve ($d = 2$) or surface ($d = 3$) that encloses a domain Ω, such that its boundary $\partial\Omega$ is smooth. We assume that Γ is an open set with respect to the induced topology on $\partial\Omega$. The normal vector ν on Γ is defined as the outward normal vector to Ω (see Fig. 1). To define traces and normal derivatives of functions on Γ, we use the following notation for all $x \in \Gamma$:

$$f^{\pm}(x) = \lim_{h \to 0^+} f(x \pm h\nu(x)) \quad \text{and} \quad \partial^{\pm}_{\nu} f(x) = \lim_{h \to 0^+} \nu(x).\nabla f(x \pm h\nu(x)).$$

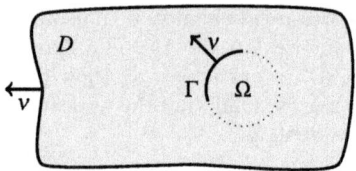

Fig. 1. Example of setting in $\mathbb{R}^2$.

We shall also work with the jump functions

$$[f] := f^+ - f^- \quad \text{and} \quad \left[\frac{\partial f}{\partial \nu}\right] := \partial^+_{\nu} f - \partial^-_{\nu} f.$$

We assume that the propagation of waves in time harmonic regime in the reference medium is governed by the Helmholtz equation $\Delta u + k^2 u = 0$ in $\mathbb{R}^d$ where Δ stands for the Laplace operator of $\mathbb{R}^d$ and where k is the wave number. We assume that the cracks are embedded in a local perturbation of the reference medium. To model this perturbation, we introduce $n \in L^\infty(\mathbb{R}^d)$ a complex valued coefficient (the refractive index of the medium) such that $n = 1$ in $\mathbb{R}^d \setminus \overline{D}$ and $n \neq 1$ in D. Here $D \subset \mathbb{R}^d$ is a bounded domain with Lipschitz boundary ∂D such that $\mathbb{R}^d \setminus D$ is connected. We assume that $\Im m\,(n) \geq 0$ in $\mathbb{R}^d$ and that $\Gamma \subset D$. The scattering of the incident plane wave $u_i(\theta, \cdot) := e^{ik\theta \cdot x}$ of direction of propagation $\theta \in \mathbb{S}^{d-1}$ by the medium is described by the problem

$$\left|\begin{array}{l} \text{Find } u = u_i + u_s \text{ such that} \\ \Delta u + k^2 n\, u = 0 \text{ in } \mathbb{R}^d \setminus \Gamma \\ \qquad \partial_\nu^\pm u = 0 \text{ on } \Gamma \\ \lim_{r \to +\infty} r^{\frac{d-1}{2}} \left(\dfrac{\partial u_s}{\partial r} - iku_s \right) = 0, \end{array}\right. \tag{1}$$

with $u_i = u_i(\theta, \cdot)$. The last line of (1), where $r = |x|$, is the Sommerfeld radiation condition which selects the outgoing scattered field and which is assumed to hold uniformly with respect to $\hat{x} = x/|x| \in \mathbb{S}^{d-1}$. For all $k > 0$, Problem (1) has a unique solution u belonging to $H^1(\mathscr{O} \setminus \Gamma)$ for all bounded domain $\mathscr{O} \subset \mathbb{R}^d$. The scattered field $u_s(\theta, \cdot)$ has the expansion

$$u_s(\theta, x) = \eta_d e^{ikr} r^{-\frac{d-1}{2}} \left(u_s^\infty(\theta, \hat{x}) + O(1/r) \right), \tag{2}$$

as $r \to +\infty$, uniformly in $\hat{x} = x/|x| \in \mathbb{S}^{d-1}$. In (2) the constant η_d is given by $\eta_d = e^{i\frac{\pi}{4}}/\sqrt{8\pi k}$ for $d = 2$ and by $= 1/(4\pi)$ for $d = 3$. The function $u_s^\infty(\theta, \cdot) : \mathbb{S}^{d-1} \to \mathbb{C}$, is called the far field pattern associated with $u_i(\theta, \cdot)$. From the far field pattern, we can define the far field operator $F : L^2(\mathbb{S}^{d-1}) \to L^2(\mathbb{S}^{d-1})$ such that

$$(Fg)(\hat{x}) = \int_{\mathbb{S}^{d-1}} g(\theta)\, u_s^\infty(\theta, \hat{x})\, ds(\theta). \tag{3}$$

By linearity, the function Fg corresponds to the far field pattern of the scattered field in (1) with

$$u_i = v_g := \int_{\mathbb{S}^{d-1}} g(\theta) e^{ik\theta \cdot x}\, ds(\theta) \qquad \text{(Herglotz wave function)}. \tag{4}$$

3   Factorization of the Far Field Operator

In this section we explain how to factorize the far field operator F defined in (3). From the Green representation theorem, computing the asymptotic behaviour of the Green's function as $r \to +\infty$ gives

$$u_s^\infty(\hat{x}) = \left(k^2 \int_D (n(y) - 1)u(y)e^{-ik\hat{x}y}dy + \int_\Gamma [u(y)]\partial_{\nu(y)}^+ e^{-ik\hat{x}y}\, ds(y) \right) \tag{5}$$

for the far field pattern of u_s in (2). A first step towards the factorization of F is to define the Herglotz operator $H : L^2(\mathbb{S}^{d-1}) \to L^2(D) \times L^2(\Gamma)$ such that

$$Hg = (v_{g|D}, \partial_\nu^+ v_{g|\Gamma}). \tag{6}$$

We give in Proposition 1 below a characterization of the closure of the range of H. Set

$$\mathscr{H} = \left\{ v \in L^2(D) \,|\, \Delta v + k^2 v = 0 \text{ in } D \right\}. \tag{7}$$

and define the map $\Psi : \mathscr{H} \to L^2(D) \times L^2(\Gamma)$ such that

$$\Psi v = (v_{|D}, \partial_\nu^+ v_{|\Gamma}). \tag{8}$$

Proposition 1. *The operator $H : L^2(\mathbb{S}^{d-1}) \to L^2(D) \times L^2(\Gamma)$ defined in (6) is injective and $\overline{R(H)} = \Psi(\mathscr{H})$.*

Proof. The proof of the injectivity of H follows a classical argument based on the Jacobi Anger expansion (apply [7, Lemma 2.1]). To establish the second part of the claim, first we note that v_g (defined in (4)) belongs to $\mathscr{H}$ so that $R(H) \subset \Psi(\mathscr{H})$. On the other hand, classical results of interior regularity ensure that is some constant $C > 0$ such that $\|\partial_\nu v\|_{L^2(\Gamma)} \leq C\|v\|_{L^2(D)}$ for all $v \in \mathscr{H}$. This in addition to $\|\Psi v\|_{L^2(D) \times L^2(\Gamma)} \geq \|v\|_{L^2(D)}$ allows one to show that $\Psi(\mathscr{H})$ is a closed subspace of $L^2(D) \times L^2(\Gamma)$. The regularity result implies that $\Psi : (\mathscr{H}, \|\cdot\|_{L^2(D)}) \to L^2(D) \times L^2(\Gamma)$ is continuous. Since the set of Herglotz wave functions is dense in $(\mathscr{H}, \|\cdot\|_{L^2(D)})$, we deduce that $\overline{R(H)} = \Psi(\mathscr{H})$. □

Next we define the operator $G : \overline{R(H)} \to L^2(\mathbb{S}^{d-1})$ such that

$$G(v, \partial_\nu^+ v) = u_s^\infty, \tag{9}$$

where u_s^∞ is the far field pattern of u_s, the outgoing scattered field which satisfies

$$\left| \begin{array}{ll} \Delta u_s + k^2 n\, u_s = k^2(1-n)v & \text{in } \mathbb{R}^d \setminus \Gamma \\ \partial_\nu^\pm u_s = -\partial_\nu^\pm v & \text{on } \Gamma. \end{array} \right. \tag{10}$$

Note that if $(v, \partial_\nu^+ v) \in \overline{R(H)}$ then interior regularity implies $\partial_\nu^+ v = \partial_\nu^- v$ on Γ. We also define the map $T : L^2(D) \times L^2(\Gamma) \to L^2(D) \times L^2(\Gamma)$ such that

$$T(v, \partial_\nu^+ v) = (k^2(n-1)(v + u_s), [v + u_s]). \tag{11}$$

Clearly we have $F = GH$. And one can check using (5) that $G = H^*T$ so that F admits the factorisation

$$F = H^*TH. \tag{12}$$

The justification of the techniques we propose below to recover the cracks will depend on the properties of the operators G, T. And the latter are related to the solvability of the so-called interior transmission problem which in our situation states as follows: given $f \in H^{3/2}(\partial D), g \in H^{1/2}(\partial D)$

$$\left| \begin{array}{ll} \text{Find } (u, v) \in L^2(D) \times L^2(D) \text{ such that} & \\ w := u - v \in \{\varphi \in H^1(D \setminus \Gamma) \,|\, \Delta\varphi \in L^2(D \setminus \Gamma)\} & \\ \Delta u + k^2 n\, u = 0 \text{ in } D \setminus \Gamma & u - v = f \text{ on } \partial D \\ \Delta v + k^2 v = 0 \text{ in } D & \partial_\nu u - \partial_\nu v = g \text{ on } \partial D \\ \partial_\nu^\pm u = 0 \text{ on } \Gamma. & \end{array} \right. \tag{13}$$

We shall say that $k > 0$ is a transmission eigenvalue if (13) with $f = g = 0$ admits a non zero solution. One can show for example that if the coefficient n is real and satisfies $1 < n_* < n < n^*$ for some constants n_*, n^*, then the set of transmission eigenvalues is discrete without accumulation point and that Problem (13) is uniquely solvable if and only if k is not a transmission eigenvalue (this will be part of a future work). We shall say that (13) is well-posed if it admits a unique solution for all $f \in H^{3/2}(\partial D), g \in H^{1/2}(\partial D)$.

Proposition 2. *Assume that $k > 0$ is not a transmission eigenvalue. Then the operator $G : \overline{R(H)} \rightarrow L^2(\mathbb{S}^{d-1})$ is compact, injective with dense range.*

Proof. First we show the injectivity of G. Let $V = (v, \partial_\nu^+ v) \in \overline{R(H)}$ such that $GV = 0$. Then from the Rellich lemma, the solution u_s of (10) is zero in $\mathbb{R}^d \setminus \overline{D}$. Therefore, if we define $u = v + u_s$, then the pair (u, v) satisfies the interior transmission problem (13) with $f = g = 0$. Since we assumed that $k > 0$ is not a transmission eigenvalue, we deduce that $v = 0$ and so $V = 0$.

Now we focus our attention on the denseness of the range of G. First we establish an identity of symmetry. Let $V_1 = (v_1, \partial_\nu^+ v_1)$, $V_2 = (v_2, \partial_\nu^+ v_2) \in \overline{R(H)}$. Denote w_1, w_2 the corresponding solutions to Problem (10). In particular we have

$$\Delta w_1 + k^2 n w_1 = k^2(1-n)v_1, \quad \Delta w_2 + k^2 n w_2 = k^2(1-n)v_2 \quad \text{in } \mathbb{R}^d \setminus \Gamma. \quad (14)$$

Multiplying the first equation by w_2 and the second by w_1, integrating by parts the difference over B_R, the open ball of radius R centered at O, we obtain

$$k^2 \int_D (n-1)(v_1 w_2 - v_2 w_1)\, dx$$
$$= \int_{\partial B_R} (\partial_\nu w_1 w_2 - w_1 \partial_\nu w_2)\, ds + \int_\Gamma ([w_2]\partial_\nu^+ v_1 - [w_1]\partial_\nu^+ v_2)\, ds.$$

Taking the limit as $R \rightarrow +\infty$ and using that $\lim_{R \rightarrow +\infty} \int_{\partial B_R} (\partial_\nu w_1 w_2 - w_1 \partial_\nu w_2)\, ds = 0$ (w_1 and w_2 satisfy the radiation condition), we find the identity

$$k^2 \int_D (n-1)v_1 w_2\, dx + \int_\Gamma \partial_\nu^+ v_1 [w_2]\, ds = k^2 \int_D (n-1)v_2 w_1\, dx + \int_\Gamma \partial_\nu^+ v_2 [w_1]\, ds. \quad (15)$$

Using (15), we deduce that for ϕ, $g \in L^2(\mathbb{S}^{d-1})$, we have

$$\langle G(H\phi), \overline{g} \rangle_{L^2(\mathbb{S}^{d-1})}$$
$$= k^2 \int_D (n-1)(H\phi + u_s(\phi))Hg\, dx + \int_\Gamma [H\phi + u_s(\phi)]\partial_\nu^+(Hg)\, ds$$
$$= k^2 \int_D (n-1)(Hg + u_s(g))H\phi\, dx + \int_\Gamma [Hg + u_s(g)]\partial_\nu^+(H\phi)\, ds$$
$$= \langle G(Hg), \overline{\phi} \rangle_{L^2(\mathbb{S}^{d-1})}.$$

Therefore if $\overline{g} \in R(G)^\perp$ then $G(Hg) = 0$. The injectivity of G and H imply that $g = 0$ which shows that G has dense range.
Finally, using again the estimate $\|\partial_\nu v\|_{L^2(\Gamma)} \leq C\|v\|_{L^2(D)}$ for all $v \in \mathcal{H}$, results

of interior regularity and the definition of H (see (6)), one can check that $H : L^2(\mathbb{S}^{d-1}) \rightarrow L^2(D) \times L^2(\Gamma)$ is compact. Since $G = H^*T$ and T is continuous, we deduce that $G : L^2(D) \times L^2(\Gamma) \rightarrow L^2(\mathbb{S}^{d-1})$ is compact. $\square$

Proposition 3. *For all* $V = (v, \partial_\nu^+ v) \in \overline{R(H)}$, *we have the energy identity*

$$\Im m \left(\langle TV, V \rangle_{L^2(D) \times L^2(\Gamma)} \right) = k^2 \int_D \Im m\,(n)|u_s + v|^2\,dx + k\|GV\|^2_{L^2(\mathbb{S}^{d-1})}, \quad (16)$$

where u_s *denotes the solution of (10). As a consequence if* $\Im m\,(n) \geq 0$ *a.e. in* D *and if* k *is not a transmission eigenvalue of (13), then* T *is injective.*

Proof. Multiplying by $\overline{u_s}$ the equation $\Delta u_s + k^2 u_s = -k^2(n-1)(u_s + v)$ and integrating by parts over the ball B_R, we obtain

$$-k^2 \int_D (n-1)(u_s + v)\overline{u_s}\,dx =$$
$$- \int_{B_R} |\nabla u_s|^2 - k^2|u_s|^2\,dx + \int_{\partial B_R} \partial_\nu u_s \overline{u_s}\,ds - \int_\Gamma \partial_\nu^+ u_s [\overline{u_s}]\,ds. \quad (17)$$

Using (17), then we find

$$\langle TV, V \rangle_{L^2(D) \times L^2(\Gamma)} = k^2 \int_D (n-1)|u_s + v|^2\,dx - \int_{B_R} |\nabla u_s|^2 - k^2|u_s|^2\,dx$$
$$+ \int_\Gamma [v + u_s]\partial_\nu^+ \overline{v}\,ds - \int_\Gamma \partial_\nu^+ u_s [\overline{u_s}]\,ds + \int_{\partial B_R} \partial_\nu u_s\,\overline{u_s}\,ds.$$

Since $\partial_\nu^+ u_s = -\partial_\nu^+ v$ and $[v] = 0$ (interior regularity) on Γ, we deduce

$$\langle TV, V \rangle_{L^2(D) \times L^2(\Gamma)} = k^2 \int_D (n-1)|u_s + v|^2\,dx - \int_{B_R} |\nabla u_s|^2 - k^2|u_s|^2\,dx$$
$$-2\Re e \left(\int_\Gamma [u_s]\partial_\nu^+ \overline{u_s}\,ds \right) + \int_{\partial B_R} \partial_\nu u_s \overline{u_s}\,ds. \quad (18)$$

The radiation condition (see (1)) implies $\lim_{R \to \infty} \int_{\partial B_R} \partial_\nu u_s \overline{u_s}\,ds = ik \int_{\mathbb{S}^{d-1}} |u_s^\infty|^2 d\theta = ik\|GV\|^2_{L^2(\mathbb{S}^{d-1})}$. As a consequence, taking the imaginary part of (18) and letting R goes to infinity, we get identity (16). Now if $TV = 0$ and if $\Im m\,(n) \geq 0$ a.e. in D, then (16) gives $GV = 0$. Since G is injective when k is not a transmission eigenvalue of (13) (Proposition 2), we deduce that T is injective. $\square$

4   Generalized Linear Sampling Method and Differential Linear Sampling Method

For $z \in \mathbb{R}^d$, we denote by $\Phi(.,z)$ the outgoing fundamental solution of the homogeneous Helmoltz equation such that

$$\Phi(x, z) = \frac{i}{4}H_0^{(1)}(k|x - z|) \text{ if } d = 2 \quad \text{and} \quad \frac{e^{ik|x-z|}}{4\pi|x - z|} \text{ if } d = 3. \quad (19)$$

Here $H_0^{(1)}$ stands for the Hankel function of first kind of order zero. The far field of $\Phi(.,z)$ is $\phi_z(\widehat{x}) = e^{-ikz.\widehat{x}}$. The GLSM uses the following theorem whose proof is classical [7].

Theorem 1. *Assume that the interior transmission problem* (13) *is well-posed. Then*

$$z \in D \qquad \text{if and only if} \qquad \phi_z \in R(G).$$

The particularity of the GLSM is to build an approximate solution $(Fg \simeq \phi_z)$ to the far field equation by minimizing the functional $J^\alpha(\phi_z, .) : L^2(\mathbb{S}^{d-1}) \to \mathbb{R}$ defined by

$$J^\alpha(\phi_z, g) = \alpha\langle F^\sharp g, g\rangle_{L^2(\mathbb{S}^{d-1})} + \|Fg - \phi_z\|^2_{L^2(\mathbb{S}^{d-1})}, \quad \forall g \in L^2(\mathbb{S}^{d-1}), \quad (20)$$

where $F^\sharp := |\frac{1}{2}(F + F^*)| + |\frac{1}{2i}(F - F^*)|$.

Theorem 2 (GLSM). *Assume that the interior transmission problem* (13) *is well-posed, that the index n satisfies $[\Im m\,(n) \geq 0,\ \Re e\,(n-1) \geq n_*$ a.e. in $D]$ or $[\Im m\,(n) \geq 0,\ \Re e\,(1-n) \geq n_*$ a.e. in $D]$ for some constant $n_* > 0$. Let $g_z^\alpha \in L^2(\mathbb{S}^{d-1})$ be a minimizing sequence of $J^\alpha(\phi_z, .)$ such that*

$$J^\alpha(\phi_z, g_z^\alpha) \leq \inf_g J^\alpha(\phi_z, g) + p(\alpha), \quad (21)$$

where $\lim_{\alpha\to 0} \alpha^{-1}p(\alpha) = 0$. Then

- $z \in D$ *if and only if* $\lim_{\alpha\to 0}\langle F^\sharp g_z^\alpha, g_z^\alpha\rangle_{L^2(\mathbb{S}^{d-1})} < +\infty.$
- *If $z \in D$ then there exists $h \in \overline{R(H)}$ such that $\phi_z = Gh$ and Hg_z^α converges strongly to h as $\alpha \to 0$.*

Thus the GLSM, justified by this theorem, offers a way to recover D, that is to identify the perturbation in the reference background. Note that the GLSM, contrary to the LSM, provides an exact characterization of D. However it does not give any information on the location of the crack Γ.

Proof. We establish this theorem by applying the abstract result of [7, Theorem 2.10]. The latter requires that the following properties hold.

(i) $F = GH = H^*TH$ is injective with dense range and G is compact.
(ii) $F^\sharp$ factorizes as $F^\sharp = H^*T^\sharp H$ where $T^\sharp$ satisfies the coercivity property

$$\exists \mu > 0,\ \forall V \in \overline{R(H)}, \qquad |\langle T^\sharp V, V\rangle_{L^2(D)\times L^2(\Gamma)}| \geq \mu\|V\|^2_{L^2(D)\times L^2(\Gamma)}; \quad (22)$$

(iii) $V \mapsto |\langle T^\sharp V, V\rangle_{L^2(D)\times L^2(\Gamma)}|^{1/2}$ is uniformly convex on $\overline{R(H)}$.

Item (i) is a consequence of Propositions 1, 2 and 3. Moreover, we deduce (iii) from (ii) and from the fact that $\langle F^\sharp g, g\rangle_{L^2(D)\times L^2(\Gamma)} = \|(F^\sharp)^{1/2}g\|^2_{L^2(\mathbb{S}^{d-1})}$ (see e.g. [7]). Therefore, it remains to show ii). To proceed, we use [7, Theorem 2.31] which guarantees that it is true if :

- T injective on $\overline{R(H)}$;
- $\Im m(\langle TV, V \rangle_{L^2(D) \times L^2(\Gamma)}) \geq 0$ for all $V \in \overline{R(H)}$;
- $\Re e(T)$ decomposes as $\Re e(T) = T_0 + C$ where T_0 satisfies (22) and where C is compact on $\overline{R(H)}$.

The first two items have been proved in Proposition 3. Let us focus our attention on the last one. By definition, we have $TV = (k^2(n-1)(v+u_s), [v+u_s])$. Set $\tilde{C}V = (k^2(n-1)u_s, [v+u_s] - \partial_\nu^+ v|_\Gamma)$. Using results of interior regularity, one can check that $C = \Re e(\tilde{C})$ is compact. Now, define $T_0 := \Re e(T) - C = (k^2\Re e(n-1)v, \partial_\nu^+ v|_\Gamma)$. Clearly one has $|\langle T_0 V, V \rangle_{L^2(D) \times L^2(\Gamma)}| \geq n_* \|V\|^2_{L^2(D) \times L^2(\Gamma)}$ when $\Re e(n-1) \geq n_*$. The case $\Re e(1-n) \geq n_*$ can be dealt in a similar way. □

When one has only acces to a noisy version F^δ of F, then $F^{\sharp,\delta}$ might not have the required factorization and the cost function (21) must be regularized. For this aspect, we refer the reader to [2, Sect. 5.2].

We now give the theoretical foundation of the DLSM which will allow us to localize the position of the crack Γ. The DLSM relies on the comparison of the solutions of the following interior transmission problems (without and with cracks).

$$\mathscr{P}(D) \begin{vmatrix} \Delta u_0 + k^2 n u_0 = 0 & \text{in } D \\ \Delta v_0 + k^2 v_0 = 0 & \text{in } D \\ u_0 - v_0 = \Phi_z & \text{on } \partial D \\ \partial_\nu u_0 - \partial_\nu v_0 = \partial_\nu \Phi_z & \text{on } \partial D, \end{vmatrix} \qquad \mathscr{P}_\Gamma(D) \begin{vmatrix} \Delta u + k^2 n u = 0 & \text{in } D \\ \Delta v + k^2 v = 0 & \text{in } D \\ \partial_\nu^\pm u = 0 & \text{on } \Gamma \\ u - v = \Phi_z & \text{on } \partial D \\ \partial_\nu u - \partial_\nu v = \partial_\nu \Phi_z & \text{on } \partial D, \end{vmatrix}$$

$$(23)$$

where u_0, v_0, u, $v \in L^2(D)$, $u_0 - v_0 \in H^2(D)$ and $u - v \in H^1(D \setminus \Gamma)$ is such that $\Delta(u - v) \in L^2(D \setminus \Gamma)$. We split the domain D into two kinds of connected components (see Fig. 2): The ones containing cracks are listed by $(D_\Gamma^j)_j$; others are listed by $(D_0^j)_j$. And we set $D_\Gamma := \cup_j D_\Gamma^j$ and $D_0 := \cup_j D_0^j$ so that $D = D_\Gamma \cup D_0$.

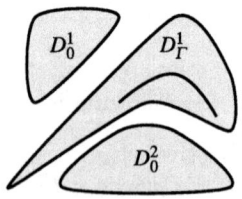

Fig. 2. We split D into two families of connected components.

Theorem 3. *Assume that Γ is a part of the boundary of a domain Ω such that $\partial\Omega$ is analytic. Assume that n is analytic in D_Γ and does not vanish. Assume also that k is not a Neumann eigenvalue for $-n^{-1}\Delta$ in Ω and is such that both $\mathscr{P}(D)$ and $\mathscr{P}_\Gamma(D)$ (see (23)) are well-posed.*
    (i) If $z \in D_0$ then $v = v_0$ in D.   *(ii) If $z \in D_\Gamma$ then $v \neq v_0$ in D_Γ.*

Proof. (i) Let $z \in D_0$. In D_0, the equations for $\mathscr{P}(D)$ and $\mathscr{P}_\Gamma(D)$ coincide. By uniqueness of the solution for these problems, we deduce that $v = v_0$ in D_0. On the other hand, one observes that $(0, -\Phi_z)$ satisfies the equations of $\mathscr{P}(D)$ and $\mathscr{P}_\Gamma(D)$ in D_Γ. As a consequence, by uniqueness of the solution for these problems, we also have $v = v_0 = -\Phi_z$ in D_Γ.

(ii) Now let $z \in D_\Gamma$. We wish to show that $v \neq v_0$ in D_Γ. We proceed by contradiction assuming that $v = v_0$ in D_Γ. Define U such that $U = u - u_0$ in $D_\Gamma \setminus \Gamma$ and $U = 0$ in $\mathbb{R}^d \setminus D_\Gamma$. Since $U = \partial_\nu U = 0$ on ∂D_Γ, from the unique continuation principle, we find $U = 0$ in $\mathbb{R}^d \setminus \Gamma$ and so $\partial_\nu^\pm u_0 = 0$ on Γ (because $\partial_\nu^\pm u = 0$ on Γ). Furthermore the regularity of n implies that $\partial_\nu^\pm u_0$ is analytic on $\partial\Omega$ and we conclude that $\partial_\nu^\pm u_0 = 0$ on $\partial\Omega$. Since we assumed that k is not a Neumann eigenvalue for $-n^{-1}\Delta$ in Ω, we deduce that $u_0 = 0$ in Ω, and by unique continuation, $u_0 = 0$ in D_Γ. Thus we must have $v_0 = -\Phi_z$ in D_Γ which contradicts the fact that $u_0 - v_0 \in H^2(D)$. $\qquad\qquad\square$

Now we consider a first heterogeneous medium without crack with a perturbation of the reference background supported in $\overline{D}$ modeled by some index n, and a second medium with the same n but with an additional crack inside D. The corresponding far field operators are denoted respectively F_0 and F_1. Then for $j = 0, 1$, let $g_{j,z}^\alpha$ refer to the sequences introduced in the statement of Theorem 2 with $F_j^\sharp = |\frac{1}{2}(F_j + F_j^*)| + |\frac{1}{2i}(F_j - F_j^*)|$. We also set for $j = 0, 1$

$$\mathscr{A}_j^\alpha(z) = \langle F_j^\sharp g_{j,z}^\alpha, g_{j,z}^\alpha \rangle_{L^2(\mathbb{S}^{d-1})}; \quad \mathscr{D}_j^\alpha(z) = \langle F_j^\sharp (g_{1,z}^\alpha - g_{0,z}^\alpha), (g_{1,z}^\alpha - g_{0,z}^\alpha) \rangle_{L^2(\mathbb{S}^{d-1})}. \tag{24}$$

The combination of Theorems 2 and 3 leads to the following result.

Theorem 4 (DLSM). *Assume that k, n and Γ are as in Theorem 3 and that n also satisfies the assumptions of Theorem 2. Then for $j = 0$ or 1*
$$\left[z \in D_0 \right] \Rightarrow \left[\lim_{\alpha \to 0} \mathscr{D}_j^\alpha(z) = 0 \right] \quad and \quad \left[z \in D_\Gamma \right] \Rightarrow \left[0 < \lim_{\alpha \to 0} \mathscr{D}_j^\alpha(z) < +\infty \right].$$

Proof. As explained in the proof of Theorem 2, $F_1^\sharp$ admits a factorization of the form $H^* T_1^\sharp H$ where $T_1^\sharp$ is continuous and $\langle T_1^\sharp \cdot, \cdot \rangle$ is coercive. According to the study of crack-free inhomogeneous medium a same factorization stands for F_0 involving an operator $T_0^\sharp$ that have the same properties of $T_1^\sharp$. This implies (for $j = 0$ or 1) the existence of two positive constants κ and K such that

$$\kappa \|H(g_{1,z}^\alpha - g_{0,z}^\alpha)\|_{L^2(D)}^2 \le \mathscr{D}_j^\alpha(z) \le K \|H(g_{1,z}^\alpha - g_{0,z}^\alpha)\|_{L^2(D)}^2. \tag{25}$$

Now for $z \in D$, if we denote (u_0, v_0) (resp. (u_1, v_1)) the solution of $\mathscr{P}(D)$ (resp. $\mathscr{P}_\Gamma(D)$), then Theorem 2 and the GLSM for the crack-free inhomogeneous medium (see the justification in [7]) guarantee that $\lim_{\alpha \to 0} \|H(g_{1,z}^\alpha - g_{0,z}^\alpha)\| = \|H(v - v_0)\|$. Then the result follows from Theorem 3. $\qquad\qquad\square$

From Theorems 2 and 4, one can design indicators for D and D_Γ. Set for $j = 0$ or 1,

$$I^{\mathrm{GLSM}}(z) = \lim_{\alpha \to 0} \frac{1}{\mathscr{A}_1^\alpha(z)} \quad and \quad I_j^{\mathrm{DLSM}}(z) = \lim_{\alpha \to 0} \frac{1}{\mathscr{A}_0^\alpha(z)\left(1 + \frac{\mathscr{A}_0^\alpha(z)}{\mathscr{D}_j^\alpha(z)}\right)}. \tag{26}$$

For these indicators, one can show the following theorem which allows one to identify the connected components of D in which some cracks have appeared.

Corollary 1. *Under the assumptions of Theorem 4, we have for $j = 0$ or 1*

- $I^{\text{GLSM}}(z) = 0$ *in* $\mathbb{R}^d \setminus D$ *and* $I^{\text{GLSM}}(z) > 0$ *in* D.

- $I_j^{\text{DLSM}}(z) = 0$ *in* $\mathbb{R}^d \setminus D_\Gamma$ *and* $I_j^{\text{DLSM}}(z) > 0$ *in* D_Γ.

5   Numerical Results

To conclude this work, we apply the GLSM and the DLSM on simulated backgrounds. All backgrounds have the same shape D constituted of three disjoint disks of radius 0.75 and of index $n = 1.5$. They differ from one another in the distribution of cracks inside the disks. Admittedly, the straight cracks appearing in the backgrounds are not a portion of the boundary of an analytic domain. However, we expect that our algorithm remains robust when this theoretical assumption is not satisfied. For each background we generate a discretization of the far field operator F by solving numerically the direct problem for multiple incident fields $u_i(\theta_p)$ with wave number $k = 4\pi$. Then we compute the matrix $F = (u_s^\infty(\theta_p, \widehat{x_q}))_{p,q}$ for θ_p, $\widehat{x_q}$ in $\{\cos(\frac{2l\pi}{100}), \sin(\frac{2l\pi}{100}), l = 1..100\}$ (somehow we discretize $L^2(\mathbb{S}^1)$). Finally, we add random noise to the simulated F and obtain our final synthetic far field data F^δ with $F_{pq}^\delta = F_{pq}(1 + \sigma N)$. Here N is a complex random variable whose real and imaginary parts are uniformly chosen in $[-1,1]^2$. The parameter $\sigma > 0$ is chosen so that $\|F^\delta - F\| = 0.05\|F^\delta\|$.

5.1   GLSM

To handle the noise δ added on the far field data, we use a regularized version of the GLSM consisting in finding the minimizers $g_z^{\alpha,\delta}$ of the functional

$$g \mapsto J^{\alpha,\delta}(\phi_z, g) = \alpha(|\langle F^{\sharp\delta} g, g \rangle_{L^2(\mathbb{S}^{d-1})}| + \delta\|F^\delta\|\|g\|_{L^2(\mathbb{S}^2)}^2) + \|F^\delta g - \phi_z\|_{L^2(\mathbb{S}^{d-1})}^2,$$

where $F^{\sharp\delta} := |\frac{1}{2}(F^\delta + F^{\delta*})| + |\frac{1}{2i}(F^\delta - F^{\delta*})|$. We fit α to δ according to [2, Sect. 5.2]. The new relevant indicator function for the regularized GLSM is then given by

$$I_{\text{GLSM}}^{\alpha,\delta}(z) = \frac{1}{\mathscr{A}^{\alpha,\delta}(z)}$$

where $\mathscr{A}^{\alpha,\delta}(z) = \langle F^{\sharp\delta} g_z^{\alpha,\delta}, g_z^{\alpha,\delta} \rangle_{L^2(\mathbb{S}^{d-1})} + \delta\|F^\delta\|\|g_z^{\alpha,\delta}\|_{L^2(\mathbb{S}^{d-1})}^2.$

Figure 3 shows the results of GLSM indicator function $z \mapsto I_{\text{GLSM}}^{\alpha,\delta}(z)$ for two different configurations where the second one is obtained from the first one by adding a crack to the third component. The two other components contain the same crack. One observes that GLSM is capable of retrieving the domain D for each configuration. We also observe how the behavior of the indicator function is different inside the third component. This is somehow what the DLSM exploits to isolate the component where a defect appears and this is what is discussed next.

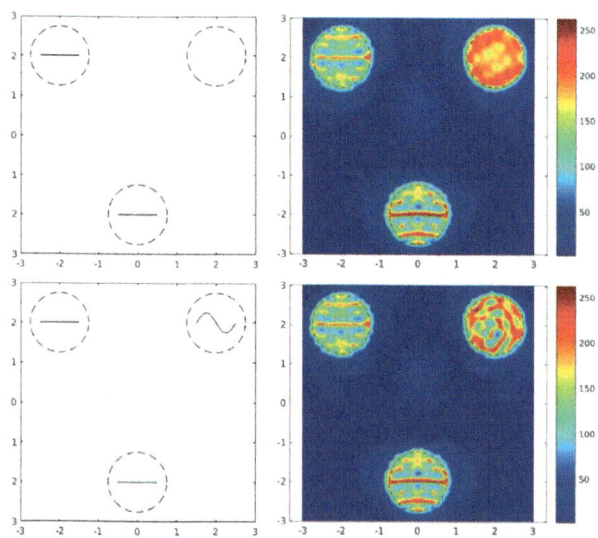

Fig. 3. Simulated backgrounds on the left and associated GLSM indicator function $z \mapsto I_{\mathrm{GLSM}}^{\alpha,\delta}(z)$ on the right.

5.2 DLSM

Given two far field data F_0^δ and F_1^δ, we respectively define $F_0^{\sharp\delta}$, $g_{0,z}^{\alpha,\delta}$, $\mathscr{A}_0^{\alpha,\delta}(z)$ and $F_1^{\sharp\delta}$, $g_{1,z}^{\alpha,\delta}$, $\mathscr{A}_1^{\alpha,\delta}(z)$ associated to each data as described in the previous paragraph. We also define

$$\mathscr{D}^{\alpha,\delta}(z) = \langle F_0^\sharp (g_{1,z}^{\alpha,\delta} - g_{0,z}^{\alpha,\delta}), (g_{1,z}^{\alpha,\delta} - g_{0,z}^{\alpha,\delta}) \rangle_{L^2(\mathbb{S}^{d-1})}.$$

Then, according to (26), the DLSM indicator is given by

$$I_{\mathrm{DLSM}}^{\alpha,\delta}(z) = \frac{1}{\mathscr{A}_0^{\alpha,\delta}(z)\left(1 + \frac{\mathscr{A}_0^{\alpha,\delta}(z)}{\mathscr{D}^{\alpha,\delta}(z)}\right)}.$$

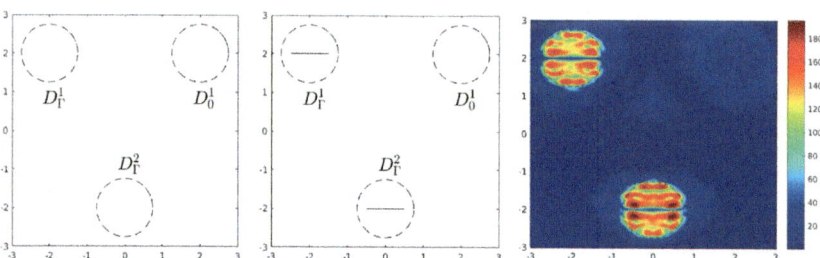

Fig. 4. A scenario for DLSM simulating the emergence of cracks in two components of a defect free background.

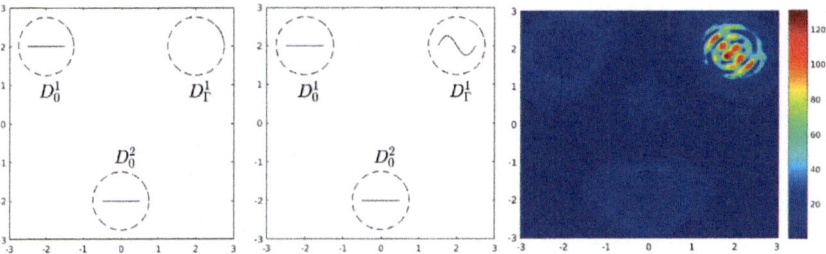

Fig. 5. A scenario for DLSM simulating the emergence of a crack in a healthy component of an already damaged background.

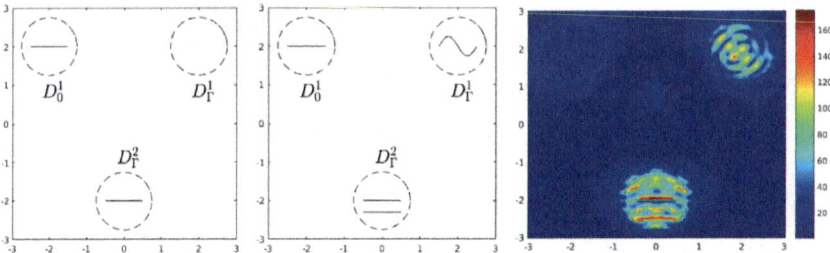

Fig. 6. A scenario for DLSM simulating the emergence of additional cracks in a healthy and a damaged components of an already damaged background.

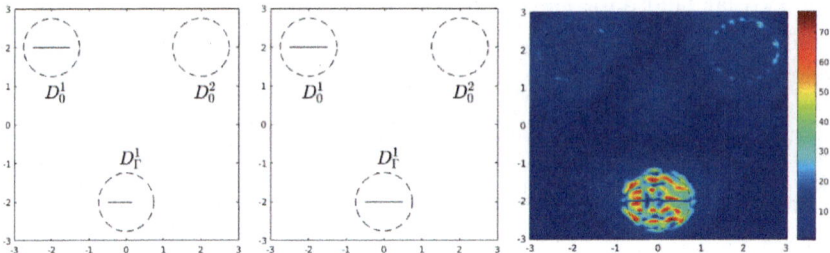

Fig. 7. A scenario for DLSM simulating the increase of the crack size in one component of an already damaged background.

The behavior of the DLSM indicator function is illustrated below for several scenarios shown in Figs. 4, 5, 6 and 7. In each figure is presented from left to right, the initial background (associated with F_0^δ), the damaged background (associated with F_1^δ) and the DLSM indicator function $z \mapsto I_{\mathrm{DLSM}}^{\alpha,\delta}(z)$. As expected, the latter allows us to identify for all scenarios the component(s) D_Γ where (additional) cracks appeared. We also remark that it slightly accentuates the border of D_0. But this effect is not explained by our theory and it does not contradict it: Our theoretical result does not stipulate that the indicator function is "uniformly" close to 0 outside D_Γ.

6 Conclusion

We analyzed the DLSM to identify emergence of cracks embedded in an unknown background and image defective components from differential measurements of far field data at a fixed frequency. The analysis is based on the justification of the GLSM for backgrounds with cracks which necessitates the study of a special interior transmission problem and the derivation of specific factorizations of the far field operator. The numerical tests on toy problems show that our method is reliable for different scenarios simulating the appearance of cracks between two measurements campaigns. This is a first step before addressing practical problems where the issues of limited aperture data and/or highly cluttered backgrounds should be solved.

References

1. Audibert, L.: Qualitative methods for heterogeneous media. Ph.D. thesis, École Doctorale Polytechnique (2015)
2. Audibert, L., Haddar, H.: A generalized formulation of the linear sampling method with exact characterization of targets in terms of farfeld measurements. Inverse Probl. **30**, 035011 (2014)
3. Audibert, L., Girard, A., Haddar, H.: Identifying defects in an unknown background using differential measurements. Inverse Probl. Imag. **9**(3), 625–643 (2015)
4. Hassen, F.B., Boukari, Y., Haddar, H.: Application of the linear sampling method to retrieve cracks with impedance boundary conditions. Rapport de recherche RR-7478, INRIA (2010)
5. Bourgeois, L., Lunéville, E.: On the use of the linear sampling method to identify cracks in elastic waveguides. Inverse Probl. **29**, 025017 (2013)
6. Cakoni, F., Colton, D.: Qualitative methods in inverse scattering theory interaction of mechanics and mathematics: an introduction. Springer (2006)
7. Cakoni, F., Colton, D., Haddar, H.: Inverse Scattering Theory and Transmission Eigenvalues. Series in Applied Mathematics (2016)
8. Colton, D., Kress, R.: Inverse Acoustic and Electromagnetic Scattering Theory. Springer, Heidelberg (1992)
9. Colton, D., Kress, R.: Integral Equation Methods in Scattering Theory. Wiley, Hoboken (1993)
10. Kirsch, A.: An Introduction to the Mathematical Theory of Inverse Problems. Springer, Heidelberg (2011)
11. Kirsch, A., Ritter, S.: A linear sampling method for inverse scattering from an open arc. Inverse Prob. **16**, 89 (2000)
12. McLean, W.: Strongly Ellyptic Systems and Boundary Integral Equations. Cambridge University Press, Cambridge (2000)

On the Research of Extra Characteristic Frequencies in a Planetary Gearbox

Oussama Graja[1]([✉]), Bacem Zghal[1], Kajetan Dziedziech[2], Fakher Chaari[1],
Adam Jablonski[2], Tomasz Barszcz[2], and Mohamed Haddar[1]

[1] Laboratory of Mechanics Modelling and Production,
National School of Engineers of Sfax, BP 1173, Sfax 3038, Tunisia
grajaoussama@gmail.com

[2] AGH University of Science and Technology, Krakow, Poland

Abstract. Gearboxes have been investigated and monitored for decades since they present one of the important transmission power systems which have been used in navy, air and automotive sectors. One of the most adopted one is the planetary gearbox since it has an important reduction ratio within compact space. The dynamic behaviour of a such one is very complicated because it possesses several gears in mesh and differs from other types of gearboxes by the fact that planet gears can occupy different positions in one period carrier rotation which leads to an important influence on the overall vibration signal acquired by a transducer mounted one the external housing. Consequently, in a measured vibration spectrum, the pass planet frequency component is identified and its energy level is considered only as the pass planet energy. However, there is another phenomenon that increases the level of the pass planet frequency component which is the disequilibrium phenomenon due to the rotation of planets. In this work, a comprehensive monitoring of a staged planetary gearbox is presented. The unbalance phenomenon is investigated in every stage. Then, an experimental validation is provided in order to support our hypothesis claiming that the disequilibrium phenomenon depends on the parity of the number of planets.

Keywords: Planetary gearbox · Characteristic frequencies

1 Introduction

Since planetary gearbox is widely used, researchers and engineers have focused on its behaviour. The same methodology is followed: they mount a transducer on the stationary gear (e.g.: ring gear in case of planetary gearbox) then they record the induced vibration as a signal during a definite time to obtain the time series. Next, they applied one or more signal processing tools in order to seek the dynamic behaviour in either healthy or damaged cases. The famous signal processing tool is the FAST FOURIER TRANSFORM (FFT) which transforms the vibration signal to a gathered frequency components indicating how many mixed signals are in the overall time series.

© Springer Nature Switzerland AG 2019
T. Fakhfakh et al. (Eds.): ICAV 2018, ACM 13, pp. 74–80, 2019.
https://doi.org/10.1007/978-3-319-94616-0_7

The methodology of analysing spectra either simulated or measured is found in many works, for instance, conducted by Chaari et al. [5], Inalpolat et al. [3,4], Feng et al. [2] and Liu et al. [1]. Their main purpose is to identify frequency components of different existing phenomena in a gearbox. Chaari et al. [5] developed an analytical model to investigate the vibration signature in a spectrum of two different tooth faults: spalling and breakage. Inalpolat and Kahraman [3] examined the side-band activity of different type of planetary gear-set in healthy case. They concluded that planetary gearbox can be divided into five groups depending on the space between planets and the phase between gear-meshes. Later on, Inalpolat and Kahraman [4] studied vibration spectra of planetary gear-set having some manufacturing errors in order to scrutinize their vibration signatures. Feng et al. [2] developed a vibration signal model of a planetary gear-set in order to deduce equations for calculation of characteristic frequencies of either faulty sun gear, planet gear or ring gear. The spectra was used to identify the signature of faults. Liu et al. [1] developed a lumped parameter model of a planetary gear-set for the purpose of the transmission path effect investigation on the overall vibration signal. To validate their work, they used an acceleration sensor to acquire vibration signal then they compared both simulated and experimental of both time series and spectra.

In every measured spectrum, frequency components are investigated in terms of localisation but not in terms of energy level. Hence, it can be possible that two phenomena which have the same characteristic frequency will overlap. This work is focused on a such problem specifically on the pass planet frequency component. In Sect. 2, a staged planetary gearbox is presented. Later, in Sect. 3, it is monitored to explore another phenomenon which has the same frequency component of the pass planet phenomenon. As a consequence, the energy level at this frequency component is equal to the energy summation of two different phenomena. Finally, in Sect. 4, a validation of the hypothesis is displayed by analysing experimental measurements.

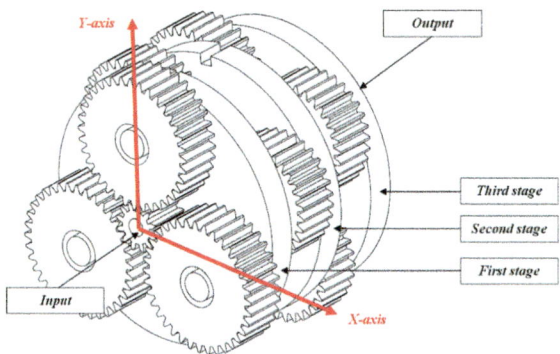

Fig. 1. 3D staged planetary gearbox

2   Presentation of the Staged Planetary Gearbox

Figure 1 presents the investigated planetary gearbox. It contains totally three stages with the same common ring which is hidden in order to display all stages.

Table 1 resumes geometric and physical characteristic of the staged planetary gearbox.

Table 1. Geometric and physical parameters of the gearbox

Components	⊘(mm)/Number	Teeth	Modulus
Sun(st1, st2, st3)	7.2/1, 12.6/1, 12.6/1	12, 21, 21	0.6
Planet(st1, st2, st3)	21.6/3, 18.6/3, 18.6/4	36, 31, 31	0.6
Common ring	50.4/1	84	0.6

3   Monitoring the Staged Planetary Gearbox

By monitoring the planetary gearbox, some snapshots are taken and presented to focus on the motion of planets of each stage. They are given in Figs. 2 and 3.

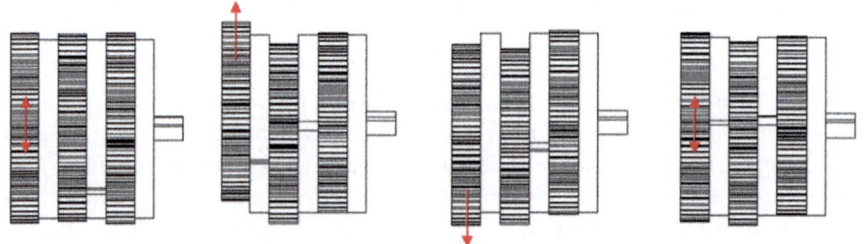

Fig. 2. Snapshots of the first stage

As shown in Fig. 2, the instantaneous positions of the first stage planet gears will create a disequilibrium where its frequency is equal to the first pass planet frequency.

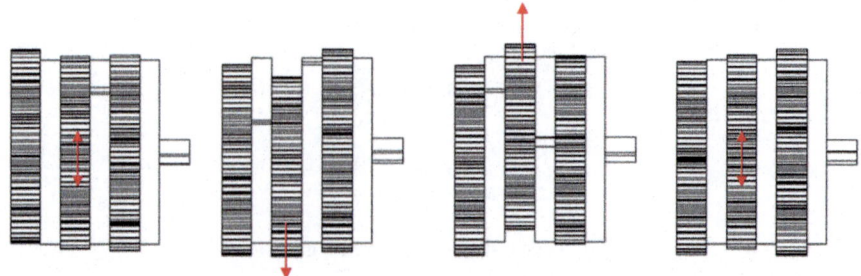

Fig. 3. Snapshots of the second stage

The same phenomenon will occur in the second stage with a frequency equal to the second pass planet frequency. However, focused on planets of the third stage, they will not cause any disequilibrium since they move with the same path from either side.

Matching those results with the kinematics of the planetary gearbox, we can conclude that:

– When the number of planets N is an odd number (e.g.: 3 planets of stage 1), the disequilibrium occurs and its frequency is equal to $f_{diseq} = N \times f_c$ with f_c is the frequency of the carrier which holds the N planets.
– If not, in other words, the number of planets N is an even number, the disequilibrium does not occur.

Consequently, there are totally two signals of the disequilibrium induced by only the first and the second stage since those latter hold odd number of planets.

As a result, the frequency component located at $N \times f_c$ does not present only the energy of pass planet but also the energy caused by the disequilibrium phenomenon. In other words, it displays the summation of both phenomena.

4   Experimental Results and Discussion

To investigate the staged planetary gearbox, a test rig in the laboratory of AGH University, presented in Fig. 4, is used.

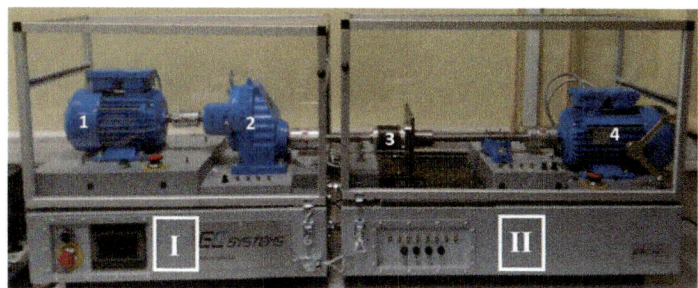

Fig. 4. Test rig

The test rig is divided into two main sides: a driving side (I) composed by a motor (1) and a parallel gearbox (2) with a reduction ratio equal 2.91:1, and a driven side (II) composed by a motor load (4) and the investigated planetary gearbox (3). The motor was operating at 1009 rpm. Passing through the parallel gearbox, the speed was reduced to 346.5 rpm which is the input shaft speed of the staged planetary gearbox. Consequently, the pass planet frequency components of the first, second and third stages are respectively 2.17 Hz, 0.43 Hz and 0.12 Hz. Since we deal with low frequency components, a strain gage is installed to measure strains on X-axis as Figure 5 shows. It was glued to the casing of the gearbox in this direction in order to measure the induced strains on X-axis.

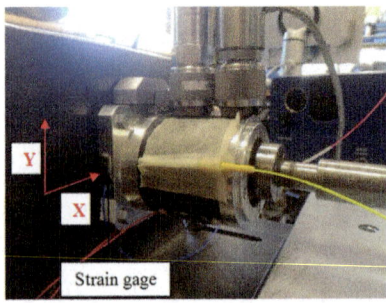

Fig. 5. Strain gage location

The measurement of both strain and velocity signals acquired by respectively the strain gage and the LAZER Vibrometer were done during 100 seconds. Hence, the frequency resolution was equal to 0.01 Hz which presents a good resolution to our case study and the sampling frequency was equal to 25 KHz which leads to a Nyquist frequency equal to 12.5 KHz, so all characteristic frequency components can be shown within this range.

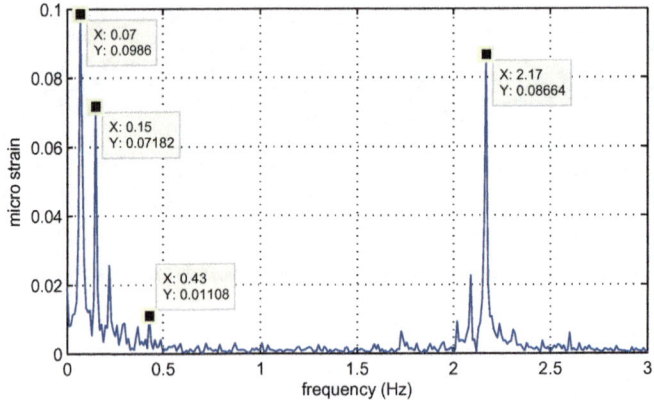

Fig. 6. Measured spectrum on X-axis

Figure 6 displays the measured spectrum. The components at 0.43 Hz and 2.17 Hz remain figurative. But, the component at 0.12 Hz does not occur. Consequently, the experimental findings deal with the hypothesis mentioned in Sect. 3 which declares that the disequilibrium frequency $f_{diseq} = N \times f_c$ might be a characteristic frequency if the number N of planets held by the carrier is an odd number, if not, it cannot be.

To emphasize the fact that the frequency component located at $N \times f_c$ does not present only the energy of pass planet, for instance $f_{pp} = N_1 \times f_{c1} = 2.17\,\text{Hz}$, another measurement on X-axis was done with using a LAZER-Vibrometer, as shown in Fig. 7, in order to measure the non relative motion on X-axis of the planetary gearbox since this transducer is not attached to it.

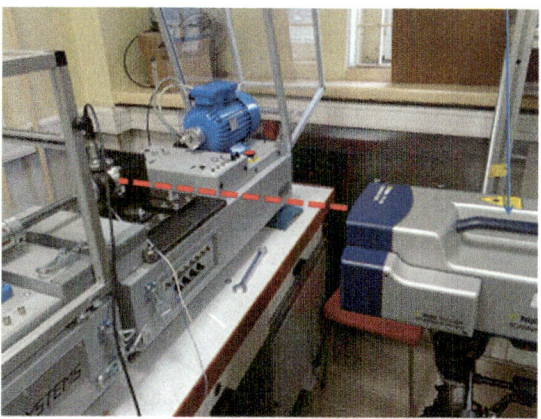

Fig. 7. Measurement done by LAZER-Vibrometer

Figure 8 shows the measured spectrum. A frequency component located at 2.16 Hz remains figurative. Therefore, the disequilibrium phenomenon of the first stage occurs with a fundamental frequency equal to 2.16 Hz.

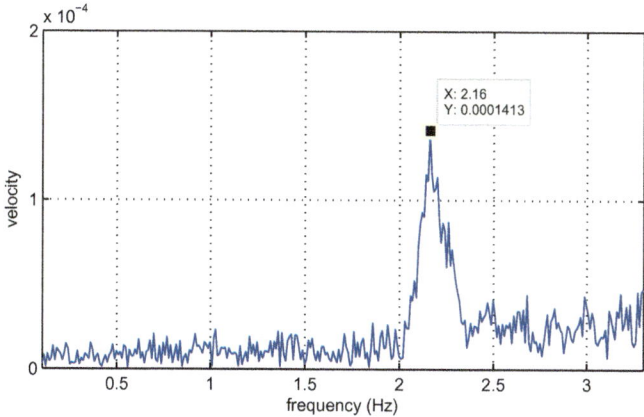

Fig. 8. Measured spectrum on X-axis

5 Conclusion

In this work, a comprehensive monitoring of a staged planetary gearbox was presented in order to scrutinise the existing of a novel phenonmenon which is the disequilibrium due to the rotation of planets around the center of the planetary gearbox. The fundamental frequency of this phenomenon is equal unfortunately to the pass planet frequency. Hence, the aim of this work is to demonstrate the overlapping between those two phenomena. For this purpose, two different sensors were used: one of them was attached and the other was not in order to capture the non relative motion of the hole gearbox. Finally, by analyzing the measured specra from both transducers, we conclued that the disequilibrium phenomenon occurs only if the number of planets is odd.

Acknowledgements. This work is partially supported by NATIONAL SCHOOL OF ENGINEERS OF SFAX (ENIS)/Laboratory of Mechanics, Modelling and Production (LA2MP) and The National Centre of Research and Development (NCRD) in Poland under the research project no. PBS3/B6/21/2015.

References

1. Liu, L., Liang, X., Zuo, M.J.: Vibration signal modeling of a planetary gear set with transmission path effect analysis. Measurement **85**, 20–31 (2016)
2. Feng, Z., Zuo, M.J.: Vibration signal models for fault diagnosis of planetary gearboxes. J. Sound Vib. **331**, 4919–4939 (2012)
3. Inalpolat, M., Kahraman, A.: A theoretical and experimental investigation of modulation sidebands of planetary gear sets. J. Sound Vib. **323**, 677–696 (2009)
4. Inalpolat, M., Kahraman, A.: A dynamic model to predict modulation sidebands of a planetary gear set having manufacturing errors. J. Sound Vib. **329**, 371–393 (2010)
5. Chaari, F., Baccar, W., Abbes, M.S., Haddar, M.: Effect of spalling or tooth breakage on gearmesh stiffness and dynamic response of a one-stage spur gear transmission. Eur. J. Mech. A/Solids **27**, 691–705 (2008)

Mutiphysics Systems Dynamics

Analytical Based Approach for Vibration Analysis in Modelica: Application to the Bridge Crane System

Ghazoi Hamza[1]([⊠]), Moncef Hammadi[2], Maher Barkallah[1],
Jean-Yves Choley[2], Alain Riviere[2], Jamel Louati[1],
and Mohamed Haddar[1]

[1] Mechanics Modeling and Production Research Laboratory (LA2MP),
National School of Engineers of Sfax (ENIS), University of Sfax,
B.P. 1173, 3038 Sfax, Tunisia
hamza.ghazoi@gmail.com, bark_maher@yahoo.fr,
louati.ttg@gnet.tn, mohamed.haddar@enis.rnu.tn
[2] QUARTZ EA 7393, SUPMECA, 3 rue Fernand Hainaut,
93407 Saint-Ouen Cedex, France
{moncef.hammadi,jean-yves.choley,
alain.riviere}@supmeca.fr

Abstract. Most engineering system, machines and products have moving parts and in order to achieve a desired performance, they require the manipulation of their mechanical or dynamic behavior from the early stage of design. Also, the dynamic interaction between the moving object and the structure should be properly considered at this level. The objective of the presented paper is to propose a new methodology for the pre-design of a mechatronic system, considering the vibrational behavior using the object oriented modeling language Modelica with Dymola environment. In fact, we study the dynamic behavior of a supporting flexible beam structure (simply supported at both ends) traversed by moving masses at variable speeds, based on the object-oriented modelling paradigm developed in Modelica. An analytical approach is adopted, providing a compromise between the results accuracy and the computation time. To illustrate the methodology, the bridge crane system is used as a supporting study. This machine is commonly used in industrial facilities. The effects of varying the different parameters on the dynamic response of the system are investigated. This methodology would be useful for a designer to have an overview about the system response and the interaction between the different subcomponents in the conceptual design phase.

Keywords: Conceptual design · Vibration mechatronic system
Beam structure · Modelica

1 Introduction

Mechatronic systems play an important role in different types of industry such as transportation, automotive, aerospace, etc. The design of a mechatronic system requires a high degree of integration (Hammadi et al. 2012). Therefore, the mechatronic system

© Springer Nature Switzerland AG 2019
T. Fakhfakh et al. (Eds.): ICAV 2018, ACM 13, pp. 83–91, 2019.
https://doi.org/10.1007/978-3-319-94616-0_8

is frequently divided into simpler subsystems or subcomponents and assigned to different design teams. During the mechatronic system design process, the collaboration between the different domains plays a key role.

The preliminary design or the conceptual design is an important stage in the development of a mechatronic system (Hamza et al. 2017). The target of this step is to specify the system components before the detailed design of each of them. To take into account the different design constraints, it is necessary to have models as simple as possible to reduce the number of parameters to be provided, and as fast as possible to minimize the calculation time (Hamza et al. 2014; Hamza et al. 2015a, b).

Products are becoming more complex due to the integration of different components form different engineering domains such as electrical, mechanical, hydraulics, control, etc. An integrated environment is essential to closely engage the different domains together (Hammadi et al. 2014; Hammadi and Choley 2015).

Mechatronic systems are often subjected to different phenomena such as heat conduction, convection, vibration, etc. For such phenomena like vibrations in flexible mechanical structures, models are generally of the form of partial differential equation (PDE). These problems cannot be solved in all cases analytically so we resort to numerical methods such as the finite element method (FEM) which provides approximate solutions.

Vibrations analysis of structures such as plates, shells and beams have been of general interest to the scientific and engineering communities. These structures have many applications in almost every industry. Engineering structures must be strong, safe and economical, materials should be used efficiently (Mehmood et al. 2014).

Flexible structures in industry are common. Flexibility is not embedded in most systems from the early stage of design. In an earlier paper, the authors (Hamza et al. 2015a, b) proposed a new pre-design method applied to a mechatronic system taking into account the vibrational effect. In fact, the authors studied the dynamic interaction between components (motors and electronics cards) located on a simply supported plate using an analytical approach with Dymola.

Elastic structures, such as plates and beams under the action of moving loads are one of the major research topic in structural engineering fields and civil. In the last century, moving load problem has been the subject of many research efforts. Two approaches are generally adopted to build a model, that is, an analytical method or a finite element (FE) method. Analytical approach is used to simple cases of structure; however, the FEM is adopted generally for complex structures. For instance, Awodola (2014) studied the dynamic vibration of a simply supported plate resting on Winkler elastic foundation with stiffness variation, carrying moving masses using the technique of variable separation.

The bridge crane system is a typical structure under moving loads in mechanical engineering. The bridge crane is used to transfer the heavy payload from one location to another. Gašić et al. (2011) studied the dynamic response of the bridge crane system using the direct integration method. The effect of selected parameters such as the trolley speed and mass are analyzed.

In the present study, the objective is to investigate a new pre-design method based on an analytical approach, taking into consideration the effect of vibration. More precisely, we focus on the effect of successive moving loads traversing a simply supported beam. This methodology is illustrated for the bridge crane system.

The paper is organized as follows: in Sect. 2 the mathematical formulation of the system is presented, in Sect. 3 the Modelica/Dymola implementation is analyzed. Selected simulations results are presented in Sect. 4. The conclusion remarks of our paper are given in Sect. 5.

2   Model Formulation of the Beam Vibration

The system under consideration consists of a simply supported beam of length L carrying an arbitrary number (say N) of concentrated moving masses M_i, moving with constant velocities v_i (i = 1, 2, 3....N) from left to right. The model is depicted in Fig. 1. The beam obeys the Euler Bernoulli theory. The beam is supposed to undergo small deflections. The moving masses are always in contact with the beam and do not generate friction force. The mathematical formulation of the beam vibration has been inspired from the model developed in reference (Stanišić and Hardin 1969).

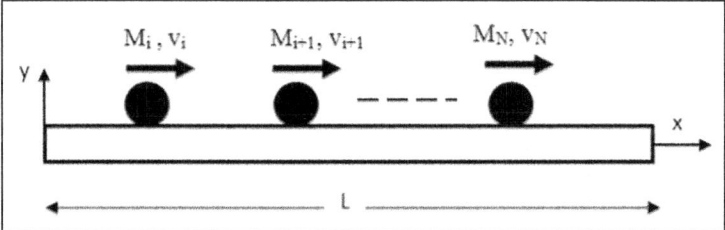

Fig. 1. The beam model subjected to moving concentrated masses

The governing equation of motion of an Euler-Bernoulli beam subjected to moving masses can be written as:

$$EI\frac{\partial^4 w(x,t)}{\partial x^4} + \left[\rho S + \sum_{i=1}^{N} M_i\delta(x - x_i)\right]\frac{\partial^2 w(x,t)}{\partial t^2} = g\sum_{i=1}^{N} M_i\delta(x - v_it) \qquad (1)$$

In this equation, w is the transverse deflection of the beam, EI is the flexural rigidity of the beam, ρ is the beam mass density, S is the cross sectional area of the beam, g is the acceleration of gravity. The beam is considered to be square in cross section.

$$\delta(x - v_it) \text{ is the Dirac Delta function}$$
$$\delta(x - v_it) = 0 \text{ for } x \neq v_it \qquad (2)$$

The boundary conditions of the beam are:

$$W(0) = 0, \ W(l) = 0$$
$$\frac{d^2W}{dx^2}(0) = 0, \frac{d^2W}{dx^2}(l) = 0 \tag{3}$$

The beam is simply supported at both ends; the Eigen frequency for m^{th} mode is given by:

$$\omega_m = \frac{m^2\pi^2}{l^2}\sqrt{\frac{EI}{\rho A}} \tag{4}$$

In addition, $\int_0^L \delta(x - v_i t)dx = 1 \tag{5}$

Applying the Fourier finite sine transform,

$$Z(m,t) = \int_0^L w(x,t)\sin\frac{m\pi x}{L}dx \tag{6}$$

The transformed equation of the problem is,

$$Z_{tt}(m,t) + \omega_m^2 Z(m,t) + \frac{1}{\rho SL}\sum_{i=1}^N M_i\left[Z_{tt}(m,t) + 2\sum_{k=1}^{\infty} Z_{tt}(k,t)\sin\frac{k\pi v_i t}{L}\sin\frac{m\pi v_i t}{L}\right]$$
$$= \frac{g}{\rho S}\sum_{i=1}^N M_i\sin\frac{m\pi v_i t}{L} \tag{7}$$

We consider only one mass M moving with velocity v.
If we consider only the linear inertia term, Eq. 7 becomes:

$$Z_{tt}(m,t) + \frac{\omega_m^2}{(1+R)}Z(m,t) = \frac{P}{(1+R)}\sin\frac{m\pi vt}{L}$$
$$R = \frac{M}{\rho SL} \tag{8}$$

Then,

$$Z(m,t) = \frac{p}{\omega_m^2\left[1 - \left[\frac{m\pi v\sqrt{1+R}}{L\omega_m}\right]^2\right]}\left[\sin\frac{m\pi vt}{L} - \frac{m\pi v\sqrt{(1+R)}}{L\omega_m}\sin\frac{\omega_m t}{\sqrt{(1+R)}}\right] \tag{9}$$

Defining,

$$P = \frac{Mg}{\rho S} \tag{10}$$

The expression of the beam deflection $W(x,\ t)$ can be written as in the following,

$$w(x,t) = \frac{2P}{L} \sum_{m=1}^{\infty} \frac{\left[\sin\frac{m\pi vt}{L} - \frac{m\pi v\sqrt{1+R}}{L\omega_m} \sin\frac{\omega_m}{\sqrt{(1+R)}}t \right]}{\omega_m^2 \left(1 - \left(\frac{m\pi v\sqrt{(1+R)}}{L\omega_m} \right)^2 \right)} \sin\frac{m\pi x}{L} \tag{11}$$

3 System Description

The bridge crane model system is composed of two main parts (Fig. 2) which are the moving system (the trolley) and the framework (structure). This example will serve to illustrate the method. We consider that the beam portion of the crane was simply supported and that it can be modelled adequately using the Euler-Bernoulli beam theory. The beam is assumed to undergo small deflections. The crane and the payload are modeled as a concentrated moving load.

Fig. 2. Bridge crane system (Gašić et al. 2011)

4 Modelica System Modeling

In this paper, the bridge crane system model has been modelled using Modelica language. In fact, based on the analytical model presented in Sect. 2, we have created in Modelica/Dymola two components which are the beam model and the carriage model.

Figure 3 shows the top-level structure of the bridge crane system. The system model is composed by two main components which are the beam (framework) and the carriage (trolley and payload). The beam model parameters include beam specific design data such as the beam length and the Young modulus, etc. The carriage model is represented as a concentrated mass moving at a specified speed. To connect these new models built in Dymola, we have developed a new connector. The code of the new connector is the following:

```
connector C1
Modelica.SIunits.Position s;
Modelica.SIunits.Force F;
 a
end C1;
```

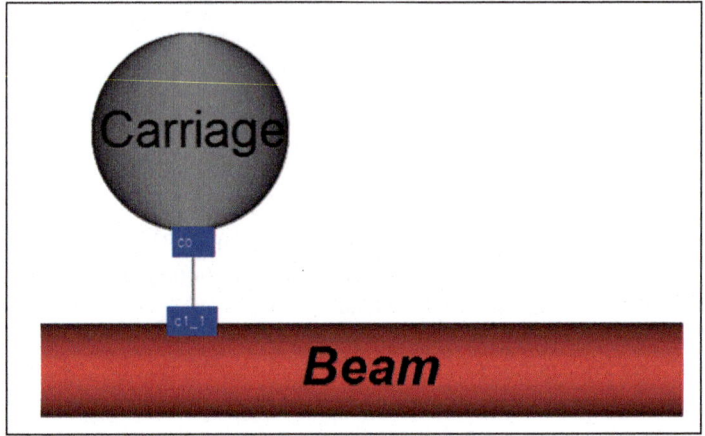

Fig. 3. Bridge crane model developed in Modelica

This early mechatronic configuration allows us to estimate the vibrational behavior of the bridge crane system. It represents a first attempt in order to increase the understanding of the cranes dynamics due to the moving masses.

5 Parametric Study

Using the proposed method of system analysis, we can analyze the influence of the system parameters on its characteristics and the effects of some selected parameters on the dynamic flexibility. The parameters used in the simulation are listed in Table 1.

Table 1. System parameters

Parameter	Value
Young modulus	$E = 2.1 \times 10^{11}$ N/m^2
Mass density	$\rho = 7850$ kg/m^3
Beam length	$L = 40$ m
Cross section	$A = 0.04$ m^2
Moment of inertia	$I = 0.00667$ m^4
Speed	$v = 5$ m/s
Load mass	$M = 10,000$ kg

The deflection for different points on the length of the beam is shown in Fig. 4. It can be seen that the three curves have the same oscillation frequency and they are in phase, moreover the beam displacements dependent on the point position and the maximum deflection is in the mid-point.

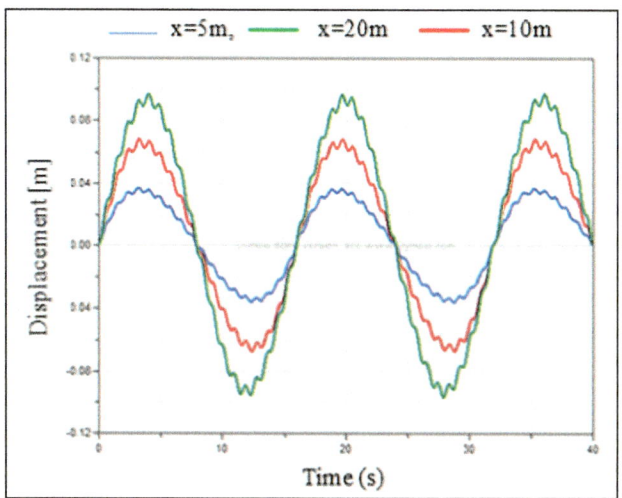

Fig. 4. Beam deflections for different points on the length of the beam ($M = 10^4$ kg, $v = 5$ m/s)

Figure 5 depicts the beam deflection for different values of the carriage speed. As expected, the three curves are not in phase and the beam deflection appears to increase slightly with increasing speed. Then, the beam deflection for a given set of trolley and payload mass is dependent upon the trolley speed. Fluctuations depend on the value of the ratio $T1/\tau$. The symbol T_1 denotes the fundamental period of the beam and $\tau = L/v$ represents the travel time of the load from left end to the right end of the beam.

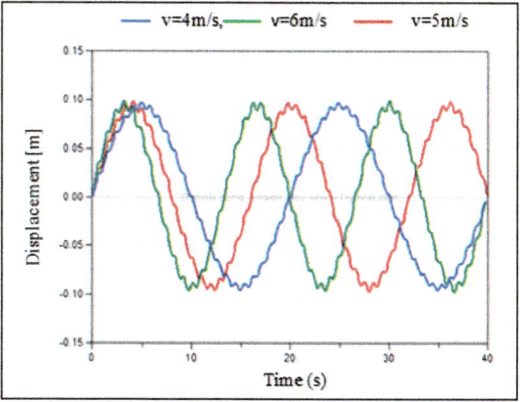

Fig. 5. Beam deflections for different value of the trolley speed

The result of the investigation of the influence of the mass on the dynamic response of the beam is illustrated in Fig. 6. It is cleared that the amplitude of deflection increases due to the increase of the load mass.

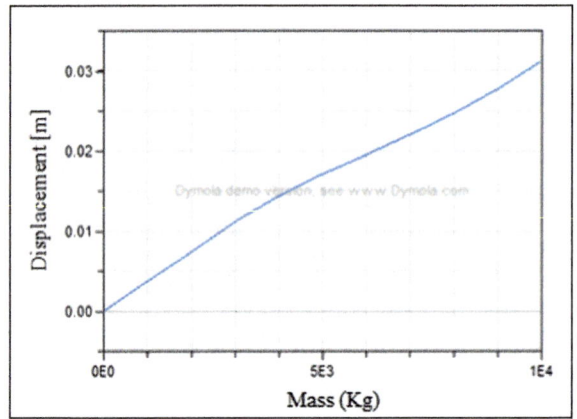

Fig. 6. Beam deflection according to the load mass

Figure 7 shows the effect of the speed of the moving mass on the amplitude of deflection of the beam. As it can be observed, the amplitude of deflection increases as the speed of the moving mass is increased while the other parameters are fixed.

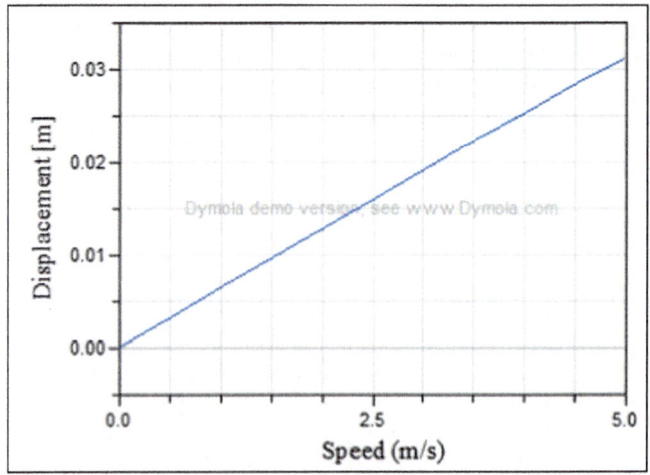

Fig. 7. Beam deflection according to the trolley speed

6   Conclusion

We propose a new method to model and simulate a mechatronic system, considering the vibrational behavior with Modelica/Dymola. In fact, we investigate the dynamic phenomena produced by the interaction between a supporting flexible structure (simply supported beam) and moving masses. Our methodology is based on an analytical method, which decrease the need for costly finite element analysis. This approach was applied to the bridge crane system. Having a global system model in the conceptual design level is of great interest since it captures the interaction between the different components. In order to capture the pertinent characteristics, various plots of the beam deflections are presented and discussed.

References

Hammadi, M., Choley, J.Y., Penas, O., Riviere, A., Louati, J., Haddar, M.: A new multi-criteria indicator for mechatronic system performance evaluation in preliminary design level. In: 2012 9th France-Japan and 7th Europe-Asia Congress on Mechatronics (MECATRONICS) and 2012 13th International Workshop on Research and Education in Mechatronics (REM), pp. 409–416. IEEE (2012)

Hamza, G., Hammadi, M., Barkallah, M., Choley, J.Y., Riviere, A., Louati, J., Haddar, M.: Conceptual design methodology for the preliminary study of a mechatronic system: application to wind turbine system. Mech. Ind. **18**(4), 413 (2017)

Hamza, G., Barkallah, M., Louati, J., Haddar, M., Hammadi, M., Choley, J.Y., Riviere, A.: Analytical approach for the integrated preliminary analysis of mechatronic systems subjected to vibration. In: 2014 10th France-Japan/8th Europe-Asia Congress on Mecatronics (MECATRONICS), pp. 151–155. IEEE (2014)

Hamza, G., Choley, J.Y., Hammadi, M., Barkallah, M., Louati, J., Riviere, A., Haddar, M.: Pre-dimensioning of the dynamic properties of the wind turbine system using analytical approach. In: Design and Modeling of Mechanical Systems-II, pp. 179–188. Springer (2015a)

Hammadi, M., Choley, J.Y., Mhenni, F.: A multi-agent methodology for multi-level modeling of mechatronic systems. Adv. Eng. Inform. **28**(3), 208–217 (2014)

Hammadi, M., Choley, J.Y.: Parametric compact modelling of dynamical systems using meshfree method with multi-port technique. Int. J. Dyn. Syst. Differ. Equ. **5**(3), 206–219 (2015)

Hamza, G., Choley, J.Y., Hammadi, M., Riviere, A., Barkallah, M., Louati, J., Haddar, M.: Pre-designing of a mechatronic system using an analytical approach with Dymola. JTAM **53**(3), 697–710 (2015b)

Mehmood, A., Khan, A.A., Mehdi, H.: Vibration analysis of beam subjected to moving loads using finite element method. IOSR J. Eng. (IOSRJEN) **4**(05), 07–17 (2014)

Stanišić, M.M., Hardin, J.C.: On the response of beams to an arbitrary number of concentrated moving masses. J. Frankl. Inst. **287**(2), 115–123 (1969)

Awodola, T.O.: Flexural motions under moving concentrated masses of elastically supported rectangular plates resting on variable winkler elastic foundation. Latin Am. J. Solids Struct. **11**(9), 1515–1540 (2014)

Gašić, V., Zrnić, N., Obradović, A., Bošnjak, S.: Consideration of moving oscillator problem in dynamic responses of bridge cranes. FME Trans. **39**(1), 17–24 (2011)

Dynamic Relaxation Coordination Based Collaborative Optimization for Optimal Design of Multi-physics Systems

Hamda Chagraoui[(✉)] and Mohamed Soula

Laboratory of Applied Mechanics and Engineering, ENIT,
Department of Mechanical Engineering ENSIT-Tunis,
University of Tunisia, Tunis, Tunisia
chagraoui_hamda@yahoo.fr, Soulamed2003@yahoo.fr

Abstract. To solve problems of higher computational burden in standard collaborative optimization (CO) approach during the processing of design problem of the multi-physics systems with multiples disciplines, a Dynamic Relaxation Coordination based Collaborative Optimization (DRC-CO) method is presented. The main concept of DRC-CO method is to decompose the global design problem into one optimization problem at the system level and several autonomous sub-problems at disciplinary level. At the system level, the dynamic relaxation coordination aims to solve the inconsistency between all disciplines, which leads the optimization process converging to the feasible optimum efficiently. To demonstrate the efficiency and accuracy of the proposed DRC-CO method, a safety isolation transformer is considered. The obtained results of the engineering multi-physics system show the effectiveness of the proposed DRC-CO process compared to Single Level Optimization (SLO) and standard CO methods. The obtained optimal configuration of the safety isolation transformer in terms of total mass using DRC-CO method (2.30 kg) is close to the result obtained from SLO method (2.31 kg) with an absolute percentage error is less than 0.5%. Moreover, our approach requires 3 system iterations to find realizable designs. However, an important number of disciplinary design problems were evaluated at the disciplinary level optimizer.

Keywords: Multi-physics system · Improved collaborative optimization
Dynamic relaxation coordination · Single level optimization
Safety isolation transformer

1 Introduction

Multidisciplinary design optimization (MDO) is an effective method for solving large-scale and complex engineering systems that involve an important number of disciplines, such as aerospace design problems. In recent years, MDO method has broken into other fields such as automotive, mechanical, electrical engineering and electromagnetic devices. For example, the electromagnetic devices often involve an important number of disciplines, such as mechanical, electrical, thermal, which are fully related to each other by interdisciplinary interactions.

© Springer Nature Switzerland AG 2019
T. Fakhfakh et al. (Eds.): ICAV 2018, ACM 13, pp. 92–100, 2019.
https://doi.org/10.1007/978-3-319-94616-0_9

Many researchers have shown in their archival articles (Chagraoui and Soula 2017; Chagraoui et al. 2016; Xia et al. 2016; Balesdent et al. 2012) the interest in the use Multidisciplinary Design Optimization (MDO) methods to solving the design problem of complex coupled system. The MDO methods can be classified into two types: single-level and multi-level architectures (Balesdent et al. 2012).

Single level architectures employ only an optimizer for solving the entire design problems. The single level architectures aim to handle the MDO problem by casting it as a single optimization problem, which is simple to be implemented for less complex MDO problems. But for complex and/or multi-physics systems where each discipline works independently of one another, these approaches may encounter a big challenge in integrating all the disciplines simultaneous.

The single level architectures are summarized in (Balesdent et al. 2012) and which comprise: all at once, individual discipline feasible and multiple disciplines feasible. One big disadvantage of these mono-level optimization methods to a design problem of the complex system that, it is time-consuming or difficult to rapidly evaluate trade-offs between all disciplines using the traditional optimization approach.

However, multi-level approaches solve the MDO problem by decomposing it into a number of a disciplinary design problem. Collaborative optimization (Braun and Kroo 1996), concurrent subspace optimization (Sobieszczanski-Sobieski et al. 1998), bi-level integrated system synthesis (Sobieszczanski-Sobieski et al. 2000) and analytical target cascading (Kim et al. 2003) are four typical MDO methods. Among these methods, the Collaborative optimization (CO) approach requires less information exchange between the various disciplines and permits more flexibility in disciplinary optimization (Weiwei et al. 2013). This characteristic of the original CO method has attracted interests between researchers in refining the CO approach.

The original CO formulation was intended to hierarchically decompose the original optimization problem into a system level optimization problem and a number of independent sub-problems at disciplinary level. The goal of the system optimizer is to minimize the system objective function while making the compatibility constraint of each discipline zero. The aim of each discipline is to minimize its compatibility constraint which is defined by the discrepancy between the system level variables and the values obtained from disciplinary level while satisfying the local constraints.

There are a number of challenges associated with CO, that limit its performance in practical MDO problems. These challenges usually include (i) system level consistency equality constraints, (ii) outside the feasible region of the original optimization problem, (iii) easily trapped into local optima, (iv) low convergence rate. These issues result in a very high overall computational and poor performance in practice.

In order to deal with the problems above, a dynamic relaxation coordination based collaborative optimization (DRC-CO) is presented to improve the standard CO method in order to address the above-mentioned difficulties. The proposed DRC-CO method adopted the multi-level CO architecture using SQP algorithm as an optimizer for each design problem at system and discipline level optimizer to find the optimal configuration of the safety isolation transformer in terms of total mass.

The outline of the present work is organized as follows: In Sect. 2, the DRC-CO method is introduced. In Sect. 3, a safety isolation transformer optimization problem is considered to show the ability of the proposed approach. Finally, some relevant conclusions are presented in Sect. 4.

2 Dynamic Relaxation Coordination Based Collaborative Optimization: DRC-CO

The basic idea in the proposed DRC-CO approach is to divide the system optimization problem into one optimization problem at the system level and several independent optimization sub-problems disciplinary level as shown in Fig. 1. At the system level optimizer, a dynamic relaxation coordination is imposed for improving the efficiency and convergence rate of DRC-CO approach.

As shown in Fig. 1, only two disciplines are considered to show the proposed approach, but the proposed DRC-CO approach is applicable to systems with more than two disciplines.

In the DRC-CO approach in Fig. 1, it can be seen that the MDO problem is hierarchically decomposed into a number of disciplinary optimization problems. A system level optimizer aims to find the system design variables x_{sh}^s, denoted shared variables, by optimizing the system's objective functions f_s while satisfying the system's constraints g_s and the compatibility constraints (represented by $L_2 Norm$: $\left\| 1 - \left(y_{12}/y_{12}^* \right) \right\|_2^2 \le \epsilon_1$ and $\left\| 1 - \left(y_{21}/y_{21}^* \right) \right\|_2^2 \le \epsilon_2$). The aim of the two-compatibility constraints is to ensure the agreement between the two disciplines as following, the deviation of two system coupling variables y_{12}^s and y_{21}^s is matched with the optimal disciplinary coupling variable of discipline 1 and discipline 2 y_{12}^* and y_{21}^*, respectively. These compatibility constraints are constrained not exceed a dynamic relaxation factor (Wang et al. 2017) ε_1 and ε_2.

The dynamic relaxation factors ε_1 and ε_2 are computed by using the inconsistency information between the two disciplines as following:

$$\epsilon_1 = \lambda \times \Delta_1; \quad \epsilon_2 = \lambda \times \Delta_2 \quad 0.5 \le \lambda \le 1$$

where

$$\Delta_1 = \sqrt{\left\| 1 - (y_{12}/y_{12}^*) \right\|_2^2}; \quad \Delta_2 = \sqrt{\left\| 1 - (y_{21}/y_{21}^*) \right\|_2^2}$$

(1)

The inconsistency between disciplines gradually decreases along with the DRC-CO process and its value is close to 0 when the final convergence comes. Thus, the feasibility and robustness of optimal solution at system level optimization achieved.

The disciplinary optimal design local variables (x_1^*, x_2^{s*}) and coupling variables (y_{12}^*, y_{21}^*) are obtained at disciplinary level and used as parameters (fixed values) at system level optimizer. The subscript **12** of a coupling variable y_{12} defines the coupling variable is computed and output from discipline **1** and is input and used in discipline **2**. Likewise, in the firstly disciplines level optimizer **1**, O_1 and g_1 represent the disciplinary design objective and disciplinary constraints, respectively. In this discipline, the discipline's design variables X_1 include local variables x_1 and shared variables x_{sh}^1. y_{12} denotes the coupling variables which are defined as functions (Y_{12}) of X_1. Each disciplinary optimization problem of the DRC-CO architecture acts on its associated, local, design variables in order to find an agreement with the other discipline upon the coupling variables, while satisfying disciplinary constraints.

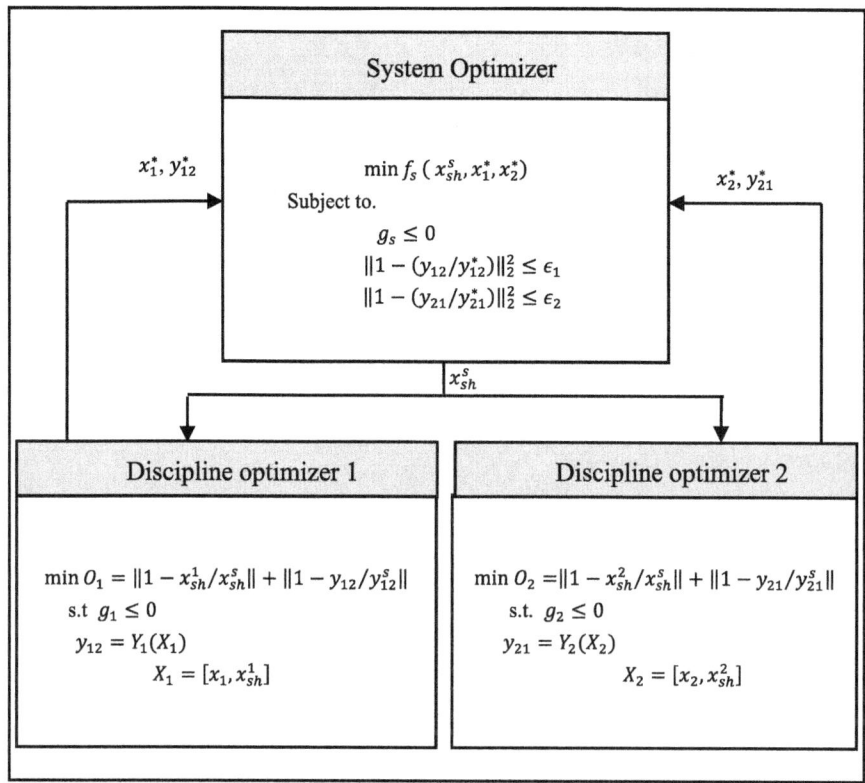

Fig. 1. Proposed DRC-CO approach

3 Safety Isolation Transformer Benchmark

3.1 Safety Isolation Transformer Model

The engineering design problem to be optimized is a low-voltage single-phase safety isolation transformer: its physical representation and design variables are shown in Fig. 2. The transformer design problem is taken from (Berbecea 2012) and which treated as a MDO benchmark. This device is a complex system that involves three fully coupled disciplines: a full-load electromagnetic discipline (EM_L), a no-load electromagnetic discipline (EM_0) and a thermal discipline (TH). The problem in hand has seven design variables are: four geometrical design variables of the transformer's iron core (a, b, c, d), two variables for the copper wire section of the primary and secondary windings denoted by S_1 and S_2 respectively, and one variable which represents the number of primary turns n_1.

The single level design problem of the transformer to solve aims to find an optimal configuration in terms of total mass of the studied device while satisfying seven physical constraints. The design problem of the transformer is stated as follow:

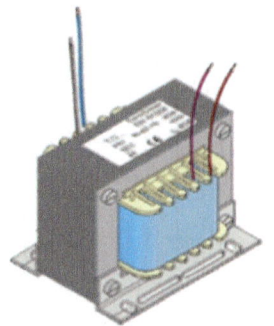   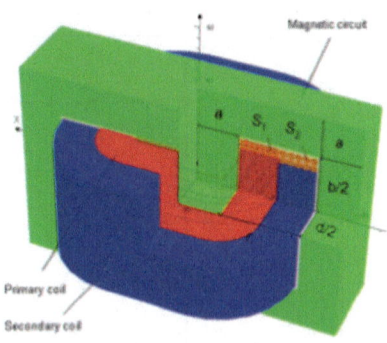

a)   **Physical representation**   b) **Geometrical variables representation**

Fig. 2. Safety isolation transformer to be optimized (Berbecea 2012).

$$Find:\quad x = [a, b, c, d, n_1, S_1, S_2]$$

$$\min:\quad f(x) = \text{mass}(x)$$

$$s.t.\quad T_{co} - 120 \leq 0 \quad T_{ir} - 120 \leq 0 \quad f_{f1} - 0.5 \leq 0 \quad 0.8 - \eta \leq 0$$

$$I_{10}/I_1 - 0.1 \leq 0 \quad \Delta V_2/V_2 - 120 \leq 0 \quad f_{f2} - 0.5 \leq 0$$

$$with:$$

$$a \in [3, 30]\,\text{mm} \quad b \in [14, 95]\,\text{mm} \quad c \in [6, 40]\,\text{mm} \quad d \in [10, 80]\text{mm}$$

$$S_1 \in [0.15, 19]\,\text{mm}^2 \quad S_2 \in [0.15, 19]\text{mm}^2 \quad n_1 \in [200, 1200] \tag{2}$$

3.2   Multi-level Optimization of the Transformer Using DRC-CO

The main purpose of this section is to practice the proposed DRC-CO multi-disciplinary method to handle the design problem of transformer device, see Eq. (1). Thus, the analytical model of the safety isolation transformer has been used within the DRC-CO optimization process. The representation of the multidisciplinary coupling model of the transformer is presented in Fig. 3. The design problem of the safety isolation transformer's device involves three coupled disciplines which are, a EM_L, a EM_0 and a TH as shown in Fig. 3. Each of these disciplines requires the same shared design variables $x = [a, b, c, d, n_1, S_1, S_2]$ between all disciplines and output of other discipline, coupling variables. The EM_0 discipline needs the output of EM_L discipline (the secondary voltage V_2 and primary current I_1) to compute the secondary voltage drop ΔV_2 and the magnetizing current I_{10}. The EM_L discipline employs the output of EM_0 discipline (voltage drop ΔV_2 and the magnetizing current I_{10}) and the output TH discipline (the windings core temperature T_{co}) to compute for the Joules and iron losses L_{co} and L_{ir}, respectively. Finally, the TH discipline aims to computes the values for the windings and magnetic core temperature T_{co} and T_{ir} respectively, based on the output values of EM_L discipline (Joules and iron losses L_{co} and L_{ir}, respectively).

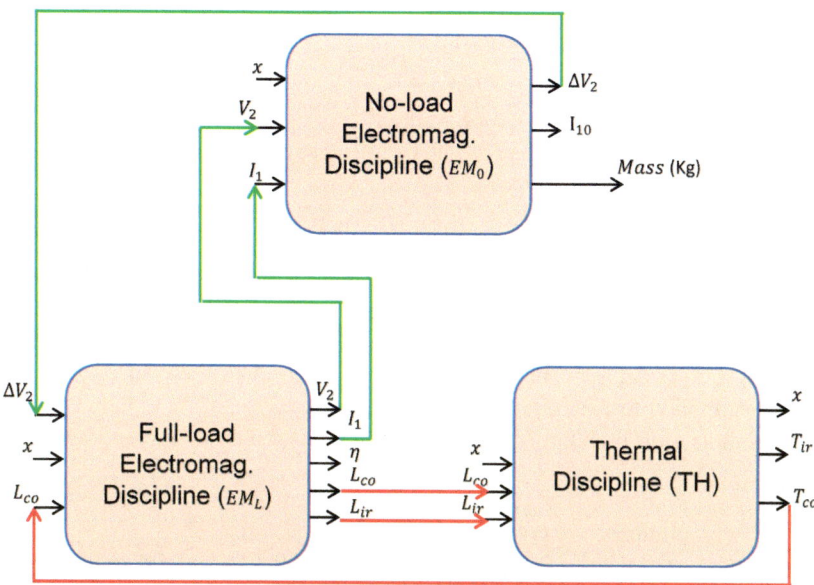

Fig. 3. Multidisciplinary isolation transformer representation

The optimization problem of the transformer, Eq. (2), can be decomposed into one optimization problem at the system level in Eq. (3), and three optimization problems at disciplinary level, see Eqs. (4)–(6), and each level has its associated optimizer using SQP algorithm as an optimizer.

The optimization problem at the system level to solve is then:

$$
\begin{aligned}
&Find: &&x_{sh}^s = [a, b, c, d, n_1, S_1, S_2]\\
&minimize: &&f(x_{sh}^s) = \mathrm{mass}(x_{sh}^s)\\
&s.t. &&\|1 - (y^s/y^*)\|_2^2 \le \varepsilon\\
& &&y^* = \left[y_{12}^*, y_{21}^*, y_{23}^*, y_{31}^*\right]\\
& &&y^s = \left[y_{12}^s, y_{21}^s, y_{23}^s, y_{31}^s\right]\\
& &&y_{12}^s = [\Delta V_2, I_{10}]\ ,\ y_{21}^s = [V_2, I_1],\ y_{23}^s = [L_{co}, L_{ir}]\ ,\ y_{31}^s = [T_{co}]\\
& &&x_{sh}^s = [a, b, c, d, n_1, S_1, S_2]
\end{aligned}
\tag{3}
$$

where x_{sh}^s represents the vector of shared variables of the transformer, y^s represents the vector of the coupling variables. y^* represents the vector of optimal coupling variables obtained at the disciplinary level optimizer, $\|y^s - y^*\|_2^2$ represents the compatibility constraints which are always satisfied for any feasible solution from the disciplinary level optimizer. In practice, the compatibility constraints $\|\cdot\|_2^2$ are satisfied within an acceptable dynamic relaxation factor ϵ of three disciplines, with its value computed according to Eq. (1).

The optimization problem of no-load EM discipline optimizer is given:

$$
\begin{aligned}
&Find: && x_{sh}^1,\ y_{12} \\
&\text{minimize:} && O_1 = \left\|1 - x_{sh}^1/x_{sh}^S\right\|_2^2 + \left\|1 - y_{12}/y_{12}^S\right\|_2^2 \\
&\text{s.t.} && I_{10}/I_1 - 0.1 \le 0 \quad \Delta V_2/V_2 - 0.1 \le 0 \\
& && y_{12} = [\Delta V_2, I_{10}] \\
& && x_{sh}^1 = [a, b, c, d, n_1, S_1, S_2]
\end{aligned}
\tag{4}
$$

The no-load EM discipline optimizer intended to match the values of the shared variables and coupling variables received from system level optimizer while satisfying the disciplinary design constraints of the no-load EM discipline in terms of magnetizing current value I_{10} and secondary voltage drop value ΔV_2, respectively. A similar treatment is applied to the EM_L and TH disciplines.

The optimization problem of the full-load EM discipline optimizer is given:

$$
\begin{aligned}
&Find: && x_{sh}^2, y_{21}, y_{23} \\
&\text{minimize:} && O_2 = \left\|1 - x_{sh}^2/x_{sh}^S\right\| + \left\|1 - y_{21}/y_{21}^S\right\| \\
&\text{s.t.} && 0.8 - \eta \le 0 \quad f_{f1} - 0.5 \le 0 \quad f_{f2} - 0.5 \le 0 \\
& && y_{21} = [V_2, I_1] \quad y_{23} = [L_{co}, L_{ir}] \\
& && x_{sh}^2 = [a, b, c, d, n_1, S_1, S_2]
\end{aligned}
\tag{5}
$$

The full-load discipline optimizer in charge with respecting the filling factors of the two windings, primary and secondary, f_{f1} and f_{f2}, respectively.

The TH discipline's optimization problem is given:

$$
\begin{aligned}
&Find: && x_{sh}^3, y_{31} \\
&\text{minimize:} && O_3 = \left\|1 - x_{sh}^3/x_{sh}^S\right\| + \left\|1 - y_{31}/y_{31}^S\right\| \\
&\text{s.t.} && T_{co} - 120 \le 0 \quad T_{ir} - 100 \le 0 \\
& && y_{23} = [T_{co}] \\
& && x_{sh}^3 = [a, b, c, d, n_1, S_1, S_2]
\end{aligned}
\tag{6}
$$

The TH discipline optimizer aims to match the best corresponding values received from the system level for the TH discipline variables (x_{sh}^3 and y_{s23}) while respect the maximum winding and magnetic core temperatures, T_{co} and T_{ir}, respectively.

The optimization process of the transformer design problem using the DRC-CO and SLO processes is started from initial feasible design. Obtained optimal results of the optimization process are presented in Table 1 along with the results of SLO which considered as a reference. As can be seen in Table 1, the optimization results obtained from DRC-CO method are comparable with obtained by SLO, in terms of precision.

The convergence of the proposed DRC-CO scheme requires 3 system level iterations, as shown in Fig. 4, by cons the single level optimization approach requires 16 iterations. An optimal mass value of 2.30 kg was obtained by the DRC-CO approach. This value is slightly lower to the global optimum obtained by the SLO method (2.31 kg). The accuracy optimization of the proposed DRC-CO is confirmed by the negligible percentage error which is less 0.5%, in comparison phase with SLO method.

Table 1. Optimal solution of the DRC-CO, CO multi-level and SOL

Methods	Design variables							Obj.	Constraints				
	a	b	c	d	n_1	S_1	S_2	Mass	Tco	Tir	DV_2/V_2	I_{10}/I_1	η
	mm	mm	mm	mm	–	mm^2	mm^2	Kg	°C	°C	–	–	–
SLO	12.9	50.1	16.6	43.2	640	0.32	2.31	2.31	108.8	99.9	0.07	0.1	0.89
CO (Berbecea 2012)	16.9	53.3	18.1	31	728	0.31	2.51	2.36	104.9	95.9	0.08	0.08	0.89
DRC-CO	12.7	50.9	16.7	43.8	640	0.31	2.9	2.30	109.5	99.9	0.07	0.1	0.89

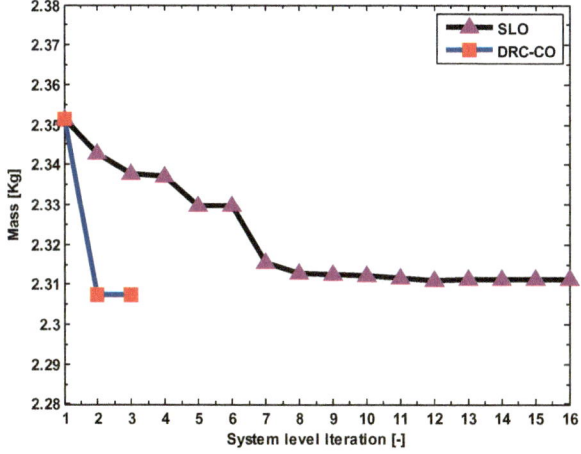

Fig. 4. System level objective evaluation from DRC-CO and SLO methods

Though, the percentage error between CO and SLO is too large, 2.16%. The slightly percentage error value between DRC-CO and SLO result the satisfaction of the interdisciplinary compatibility constraints using dynamic relaxation coordination.

4 Conclusion

In this work, Multidisciplinary Design Optimization (MDO) method of safety isolation transformer device has been carried out using DRC-CO approach. The proposed DRC-CO approach accelerates the convergence rate, using the dynamic relaxation coordination at system level optimizer, which leads the optimization process converging to the feasible and robust optimum.

The optimal solution obtained from DRC-CO scheme (*Mass* = 2.30 kg) are compared with the SLO method which provides (*Mass* = 2.31 kg). The result of the comparison indicates that the percentage of error is less than 0.5%. Therefore, the accurately of optimization process with low coordination iterations shows the efficiency of the proposed DRC-CO approach.

In summary, the proposed DRC-CO method gave other insights to treat expensive MDO engineering problems that involve multiple disciplines with several nonlinear constraints at both system and disciplines levels. The application the DRC-CO into MDO has several benefits including: (1) enabling to use the parallel computing process; (2) accelerate the evaluation time which result the reduction of the computational cost by using the dynamic relaxation coordination; (3) appropriate for solving the complex and/or multi-physics systems include several disciplines. These results encourage the application of DRC-CO to more complex MDO design problems such as aircraft wing design and automobile design. The text was modified.

References

Balesdent, M., Bérend, N., Dépincé, P., et al.: A survey of multidisciplinary design optimization methods in launch vehicle design. Struct. Multidiscip. Optim. **45**, 619–642 (2012)

Berbecea, A.C.: Multi-level approaches for optimal system design in railway applications. Dissertation, Ecole Centrale de Lille (2012)

Braun, R., Kroo, I.: Development and application of the collaborative optimization architecture in a multidisciplinary design environment. In: Alexandrov, N., Hussaini, M. (eds.) Multidisciplinary design optimization: state-of-the-art, pp. 98–116. SIAM, Philadelphia (1996)

Chagraoui, H., Soula, M.: Multidisciplinary design optimization of stiffened panels using collaborative optimization and artificial neural network. Proc. Inst. Mech. Eng. Part C: J Mech. Eng. Sci. (2017). https://doi.org/10.1177/0954406217740164

Chagraoui, H., Soula, M., Guedri, M.: A robust multi-objective and multi-physics optimization of multi-physics behavior of microstructures. J. Cent. South Univ. **23**, 3225–3238 (2016)

Kim, H.M., Michelena, N.F., Papalambros, P.Y., Jiang, T.: Target cascading in optimal system design. J. Mech. Des. **125**, 474–480 (2003)

Sobieszczanski-Sobieski, J.: Optimization by decomposition: a step from hierarchic to non-hierarchic systems. In: Proceedings of the 2nd NASA/Air Force Symposium on Recent Advances in Multidisciplinary Analysis and Optimization. Hampton, VA (1998)

Sobieszczanski-Sobieski, J., Agte, J., Sandusky, J.R.: Bi-level integrated system synthesis (BLISS). AIAA J. **38**, 164–172 (2000)

Wang, W., Gao, F., Cheng, Y., et al.: Multidisciplinary design optimization for front structure of an electric car body-in-white based on improved collaborative optimization method. Int. J. Automot. Technol. **18**, 1007–1015 (2017)

Weiwei, H., Azarm, S., Almansoori, A.: New approximation assisted multi-objective collaborative robust optimization (new AA-McRO) under interval uncertainty. Struct. Multidiscip. Optim. **47**, 19–35 (2013)

Xia, T., Li, M., Zhou, J.: A sequential robust optimization approach for multidisciplinary design optimization with uncertainty. J. Mech. Des. **138**, 111406–111410 (2016)

Electro-Mechanical System Control Based on Observers

Syrine Derbel[1,2]([✉]), Nabih Feki[2], Jean Pierre Barbot[1], Florentina Nicolau[1], Mohamed Slim Abbes[2], and Mohamed Haddar[2]

[1] Quartz Laboratory, ENSEA, 95014 Cergy, France
syrina.derbel@hotmail.fr
[2] LA2MP Laboratory, ENIS, Sfax 3038, Tunisia

Abstract. The prediction of the gear behavior is becoming major concerns in many industries. For this reason, in this article, an electro-mechanical modeling is developed in order to simulate a gear element driven by an asynchronous motor. The electrical part, which is the induction motor, is simulated by using the Kron's model while the mechanical part, which is the single stage gear element, is accounted for by a torsional model. The mechanical model that simulates the pinion-gear pair is obtained by reducing the degree of freedom of the global spur or helical gear system. The electrical and mechanical state variables are combined in order to obtain a unique differential system that describes the dynamics of the elecro-mechanical system. The global coupled electromechanical model can be characterized by a unique set of non-linear state equations. The contribution of this work is to apply the control based on observers in order to supervise the electrical and mechanical behavior of the electro-mechanical system from only its inputs and its measurements outputs (sensors outputs). Some simulations on pinon/motor angular speed, electromagnetic torque, currents, are presented, which illustrate the system evolution (i.e., the electrical and mechanical quantities) and the good performances of the proposed observers.

Keywords: Gears transmission · Observers · Simulation
Asynchronous motor

1 Introduction

Electro-mechanical systems such as mechanical gear transmission driven by induction motors are commonly used in many industrial applications. For this reason, many techniques and tools of control and diagnostics, such as vibration and sound signal analysis (Baydar and Ball 2001, Tan et al. 2007), analysis of stator currents (Feki 2012), have traditionally been used to supervise the gear element behavior and to detect faults (Chaari et al. 2008). These techniques are advantageous by their reduced cost and present a high reliability. However, they are sensitive to the positioning of the sensors and in some applications, they

© Springer Nature Switzerland AG 2019
T. Fakhfakh et al. (Eds.): ICAV 2018, ACM 13, pp. 101–110, 2019.
https://doi.org/10.1007/978-3-319-94616-0_10

present technical difficulties to implement sensors in rotating parts or hostile environment. Although gear monitoring by vibration signal analysis and current stator analysis are still widely used, a new method of electro-mechnical system control and monitoring based on observers will be presented in this paper.

In general, for technical and economic reasons, the state of the system is not completely accessible. Indeed, the complexity of the technical feasibility as well as prohibitive costs for the implantation of several sensors can considerably reduce the number of states measured. In this case, the state vector size is greater than the output vector size. However, under some conditions of existence, the state can be reconstructed using an observer Larroque (2008).

The paper will be organized as follows: (a) in Sect. 2, a single stage gear element is accounted for by using torsional model, (b) in Sect. 3, an electrical modeling of the induction motor is presented, (c) the electro-mechnical coupling is developed in Sect. 4, (d) the Sect. 5 is dedicated to implement the used observer and (e) the simulation results of the electro-mechanical system dynamic behavior are presented in Sect. 6.

2   Mechanical Modeling

The modeling of the mechanical part (see Fig. 1) is based only on gear element, meaning that we reduce the global model of 36 degrees of freedom (Feki et al. 2012) to 2 degrees of freedom. A driving torque C_m is applied to the pinion gear and a load torque C_r is applied to the wheel gear (attached to output shaft). The system modeling is accounted for by two degrees of freedom, which correspond to the torsional components. Using Euler-Lagrange equation, the motion equation of the torsional model of the gear element is obtained as follows (1):

$$M\ddot{x} + C\dot{x} + K(t, x)x = F, \tag{1}$$

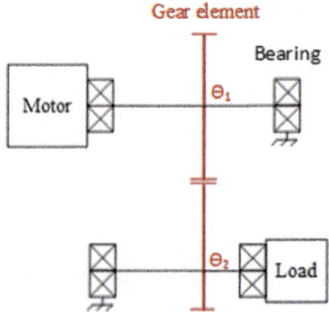

Fig. 1. Electro-mechanical system

where

- $M = diag(I_1, I_2)$: mass matrix with I_k the polar moment of inertia of the gear k,

- $K(t,x)$: stiffness matrix that depends on the state vector and on the time,
- $C = \gamma_1 M + \gamma_2 K_{moy}$: Rayleigh model damping,
- $F = [C_m \quad C_r]^t \in \mathbb{R}^2$: external forces,
- $x = [\theta_1 \quad \theta_2]^t \in \mathbb{R}^2$: two degrees of freedom vector.

The stiffness matrix can be expressed in terms of a structural vector $V(M_i)$ (Maatar and Velex 1996) by using the following form $K(t,x) = k(t)V(M_i)V(M_i)^t$, with $k(t)$ the stiffness simulated by a square waveform. Developing (1), we obtain the space representation of the gear element:

$$\begin{bmatrix} \dot{x} \\ \ddot{x} \end{bmatrix} = \begin{bmatrix} 0 & I_d \\ -M^{-1}K & -M^{-1}C \end{bmatrix} \begin{bmatrix} x \\ \dot{x} \end{bmatrix} + \begin{bmatrix} 0 \\ M^{-1}F \end{bmatrix}, \tag{2}$$

with I_d the 2nd order identity matrix.

3 Electrical Modeling

The motor is modeled using the Kron's transformation model. The principle of this transformation is to translate the three-phases quantities (abc) of the motor to two-phases quantities (see Fig. 2).

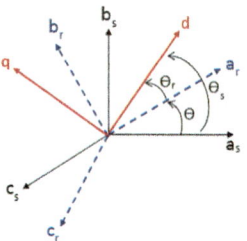

Fig. 2. Park transformation plan.

The stator variables are obtained by $\theta = \theta_s = w_s t = (a_s, d)$ while the rotor variables are calculated by $\theta = \theta_r = w_{sl} t = (a_r, d)$.

$$\begin{bmatrix} x_d \\ x_q \\ x_0 \end{bmatrix} = T_{2/3}(\theta_s) \begin{bmatrix} x_a \\ x_b \\ x_c \end{bmatrix}, \tag{3}$$

$$\begin{bmatrix} x_d \\ x_q \\ x_0 \end{bmatrix} = T_{2/3}(\theta_r) \begin{bmatrix} x_a \\ x_b \\ x_c \end{bmatrix}. \tag{4}$$

The voltage equations of the three stator and rotor phases are given by (5) and (6).

$$\begin{bmatrix} v_{as} \\ v_{bs} \\ v_{cs} \end{bmatrix} = \begin{bmatrix} R_s & 0 & 0 \\ 0 & R_s & 0 \\ 0 & 0 & R_s \end{bmatrix} \begin{bmatrix} i_{as} \\ i_{bs} \\ i_{cs} \end{bmatrix} + \frac{d}{dt} \begin{bmatrix} \phi_{as} \\ \phi_{bs} \\ \phi_{cs} \end{bmatrix}, \tag{5}$$

$$\begin{bmatrix} v_{ar} \\ v_{br} \\ v_{cr} \end{bmatrix} = \begin{bmatrix} R_r & 0 & 0 \\ 0 & R_r & 0 \\ 0 & 0 & R_r \end{bmatrix} \begin{bmatrix} i_{ar} \\ i_{br} \\ i_{cr} \end{bmatrix} + \frac{d}{dt} \begin{bmatrix} \phi_{ar} \\ \phi_{br} \\ \phi_{cr} \end{bmatrix}. \tag{6}$$

The flux equations can be written as follows:

$$\left[\phi_{abcs}\right] = L_{ss}\left[i_{abcs}\right] + L_{sr}\left[i_{abcr}\right], \tag{7}$$

$$\left[\phi_{abcr}\right] = L_{rr}\left[i_{abcr}\right] + L_{rs}\left[i_{abcs}\right], \tag{8}$$

where

- $L_{ss} = \begin{bmatrix} l_s & m_s & m_s \\ m_s & l_s & m_s \\ m_s & m_s & l_s \end{bmatrix}$: stator inductance matrix,

- $L_{rr} = \begin{bmatrix} l_r & m_r & m_r \\ m_r & l_r & m_r \\ m_r & m_r & l_r \end{bmatrix}$: rotor inductance matrix,

- $L_{sr} = L_{rs}^t = m_{sr} \begin{bmatrix} \cos(\theta) & \cos(\theta + \frac{2\Pi}{3}) & \cos(\theta - \frac{2\Pi}{3}) \\ \cos(\theta + \frac{2\Pi}{3}) & \cos(\theta) & \cos(\theta + \frac{2\Pi}{3}) \\ \cos(\theta + \frac{2\Pi}{3}) & \cos(\theta - \frac{2\Pi}{3}) & \cos(\theta) \end{bmatrix}$: mutual inductance

matrix between stator and rotor,

with the constants parameters:

- l_s (respectively l_r): self-inductance of the stator (rotor),
- m_s (respectively m_r): mutual-inductance between the stator phases (the rotor phases),
- m_{sr}: the maximum value of mutual inductances between stator and rotor phases.

Applying the Kron's transformation for (5), (6), (7) and (8), we obtain:

$$\begin{cases} v_{ds} = R_s i_{ds} + \frac{d}{dt}\phi_{ds} - \omega_s\phi_{ds}, \\ v_{qs} = R_s i_{qs} + \frac{d}{dt}\phi_{qs} + \omega_s\phi_{qs}, \end{cases} \tag{9}$$

$$\begin{cases} v_{dr} = R_r i_{dr} + \frac{d}{dt}\phi_{dr} - \omega_{sl}\phi_{dr} = 0, \\ v_{qr} = R_r i_{qr} + \frac{d}{dt}\phi_{qr} + \omega_{sl}\phi_{qr} = 0, \end{cases} \tag{10}$$

$$\begin{cases} \phi_{ds} = L_s i_{ds} + L_m i_{dr}, \\ \phi_{qs} = L_s i_{qs} + L_m i_{qr}, \end{cases} \tag{11}$$

$$\begin{cases} \phi_{dr} = L_r i_{dr} + L_m i_{ds}, \\ \phi_{qr} = L_r i_{qr} + L_m i_{qs}, \end{cases} \tag{12}$$

where L_s (respectively, L_r) represents the stator synchronous inductance (respectively, rotor synchronous inductance) and L_m is the magnetizing (synchronous) inductance. The advantage of this transformation is the fact that we get a constant mutual inductance and that along an axis, the fluxes depend only to the rotor and stator currents. Relations (9) and (10) detail the electro-magnetic behavior of the asynchronous machine written in non a linear differential equations form. These equations can be described in matrix space representation by choosing the space vector $z(t)$ composed by the both stator currents and the both rotor fluxes of the motor:

$$\dot{z}(t) = Az(t) + BU(t), \tag{13}$$

with $z(t) = [i_{ds} \ i_{qs} \ \phi_{dr} \ \phi_{qr}]^t \in \mathbb{R}^4$, $U(t) = [v_{ds} \ v_{qs}]^t \in \mathbb{R}^2$: the input vector, A, B the state, input matrices given by:

$$A = \begin{bmatrix} -(\frac{1}{T_s \sigma} + \frac{1}{T_r}\frac{1-\sigma}{\sigma}) & \omega_s & \frac{1-\sigma}{\sigma}\frac{1}{L_m T_r} & \frac{1-\sigma}{\sigma}\frac{\omega_m}{L_m} \\ -\omega_s & -(\frac{1}{T_s \sigma} + \frac{1}{T_r}\frac{1-\sigma}{\sigma}) & -\frac{1-\sigma}{\sigma}\frac{\omega_m}{L_m} & \frac{1-\sigma}{\sigma}\frac{1}{L_m T_r} \\ \frac{L_m}{T_r} & 0 & -\frac{1}{T_r} & w_{sl} \\ 0 & \frac{L_m}{T_r} & -w_{sl} & -\frac{1}{T_r} \end{bmatrix}, \tag{14}$$

$$B = \begin{bmatrix} \frac{1}{\sigma L_s} & 0 \\ 0 & \frac{1}{\sigma L_s} \\ 0 & 0 \\ 0 & 0 \end{bmatrix}, \tag{15}$$

where the constants parameters are explained below:

$- \ \sigma = 1 - \frac{L_m^2}{L_s L_r}, \quad T_s = \frac{L_s}{R_s}, \quad T_r = \frac{L_r}{R_r}.$

The electromagnetic torque is represented by (16), with p the number of pole-pairs.

$$C_{em} = \frac{pL_m}{L_r}(\phi_{dr}i_{qs} - \phi_{qr}i_{ds}). \tag{16}$$

4 Electro-Mechanical Coupling

The aim of the electro-mechanical coupling is to obtain a space representation of the system with two degrees of freedom (see Fig. 1). This modeling allows to implement control methods to supervise the dynamic behavior of the system. The electro-magnetic torque given by (16) is regarded as the input of the mechanical part and the small vibrations caused by the gear element act on the speed rotation of the asynchronous motor. The coupled system leads to a first order differential system of the form:

$$\dot{\zeta}(t) = A_c(t, \zeta)\zeta(t) + Gu, \tag{17}$$

where $\zeta(t) = [i_{ds} \ i_{qs} \ \phi_{dr} \ \phi_{qr} \ \theta_1 \ \theta_2 \ \dot{\theta}_1 \ \dot{\theta}_2]^t$: the global state vector, $A_c = \begin{bmatrix} A & 0 \\ P & H \end{bmatrix}$ is the state matrix, with A the state matrix associated to the electric part expressed in (14), $P \in \mathbb{R}^{(4 \times 4)}$ is the coupling matrix between the electric and mechanical part depending to the electromagnetic torque donated by (16) and $H = \begin{bmatrix} 0 & 0 & 1 & 0 \\ 0 & 0 & 0 & 1 \\ -M^{-1}K & -M^{-1}C \end{bmatrix}$ is the matrix associated to the gear modeling,

$$G = \begin{bmatrix} \frac{1}{\sigma L_s} & 0 & 0 & 0 \\ 0 & \frac{1}{\sigma L_s} & 0 & 0 \\ 0 & 0 & 0 & 0 \\ 0 & 0 & 0 & 0 \\ 0 & 0 & 1 & 0 \\ 0 & 0 & 0 & 1 \\ 0 & 0 & & \\ & & M^{-1} & \\ 0 & 0 & & \end{bmatrix}$$ is the input matrix and $u = \begin{bmatrix} v_{ds} \\ v_{qs} \\ 0 \\ C_r \end{bmatrix}$ the input vector.

The evolution of the system is obtained by introducing the control based on observers to monitor the dynamic behavior of the electro-mechanical system.

5 Observer Form

A state observer is a control method (Perruquetti and Barbot 2002) that gives an estimate of the internal state of the real system, only from the measurements given by the sensors and the real input of the system. The observers are used in order to control the behavior of systems, to detect the faults or to identify the unknown parameters of systems (Oueder 2012). In our case, four differentiators (Ghanes et al. 2017) are used to estimate the drift of both currents of the asynchronous motor and the displacements of the gear element. Assuming that $[s_1, ..., s_8] = [i_{ds}, \dot{i}_{ds}, i_{qs}, \dot{i}_{qs}, \theta_1, \dot{\theta}_1, \theta_2, \dot{\theta}_2]$, the observer equations are written in following form:

$$\begin{cases} \dot{\hat{s}}_i = \hat{s}_{i+1} + k_1 \mu_j |e_i|^\alpha sign(e_i), \\ \dot{\hat{s}}_{i+1} = k_2 \alpha \mu_j^2 |e_i|^{2\alpha-1} sign(e_i), \\ e_i = s_i - \hat{s}_i, \end{cases} \tag{18}$$

where

- e_i, $i = \{1, 3, 5, 7\}$ are the output estimation errors,
- k_1, k_2 are constants acting on the stability of the system,
- μ_j, $j = 1, 2$ are positive constants, the first one associated to the electric model and the second one is related to the mechanical part.

6 Simulation and Results

In this simulation, a spur gear system is considered. The main characteristics of the gear are given in Table 1 and the motor parameters are shown in Table 2.

As it can be noticed, the applied observer gives a good performances (see Figs. 3, 4, 5 and 6). In these figures, the estimated states converge, in finite

Table 1. Gears parameters

Parameters	Value
Module (mm)	4
Tooth number of Pinion	21
Tooth number of wheel	31
Face width (mm)	10
Pressure angle	20

Table 2. Electric parameters

Parameters	Value
Stator resistance R_s (Ω)	9.163
Rotor resistance R_r (Ω)	5.398
Stator inductance L_s (H)	0.115
Rotor inductance L_r (H)	0.0943
Magnetizing inductance L_m (H)	0.0943
Number of pole-pairs p	1

time, to the real quantities states. Figures 3 and 4 represent the two stator currents expressed in $(dq0)$ frame. They show that the periodic recurrence of gears meshing frequency t_m is regained in the electrical states.

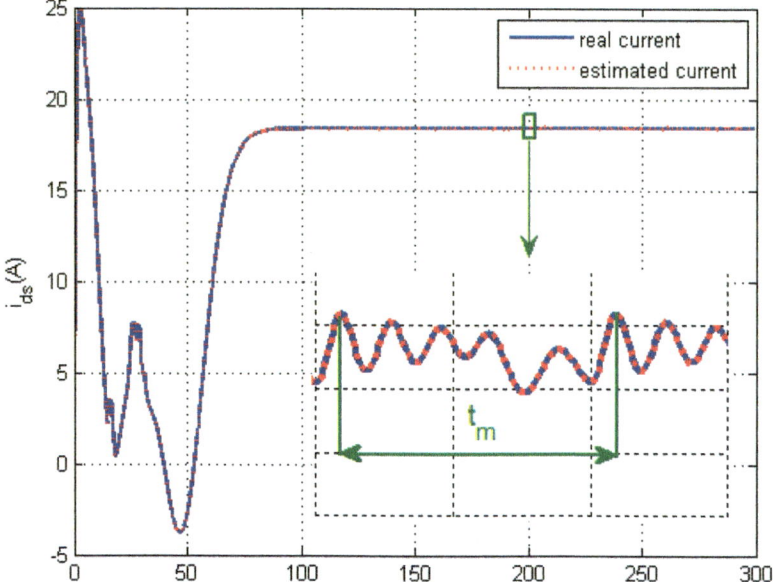

Fig. 3. State ζ_1 and its estimate

Figure 5 illustrates that the rotational speed of the pinion is of the order to 300 rad/s. This value represents the average meshing speed added to the small vibrations of the gear.

The last Figure (Fig. 6) displays the evolution of the error transmission given by the equation $R_{b1}\theta_1 + R_{b2}\theta_2$ (where R_{b1}, R_{b2} are, respectively, the base radii of the pinion and the wheel).

All results confirm the good convergence of the applied observer. This convergence has appeared in the frequency content and the amplitude of the

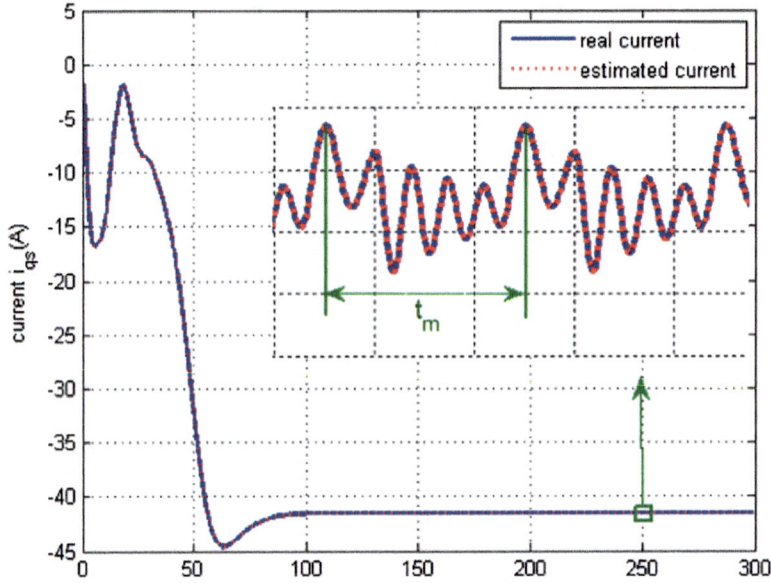

Fig. 4. State ζ_2 and its estimate

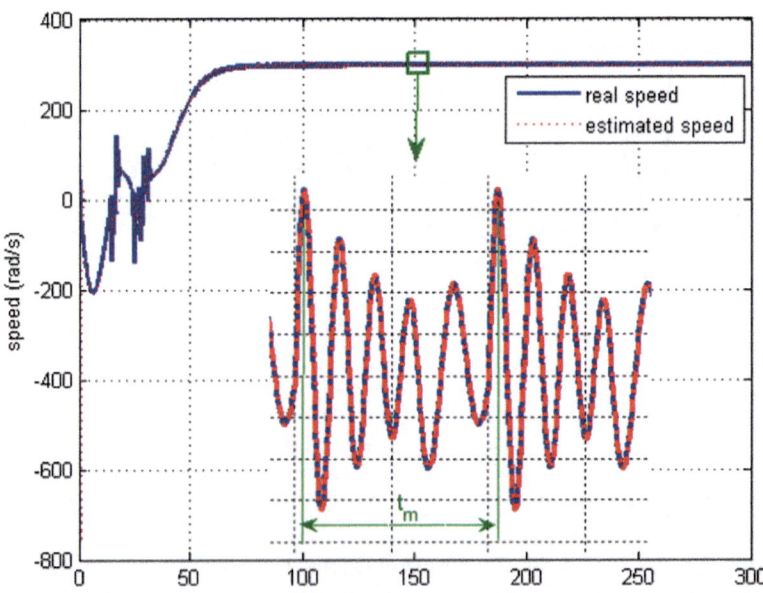

Fig. 5. State ζ_7 and its estimate

electro-mechanical system behavior signals. Meaning that the evolution of the real states and those of the estimated quantities are perfectly confused in the frequency study. The observer gives a rapid and accurate convergence for all system states.

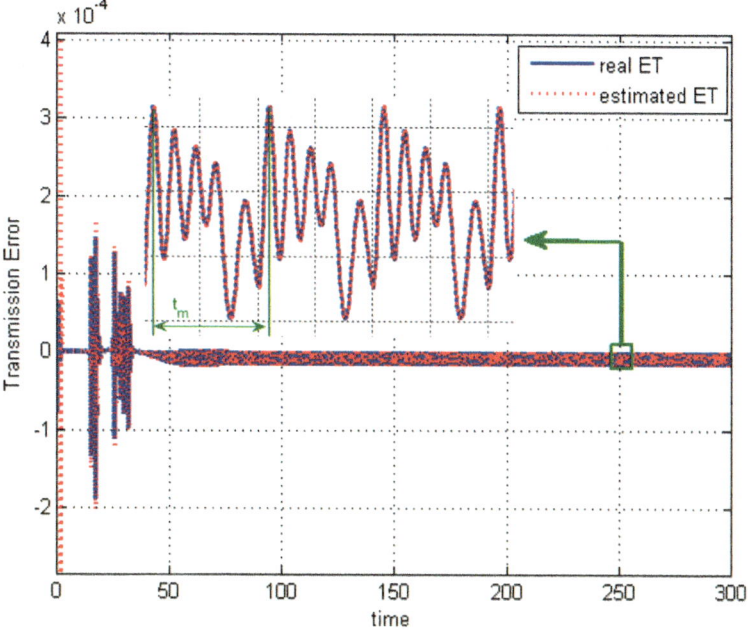

Fig. 6. Real and estimated transmission error

7   Conclusion

In this paper, an electro-mechanical coupling of the gear transmission system driven by a asynchronous motor has been studied. The monitoring of this model is obtained by using the control based on observers. Further work is in progress in order to implement other types of observers with presence of gear faults and variation of the sensors noise.

References

Baydar, N., Ball, A.: A comparative study of acoustic and vibration signals in detection of gear failures using Wigner-Ville distribution. Mech. Syst. Signal Process. **15**, 1091–1107 (2001)

Chaari, F., Baccar, W., Abbes, M.S., Haddar, M.: Effect of spalling or tooth breakage on gearmesh stiffness and dynamic response of a one-stage spur gear transmission. Eur. J. Mech.-A/Solids **27**(4), 691–705 (2008)

Feki, N.: Modelisation electro-mecanique de transmissions par engrenages : applications la detection et au suivi des avaries, Ph.D. thesis. These de doctorat dirige par Clerc, Guy et Velex, Philippe Mcanique Lyon, INSA (2012). http://www.theses.fr/2012ISAL0041/document

Feki, N., Clerc, G., Velex, P.: An integrated electro-mechanical model of motor-gear units? Applications to tooth fault detection by electric measurements. Mech. Syst. Signal Process. **29**, 377–390 (2012)

Ghanes, M., Barbot, J.-P., Fridman, L., Levant, A.: A second order sliding mode differentiator with a variable exponent. In: American Control Conference (ACC), pp. 3300–3305. IEEE (2017)

Larroque, B.: Observateurs de systémes linéaires: application à la détection et localisation de fautes. Ph.D. thesis, Institut National Polytechnique de Toulouse (2008)

Maatar, M., Velex, P.: An analytical expression for the time-varying contact length in perfect cylindrical gears: some possible applications in gear dynamics. Trans. ASME-R-J. Mech. Des. **118**(4), 586–588 (1996)

Oueder, M.: Synthese des observateurs pour les systemes non linéaires. Ph.D. thesis, Université de Caen (2012)

Perruquetti, W., Barbot, J.-P.: Sliding Mode Control in Engineering. CRC Press, Boca Raton (2002)

Tan, C.K., Irving, P., Mba, D.: A comparative experimental study on the diagnostic and prognostic capabilities of acoustics emission, vibration and spectrometric oil analysis for spur gears. Mech. Syst. Signal Process. **21**(1), 208–233 (2007)

The Design and Modeling of an Optimized Mechatronic System Using a Set Based Concurrent Engineering

Randa Ammar[1,2]($\boxtimes$), Moncef Hammadi[1], Jean-Yves Choley[1],
Maher Barkallah[2], Jamel Louati[2], and Mohamed Haddar[2]

[1] Quartz EA 7393, Laboratory/SUPMECA, 3 Rue de Fernand Hainaut,
93400 Saint-Ouen, France
Randa.ammar@supmeca.fr
[2] LA2MP Laboratory/ENIS, 3038 Sfax, Tunisia

Abstract. A mechatronic system consists in a close intersection between mechanics, electronics, control engineering and software engineering. Typically, the controller design and the system design are developed and optimized contemporaneously. However, a poorly designed mechanical system will not at any time be able to hand out a good performance by adding a good controller. Furthermore, to design complex systems, the designers have to follow the traditional point-based development model where one solution is iteratively modified until it fits the specifications. The main problem with the point-based development model lies in the several resets and modifications that return to the previous steps to satisfy the requirements that meet those of the current stage. Therefore, in this paper, we propose to use Set Based Concurrent Engineering to develop a complex system after that we propose to carry a preliminary optimization of the parametric model of the system in the preliminary design before adding the control system. This approach is shown with a simulation model using Modelica for a case study in the automotive field of an Electronic Throttle Body (ETB).

Keywords: Set Based Concurrent Engineering (SBCE) · Mechatronic system
Multi-objective optimization · Electronic Throttle Body (ETB)

1 Introduction

Evidently, the evolution of mechatronic systems was revolutionary for the industrial world because electronics has been more and more integrated in varied fields as automotive, medicine and robotics. Mechatronics expresses the close integration of mechanics, electronics, control engineering and software engineering. Hence, the design of such systems is an interdisciplinary and complex task, which involves engineers from different fields. The development activities must be coordinated and synchronized in order to save time, provide a quality product and therefore minimize its cost. Thus, to design complex systems, the designers have to follow the traditional point-based development model where one solution is iteratively modified until it fits the specifications. The main problem with the point-based development model lies in

© Springer Nature Switzerland AG 2019
T. Fakhfakh et al. (Eds.): ICAV 2018, ACM 13, pp. 111–120, 2019.
https://doi.org/10.1007/978-3-319-94616-0_11

the several resets and modifications that return to the previous steps to satisfy the requirements that meet those of the current stage. Therefore, a number of product developments have shifted from developing a single design to developing a set of possibility designs leading to the approach of the Set-Based Concurrent Engineering SBCE. This model is defined in Sobek et al. (1999) as the "Set Based Concurrent Engineering" which begins by broadly considering the set of possible solutions and gradually narrowing the set of possibilities to converge on a final solution. Thus, generally, in order to reduce the number of solutions the systems design and the controllers design are optimized simultaneously. However, a badly designed mechanical system will never be able to give a good performance by adding a good controller (Van Amerongen 2003). Therefore, it is favored to create a prefatory optimization of the physical system independently from the controller system. This will allow the designer to have an idea about the performance of the different set of possibilities and the coherence of the model and thereafter we can limit the set of solutions. So, to better understand this methodology, we apply it in an industrial case study of the Electronic Throttle Body (ETB).

2 State of the Art

The complex system design treats the integrated and the optimal design of physical systems, including actuators, sensors, embedded digital control systems and electronic components (Hammadi et al. 2012a, b). Design verification of mechatronic systems needs an integrated approach to deal with discipline interactions. Therefore, in order to evaluate the performance of complex systems in the preliminary design phases, a number of proposed and integrated research approaches (such as authors in Hammadi et al. (2012a, b)) suggested a mechatronic multi-criteria indicator using neuronal networks. Furthermore, an appropriate mechatronic system decomposition has to be developed to achieve a successful mechatronic design optimization. Therefore, to optimize the design of mechatronic systems, several multidisciplinary optimization approaches (Guzeni et al. 2014) and surrogate-based techniques (Hammadi et al. 2012a, b) are suggested for practical cases of design optimization of electric vehicles.

To design a complex system we can use a traditional (point-based) design practice or a set-based concurrent engineering design practice. "What we call Set-Based Concurrent Engineering SBCE begins by broadly considering the sets of possible solutions and gradually narrowing the set of possibilities to converge on a final solution. A wide net from the start and a gradual elimination of weaker solutions make finding the best or better solutions more likely" (Buede and Miller 2009). This approach was first seen in Toyota Motor Corporation. Although Toyota competitors considered it as an inefficient model, Toyota Motor Corporation has always been the industry leader in cost, quality and product development in lead-time (Sobek et al. 1999). Furthermore, Authors in (Ammar et al. 2017) proposed an approach to integrate SBCE in the systems engineering process to develop the architectural design of complex systems. The following figure will graphically show the Set-Based Concurrent Engineering concept (Fig. 1).

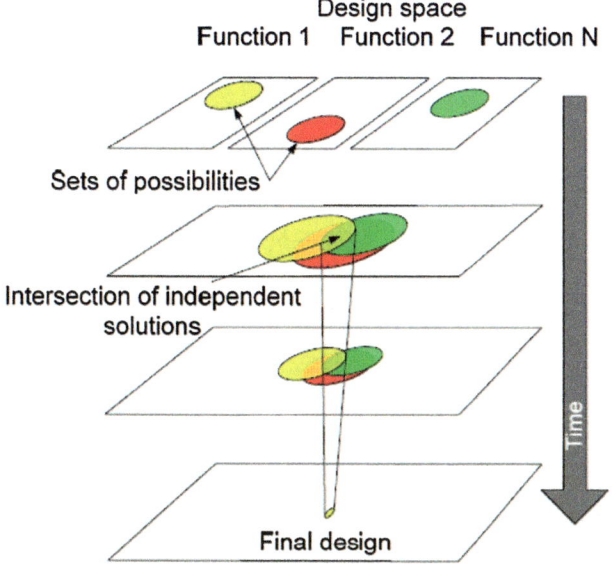

Fig. 1. The principles of Set-Based Concurrent Engineering (Raudberget 2010)

The set-based concurrent design is built on three principle phases (Raudberget 2010)

1. Map the design space:

   - Define feasible regions,
   - Explore trade-offs by designing multiple alternatives,
   - Communicate sets of possibilities.

2. Integrate by intersection:

   - Look for intersections of feasible sets,
   - Impose minimum constraint,
   - Seek conceptual robustness.

3. Establish feasibility before commitment:

   - Narrow sets gradually while increasing detail,
   - Stay within sets once committed,
   - Control by managing uncertainty at process gates.

The performance of multidisciplinary systems is motivated not only by the performance of the individual disciplines but also by their interactions and subsequently a multi-objective optimization (MDO) is necessary in the development of the complex system (Martins and Lambe 2013). MDO is an engineering field focused on the use of numerical optimization for the design of systems involving a number of disciplines or subsystems (Martins and Lambe 2013).

3 Description System

The electronic throttle body is a mechatronic device used to regulate the amount of air admitted into the cylinders of an internal combustion engine. In the beginning of their invention, the request to open the valve of the "Throttle Body" was directly controlled by a cable connected to the accelerator pedal (Fig. 2). Then, this system was developed into an Electronic Throttle Body (ETB) (Fig. 3) related to the Electronic Controller Unit (ECU) in order to control the opening and the closing of the valve.

Fig. 2. Throttle Body controlled by cable

Fig. 3. Electronic   Throttle   Body (ETB)

Then, in order to control the flow of air in the cylinders, the ETB (Electronic Throttle Body) must consist essentially of a system for opening and closing the air passage, a system for converting electrical energy into mechanical energy, a system of safety and a system to adapt the mechanical energy to reach the necessary mechanical force. The latter depends mainly on the requirement of the driver, which is translated by a pedal sensor to the electronic control unit (ECU) and on the other hand the position of the opening and closing system defined by a sensor of position allowing the ECU to know precisely its position.

4 The Set Based Concurrent Engineering Application

4.1 Define Feasible Regions

This phase aims to develop and deeply understand the sets of design possibilities for subsystems. So, the application of this phase in our case study can give a set of solutions for each subsystem as shown in Fig. 4.

As a result of this step, the set number of solution is N:

- N = 3 * 3 * 3 * 3 * 2 * 3 * 2 = 972 solutions.

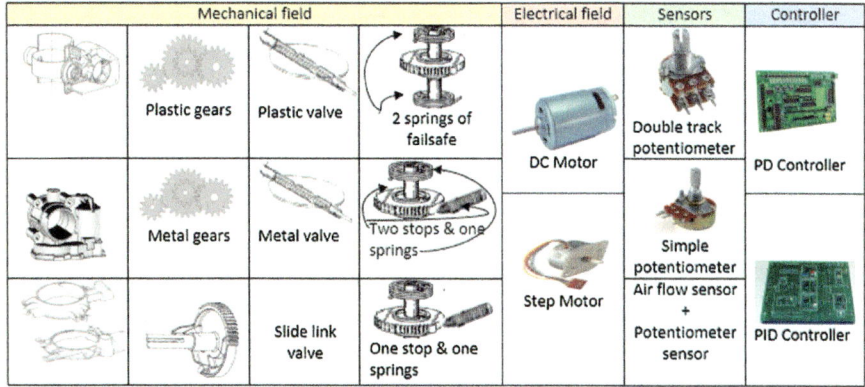

Fig. 4. Possible solutions for each subsystem

4.2 Explore Trade-Offs by Designing Multiple Alternatives

This activity can be done to some subsystems that have characteristics or performances with no effect on its integration with the other subsystems and that leads us to minimize the solution number based on customer requirements and simulations. In our case study, it can be applied to the body of the electronic throttle body ETB results and to the gears system based on the manufacturing, the cost, the performance, and the durability as shown in Figs. 5, 6, and 7.

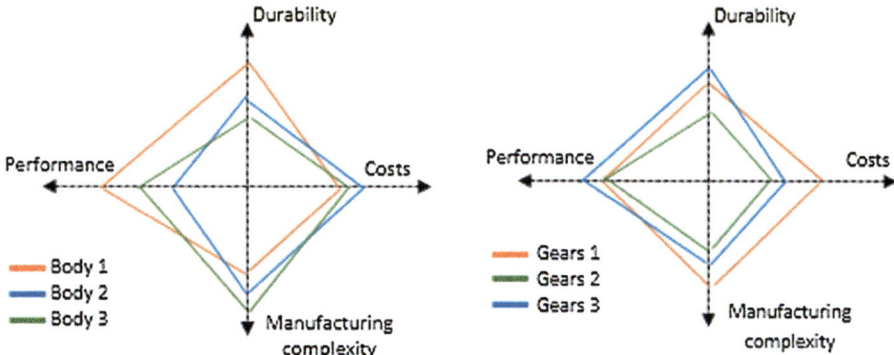

Fig. 5. Trade-off curves of the body.

Fig. 6. Trade-off curves of the regulate airflow system.

According to this step, we can eliminate solution 3 for the ETB body, we can eliminate solution 2 that corresponds to the metal gears for the gear system and we can choose the double track potentiometer and the air flow sensor for the sensors, from where the number of remaining solutions is:

- $N = 2 * 2 * 3 * 1 * 2 * 1 * 2 = 48$ solutions.

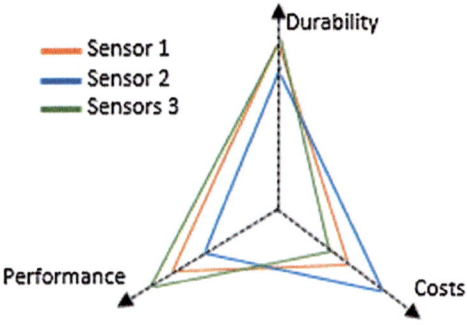

Fig. 7. Trade-off curves of the sensors

4.3   Integrate by Intersection

This phase can be considered as the set based communication principle, which ensures that the subsystem solutions are defined as feasible/compatible with all functional groups involved. In this step, firstly, we can model the possible solutions of our system by Modelica to verify the overall integration performance and reduce the set of possibilities, without forgetting our contribution which consists in a preliminary modeling of the system before adding the control system. Then, in a second step, we can make a remaining model optimization using ModelCenter to reduce the subsystem parameter intervals as well as limit the choice of subsystems.

In the following figure, you will find an example of modeling of the ETB by using the components of the Modelica library. In this model we modeled the whole system made off with a DC motor, a system of gear, a spring of failsafe system and a stop of the sector gear in combination with the gear system and the spring to limit and control the operation of the valve plate (Fig. 8).

Before connecting this model to the ModelCenter to perform a multi-objective optimization we have to assign it to the subsystem parameters as shown in the following table to verify that the subsystem integration meets our main need (Table 1).

Figure 9 shows an example realized with a valve initially positioned at 8° (limp home position) using the failsafe spring with a supply voltage 0 V. Then according to its supply with a voltage different from zero, the valve goes from 8° to 90° with a response time of 228 ms.

But, the valve opening response-time using stepper motor is too slow compared to a conventional DC electric motor which is the reason to eliminate the solution of stepper motor in electronic throttle body systems. Then the number of remaining solution is:

- N = 2 * 2 * 3 * 1 * 1 * 1 * 2 = 24 solutions.

In this step, we move to multi-objective optimization to limit the parameter interval and thereafter to ensure the correct choice of subsystems. Therefore, the specifications of our optimization are summarized as follows: (1) The opening valve time (8° to 90°) should be lower than 180 ms and (2) the closing valve time with the failsafe system (90° to 8°) must be lower than 250 ms. So, the response time specifications can occur

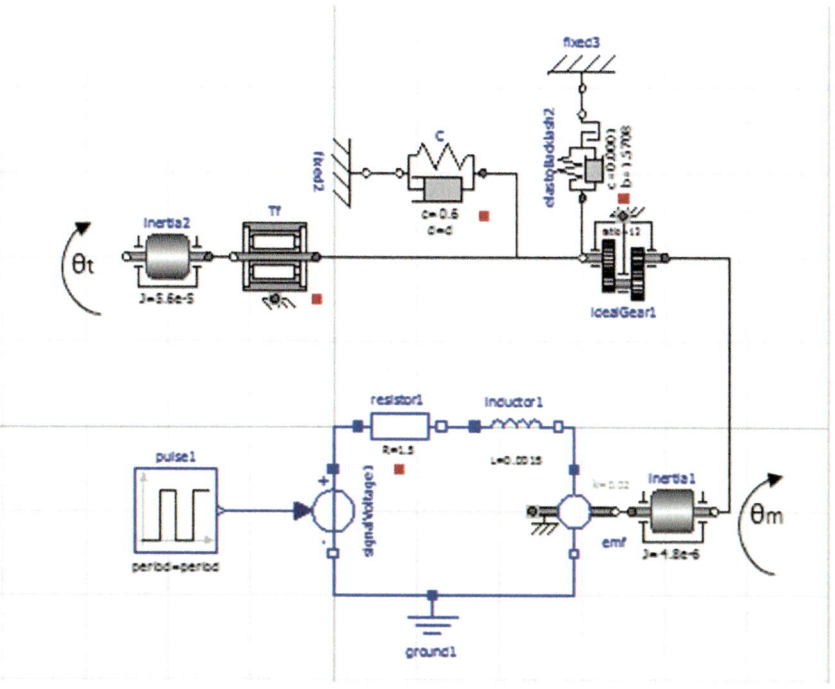

Fig. 8. Modelica Model of the ETB

Table 1. Table of design variables

Parameter	Description	Unit	Value
Motor			
Lm	Inductance	H	0.0015
Rm	Resistance	ohm	1.5
Km	Motor constant	N m/A	0.02
Ke	Back emf	N m s/rad	0.02
Jm	Inertia	$Kg\ m^2$	4.8e−6
Vm	Voltage	V	Variable
Load			
Jl	Inertia	$Kg\ m^2$	5.6e−5
C	Main spring stiffness	N m/rad	0.6
Tf	Friction Torque	N m	Variable
Ts	Springs Torque	N m	Variable
θl	Throttle angle	deg	Variable
Gear			
Ng	Ratio	–	20
Stop	Angle Interval	deg	[0, 90]

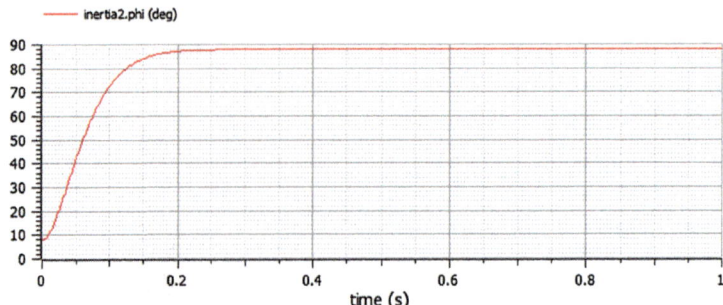

Fig. 9. Valve Opening Time

very rapidly for our application but in reality the performances will be very weak because our problem contains contradictory objectives which are defined as follows:

- If we want to minimize the closing valve time (from 90° to 8°) we must use a high stiffness spring but this will further delay the opening valve time (from 8° to 90°).

Therefore, a combination of Modelica with ModelCenter is necessary in order to do a multi-objective optimization, which is formulized in the next table (Table 2).

Table 2. Optimization constraints

	Unit	Start value	Lower bound	Upper bound
Objective to minimize				
Opening valve time	ms	228	–	180
Closing valve time	ms	1300	–	350
Problem constraints				
θ_max	deg	89.2	89	91
Design variable				
Km	N m/A	0.02	0.01	0.08
Jm	Kg m^2	4.8e−6	6e−5	6e−6
Rm	ohm	1.5	1	4
Lm	H	0.0015	0.001	0.002
Ng	–	20	5	50
C	N m/rad	0.6	0.1	1

The result of optimization is not unique since the problem is multi-objective Then we select three points from the set of Pareto solutions in order to show that there are different possibilities of technological solutions as shown in the following table (Table 3).

Now, all the solutions present such a good performance that the opening valve time can be 128 ms and the closing valve time 167 ms. It is up to the designer to trade-off between these objectives to find, without forgetting other constraints that must be considered in the decision, the components available from suppliers or components in stock.

Table 3. Optimization result

	Unit	1	2	3
Objective to minimize				
Opening valve time	ms	128	145	159
Closing valve time	ms	210	180	167
Problem constraints				
θ_max	deg	89.9999	90	89.98
Design variable				
Km	N m/A	0.0167	0.017	0.0189
Jm	Kg m^2	4.8e−6	6e−5	4.8e−6
Rm	ohm	1.5	1.59	1.12
Lm	H	0.0015	0.00199	0.00154
Ng	–	7	6	9
C1	N m/rad	0.5	0.42	0.53

4.4 Establish Feasibility Before Commitment

This phase can be considered as the principle of convergence that prescribes an aggressive elimination of inferior design sets. Then, we will increase details in order to narrow the set gradually. We should stay within sets once committed, and we must control by managing uncertainty at process.

In our case, we have 2 solutions of the body of ETB, 2 solutions for the reducer and 3 solutions of the opening and closing system. So, in this step we can add the constraints of the production system such as the availability of the machines, the production time, the cost of production and the stock to choose the right solution.

5 Conclusion

This article define an application of the SBCE (Set Based Concurrent Engineering) in the mechatronic field to design an ETB and a multi-objective optimization has integrated in the steps of the SBCE in order to get a good configuration which respects both the opening valve time and the closing valve time.

In further work, we will try to make an algorithm to reduce the number of solutions; we will also try to improve the model of ETB and to make more validations.

References

Hammadi, M., Choley, J.Y., Penas, O., Riviere, A., Louati, J., Haddar, M.: A new multi-criteria indicator for mechatronic system performance evaluation in preliminary design level. In: 2012 9th France-Japan and 7th Europe-Asia Congress on Mechatronics (MECATRONICS)/13th International Workshop on Research and Education in Mechatronics (REM) (2012)
Guizani, A., Hammadi, M., Choley, J.-Y., Soriano, T., Abbes, M.S., Haddar, M.: Multidisciplinary approach for optimizing mechatronic systems: application to the optimal design of an electric vehicle. In: 2014 IEEE/ASME International Conference on Advanced Intelligent Mechatronics (2014)

Hammadi, M., Choley, J.-Y., Penas, O., Riviere, A.: Mechatronic system optimization based on surrogate models—application to an electric vehicle. In: Proceedings of the 2nd International Conference on Simulation and Modeling Methodologies, Technologies and Applications (2012)

Buede, D.M., Miller, W.D.: The Engineering Design of Systems: Models and Methods. Wiley, London (2009)

Sobek, D.K., Ward, A.C., Liker, J.K.: Toyota's principles of setbased concurrent engineering. Sloan Manag. Rev. **40**(2), 67 (1999)

Ammar, R., Hammadi, M., Choley, J.Y., Louati, J., Barkallah, M., Haddar, M.: Architectural design of complex systems using set-based concurrent engineering. In: IEEE International Symposium on Systems Engineering (ISSE). IEEE (2017)

Raudberget, D.: Practical applications of set-based concurrent engineering in industry. J. Mech. Eng. **56**(11), 685–695 (2010)

Van Amerongen, J.: Mechatronic design. Mechatronics **13**(10), 1045–1066 (2003)

Martins, J.R.R.A., Lambe, A.B.: Multidisciplinary design optimization: a survey of architectures. AIAA J. **51**(9), 2049–2075 (2013)

Vibration Energy Localization from Nonlinear Quasi-Periodic Coupled Magnets

Zakaria Zergoune$^{(\boxtimes)}$, Najib Kacem, and Noureddine Bouhaddi

FEMTO-ST Institute, CNRS/UFC/ENSMM/UTBM,
Department of Applied Mechanics, Univ. Bourgogne Franche-Comté,
25000 Besançon, France
{zakaria.zergoune,najib.kacem,
noureddine.bouhaddi}@femto-st.fr

Abstract. The present study investigates the modeling of the vibration energy localization from a nonlinear quasi-periodic system. The periodic system consists of n moving magnets held by n elastic structures and coupled by a nonlinear magnetic force. The quasi-periodic system has been obtained by mistuning one of the n elastic structures of the system. The mistuning of the periodic system has been achieved by changing either the linear mechanical stiffness or the mass of the elastic structures. The whole system has been modeled by forced Duffing equations for each degree of freedom. The forced Duffing equations involve the geometric nonlinearity and the mechanical damping of the elastic structures and the magnetic nonlinearity of the magnetic coupling. The governing equations, modelling the quasi-periodic system, have been solved using a numerical method combining the harmonic balance method and the asymptotic numerical method. This numerical technique allows transforming the nonlinearities present in the governing equations into purely polynomial quadratic terms. The obtained results of the stiffness and mass mistuning of the quasi-periodic system have been analyzed and discussed in depth. The obtained results showed that the mistuning and the coupling coefficients have a significant effect on the oscillation amplitude of the perturbed degree of freedom.

Keywords: Energy localization · Nonlinear dynamics · Quasi-periodic system

1 Introduction

Over the last few years, energy harvesting from ambient energy has received increased attention. Several research projects have been oriented towards the design and the modeling of various harvesting systems. This trend of scavenging the ambient energy is related to the reduction of the required power supply for such microsystems and to the replacement of the battery which is limited by its life-time and requires maintenance. The harvesting approach is considering as a promising approach for innovation, miniaturization, respect for ecological issues and is part of the theme of renewable energies as well.

Diverse ambient energy sources are available in our environment and their conversion into electrical energy is a major challenge to increase the autonomy of isolated

© Springer Nature Switzerland AG 2019
T. Fakhfakh et al. (Eds.): ICAV 2018, ACM 13, pp. 121–128, 2019.
https://doi.org/10.1007/978-3-319-94616-0_12

or abandoned systems. Each environment can correspond to one or more energy sources such as sunlight, wind, thermal gradients, and mechanical vibrations. For each of these sources, one or more conversion principles exist for generating electricity. Mechanical vibration sources provide potential energy that can be scavenged for charging self-powered systems. In several researches, design of mechanical to electrical energy devices, based on different conversion mechanisms, has been attempted (El-Hami et al. 2001; Erturk and Inman 2011; Cassidy et al. 2011; Yang et al. 2014). Currently, the most existing solutions for vibration-to-electricity transduction are accomplished by electrostatic (Roundy et al. 2003; Mitcheson et al. 2004), piezoelectric (Anton and Sodano 2007), and electromagnetic applications (Yang et al. 2009).

The purpose of this study is to investigate and to analyze the modeling of a quasi-periodic system. The effects of the mistuning and nonlinearities of the proposed system are discussed. The damping factor of the quasi-periodic system was estimated experimentally by the half-power bandwidth method (Papagiannopoulos and Hatzigeorgiou 2011). The geometric and magnetic nonlinearities introduced in the model as well as the mistuning effect of the mechanical stiffness allow enlarging the bandwidth and localize the energy.

2 System Modeling

The quasi-periodic system presented in this survey was inspired by existing published works (e.g. The nonlinearity was inspired by Mann and Sims (2009), Mahmoudi et al. (2014), Ping et al. (2015), and Abed et al. (2016) while the mistuning effect and the vibration localization was inspired by Yoo et al. (2003) and Malaji and Ali (2015)). However, the main drawback of the previous harvesting systems is mainly the large mechanical damping factor. This significant damping is due to the friction of the lateral surface of the center moving magnet with the inner surface of the coil holder which affects directly the oscillation amplitude and then the harvested power.

The concept proposed in this paper uses quasi-periodic structure in order to take advantage of the multimodal approach and the vibration localization, while the mechanical and magnetic forces have been used to guide and couple the center moving magnets as well as reducing the mechanical damping factor. The considered system is composed of $n + 2$ magnets (two fixed magnets and n moving magnets). The poles of the whole magnets have been oriented to repel each other. The center moving magnets are mechanically attached to structure with a very low damping factor. The coils have been placed next to the n moving magnets. The separating distance between the $n + 2$ magnets can be tuned via threaded mechanism in order to adjust the magnetic coupling force as well as the linear resonance.

2.1 Magnetic Force

The resulting magnetic force has been estimated numerically by the 2D finite element method (Meeker 2006) while varying the gap between the magnets. Figure 1a shows the FEMM model for one degree of freedom while Fig. 2b shows the numerical estimation of the top and bottom of the magnetic force as a function of the separation distance (gap d) between two magnets (Fig. 3).

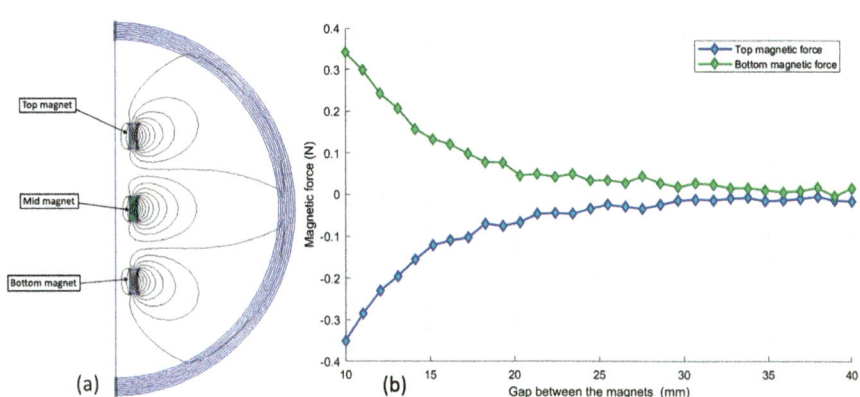

Fig. 1. Equivalent model for two moving magnets.

The numerical results of the magnetic force estimated by FEMM have been fitted for several values of gap d using a least-squares procedure. So, the total magnetic force can be identified as:

$$F^{mg}(x) = k_1^{mg}x + k_3^{mg}x^3, \tag{1}$$

where $k_1^{mg} = 2\lambda_1 + 4d\lambda_2 + 6d^2\lambda_3$ is the linear stiffness coefficient and $k_3^{mg} = 2\lambda_3$ is the cubic nonlinear stiffness coefficient in which d is the gap between the magnets. x is the displacement of the moving magnet. The FEMM result of the total magnetic force as a function of the displacement of the mid magnet and the fitting data for the gap equal to $d = 40$ mm as well as the magnetic linear stiffness k_1^{mg} deduced from the fitting FEMM data of different separating distance value.

The estimated parameters for the magnetic linear stiffness at $d = 40$ mm are $\alpha_1 = 313.71$ N m^{-1}, $\alpha_2 = -4.1e^{+3}$ N m^{-2}, and $\alpha_3 = 3.06e^{+3}$ N m^{-3}.

The accuracy of the fitted data has been checked by an overlay of the numerical data. The magnetic field B of the permanent magnets has been obtained analytically by the expression developed for ring magnets in reference (Camacho and Sosa 2013).

$$B(d) = \frac{\mu_0 M}{2} \left[\left(\frac{d}{\sqrt{d^2 + r_{out}^2}} - \frac{d - h}{\sqrt{(d - h)^2 + r_{out}^2}} \right) - \left(\frac{d}{\sqrt{d^2 + r_{int}^2}} - \frac{d - h}{\sqrt{(d - h)^2 + r_{int}^2}} \right) \right], \tag{2}$$

where d stands for the gap between two magnets, r_{int} and r_{out} are the inner and outer radius respectively.

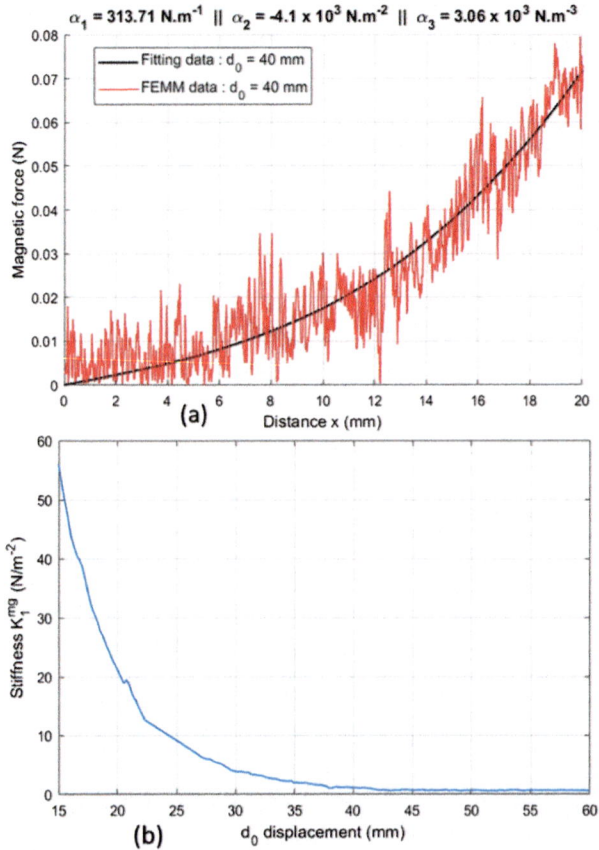

Fig. 2. (a) The FEMM result and the fitting data for $d = 40$ mm. (b) The linear stiffness k_1^{mg} estimated by fitting the FEMM data for each separating distance value.

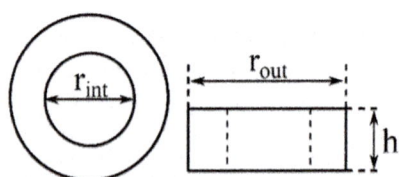

Fig. 3. Geometrical parameters of the magnet.

2.2 Governing Equations

In the present section, two center moving magnets are considered as illustrated in the equivalent mechanical and electrical model (Fig. 4). The proposed harvesting devise is

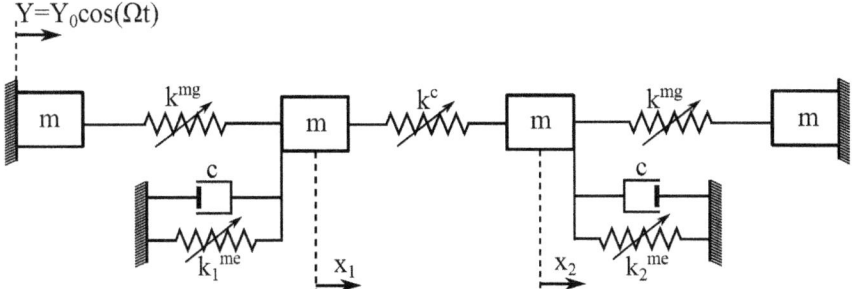

Fig. 4. Equivalent mechanical and electrical model for two moving magnets.

modeled using two forced duffing equations. So, the governing equation of the designed harvester can be written as:

$$m\ddot{x}_j + c\dot{x}_j + F_j^{me}(x) + F_j^{mg}(x) = -m\ddot{Y}; \quad with \quad j = 1, 2, \tag{3}$$

where c stands for the mechanical damping factors respectively. F_j^{me} and F_j^{mg} are the mechanical and magnetic forces for each moving magnet. $\ddot{Y}$ is the excitation acceleration of the support as shown in Fig. 4. It is assumed that the two center moving magnets have the same mass, mechanical and electrical damping.

$$\begin{cases} \ddot{x}_1 + 2\xi\omega_1\dot{x}_1 + \omega_1^2(1 + 2\beta)x_1 - \beta x_2 + \gamma x_1^3 - \beta_{NL}x_2^3 = -\ddot{Y} \\ \ddot{x}_2 + 2\xi\omega_1\dot{x}_2 + \omega_1^2(\alpha + 2\beta)x_2 - \beta x_1 + \gamma x_2^3 - \beta_{NL}x_1^3 = -\ddot{Y} \end{cases}, \tag{4}$$

$$2\xi\omega_1 = \frac{c}{m}, \beta = \frac{k_c^L}{k_1^{me}}, \beta_{NL} = \frac{k_c^{NL}}{m}, \omega_1^2 = \frac{k_1^{me}}{m}, \alpha = \frac{k_2^{me}}{k_1^{me}},$$

where α and β are the stiffness mistuning and coupling coefficients, respectively.

The solving procedure uses the classical harmonic balance method combined with the asymptotic numerical method (Cochelin and Vergez 2009). This technique allows transforming the nonlinearities present in the governing equation (Eq. 4) into purely polynomial quadratic terms.

3   Results and Discussion

In the present section, several numerical simulations have been performed in the case of two moving magnets. These simulations enable us to highlight the importance of the nonlinearity and mistuning of the designed harvesting device. The mistuning coefficients α represents the ratio of the mechanical linear stiffness of the second moving magnet to the ones of the first moving magnet. It is assumed in the present simulation that $k_{mg} = k_c$.

Figure 5 represents the frequency response for periodic and quasi-periodic structures with $\beta = 0.0083$ and an acceleration $a = 0.006 \, g$. Stiffness of the first moving

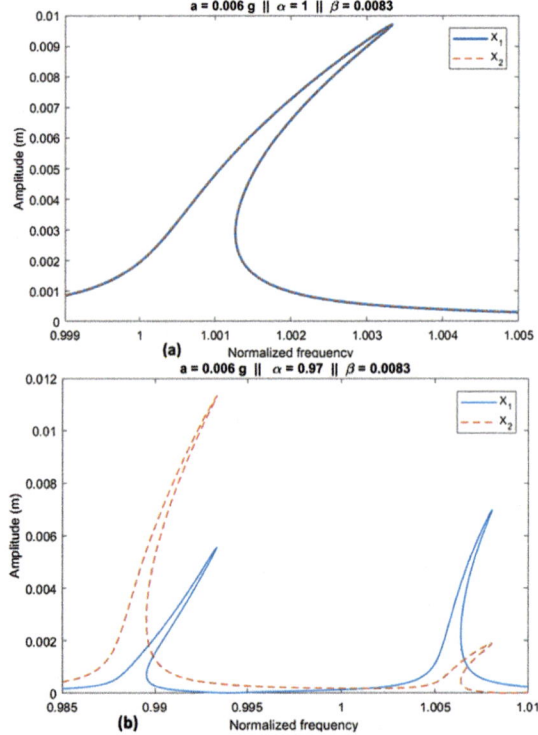

Fig. 5. Frequency response without (a) and with (b) stiffness perturbation.

magnet is taken as nominal stiffness. The mistuning was achieved by varying the stiffness of the second moving magnet. As shown in Fig. 5b, the amplitude of the perturbed dof was increased significantly with respect to the first dof. In addition, the bandwidth of the whole system was increased.

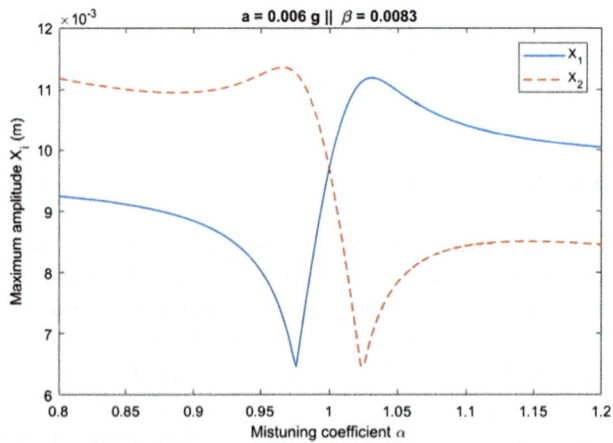

Fig. 6. Effect of the variation of the mistuning coefficient α on the maximum amplitudes.

Figure 6 shows the variation of the maximum amplitudes of the quasi-periodic system due to the variation of the mistuning coefficient α with an acceleration $a = 0.006\ g$ and $\beta = 0.0083$. As shown in this figure, when the mistuning coefficient α is less than 1, the amplitude of the perturbed dof increases with respect to the first dof. However, when $\alpha > 1$ the first dof represents an important amplitude compared to the perturbed dof.

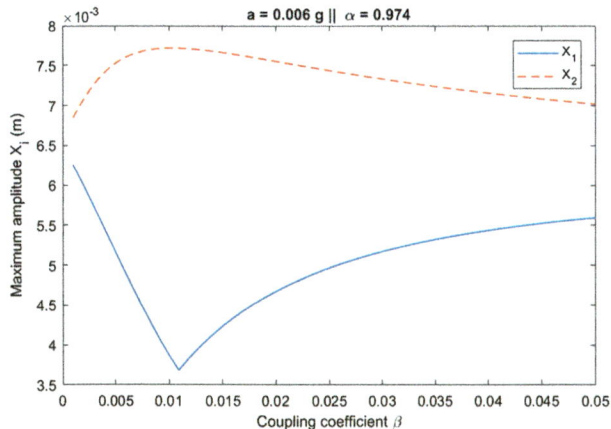

Fig. 7. Effect of the variation of the coupling coefficient β on the maximum amplitudes.

Figure 7 shows the variation of the maximum amplitudes of the present structure due to the variation of the coupling coefficient β with an acceleration $a = 0.006\ g$ and $\alpha = 0.97$. As shown in this figure, the coupling coefficient β has a significant effect on the oscillation amplitude of the proposed system.

4   Conclusion

In this paper, we studied the effect of the mistuning and coupling coefficients as well as the nonlinearity on the frequency response of a periodic structure. The obtained results show that the perturbation of one of the moving magnet, the magnetic coupling coefficient, and the nonlinearity increase the oscillation amplitude of the periodic system and enlarge the bandwidth as well. Thus, we can take advantage of these aspects to enhance the harvested power of a vibration energy harvesting mechanism. The proposed approach can be generalized to a large-scale quasi-periodic system.

Acknowledgements. This project has been performed in cooperation with the Labex ACTION program (contract ANR-11-LABX-01-01).

References

Abed, I., Kacem, N., Bouhaddi, N., Bouazizi, M.L.: Multi-modal vibration energy harvesting approach based on nonlinear oscillator arrays under magnetic levitation. Smart Mater. Struct. (2016). https://doi.org/10.1088/0964-1726/25/2/025018

Anton, S.R., Sodano, H.A.: A review of power harvesting using piezoelectric materials (2003–2006). Smart Mater. Struct. **16**, R1–R21 (2007)

Camacho, J.M., Sosa, V.: Alternative method to calculate the magnetic field of permanent magnets with azimuthal symmetry. Rev. Mex. de Fis. E **59**, 8–17 (2013)

Cassidy, I.L., Scruggs, J.T., Behrens, S.: Design of electromagnetic energy harvesters for large-scale structural vibration applications. In: Proceedings of Active and Passive Smart Structures and Integrated Systems 2011, 79770P (2011). https://doi.org/10.1117/12.880639

Cochelin, B., Vergez, C.: A high order purely frequency-based harmonic balance formulation for continuation of periodic solutions. J. Sound Vib. **324**, 243–262 (2009). https://doi.org/10.1016/j.jsv.2009.01.054

El-Hami, M., Glynne-Jones, P., White, N.M., Beeby, S., James, E., Brown, A.D., Ross, J.N.: Design and fabrication of a new vibration-based electromechanical power generator. Sens. Actuators, A **92**, 335–342 (2001)

Erturk, A., Inman, D.J.: Piezoelecric Energy Harvesting. Wiley, UK (2011). https://doi.org/10.1002/9781119991151

Mahmoudi, S., Kacem, N., Bouhaddi, N.: Enhancement of the performance of a hybrid nonlinear vibration energy harvester based on piezoelectric and electromagnetic transductions. Smart Mater. Struct. (2014). https://doi.org/10.1088/0964-1726/23/7/075024

Malaji, P.V., Ali, S.F.: Energy harvesting from near periodic structures. Vib. Eng. Technol. Mach. (2015). https://doi.org/10.1007/978-3-319-09918-7_37

Mann, B.P., Sims, N.D.: Energy harvesting from the nonlinear oscillations of magnetic levitation. J. Sound Vib. **319**, 515–530 (2009)

Meeker, D.C.: Finite Element Method Magnetics, Version 4.0.1 (2006). http://www.femm.info. Accessed 03 Dec 2006

Mitcheson, P.D., Green, T.C., Yeatman, E.M., Holmes, A.S.: Architectures for vibration-driven micropower generators. J. Microelectromech. Syst. **13**, 429–440 (2004)

Papagiannopoulos, G.A., Hatzigeorgiou, G.D.: On the use of the half-power bandwidth method to estimate damping in building structures. Soil Dyn. Earthq. Eng. **31**, 1075–1079 (2011). https://doi.org/10.1016/j.soildyn.2011.02.007

Ping, L., Shiqiao, G., Huatong, C., Lisen, W.: Theoretical analysis and experimental study for nonlinear hybrid piezoelectric and electromagnetic energy harvester (2015)

Roundy, S., Wright, P.K., Rabaey, J.: A study of low level vibrations as a power source for wireless sensor nodes. Comput. Commun. **26**, 1131–1144 (2003)

Yang, X., Wang, Y., Cao, Y., Liu, S., Zhao, Z., Dong, G.: A new hybrid piezoelectric-electromagnetic vibration-powered generator and its model and experiment research. IEEE Trans. Appl. Supercond. (2014). https://doi.org/10.1109/TASC.2013.2285944

Yang, Y.W., Tang, L.H., Li, H.Y.: Vibration energy harvesting using macro-fiber composites. Smart Mater. Struct. **18**, 115025 (2009)

Yoo, H.H., Kim, J.Y., Inman, D.J.: Vibration localization of simplified mistuned cyclic structures undertaking external harmonic force. J. Sound Vib. **261**, 859–870 (2003). https://doi.org/10.1016/S0022-460X(02)00997-5

A Numerical Parametric Analysis for the Distribution of Fins Using Phase Change Material (PCM)

Ahmed Guerine[✉] and Abdelkhalak El Hami

Laboratoire de Mécanique de Normandie LMN, INSA de Rouen Normandie,
Normandie Université, 76801 Saint Etienne du Rouvray Cedex, France
ahmedguerine@gmail.com,
abdelkhalak.elhami@insa-rouen.fr

Abstract. In this paper, the low melting point metal Phase Change Material (PCM) heat sink for coping with ultrahigh thermal shock (1 W/cm^2) is developed numerically. Sodium hydrate-based PCP is selected as the best Phase Change Material candidate from the point of view of thermal performance based on an approximate numerical analysis. Plate fin structure is investigated. The effects of fin number, heat flux, filling factor of PCM and fin width are parametrically studied; the influence of the structural material is briefly discussed. For arbitrarily given heating condition, the optimal geometric configuration of the heat sink is suggested and corresponding thermal performance is provided. The proposed low melting point metal PCM heat sink can cope with very large thermal shock with maximum device temperature, under the ambient temperature, which is extremely difficult to deal with otherwise by conventional PCMs. The conclusions drawn in this paper can serve as valuable reference for thermal design and analysis of PCM heat sink against ultra-high thermal shock. The results indicated that PCM-based heat sinks with fins are viable option for cooling plate structure with respect the number of fins, the power level of the heat source.

Keywords: Phase change material (PCM) · Heat sink · Thermal management

1 Introduction

Phase change material (PCM) cooling technique is a kind of passive cooling technique that uses phase change material as the coolant. When facing a thermal shock, PCM absorbs the heat and melts, while its temperature nearly keeps constant over the melting process, and thus prevents the power devices from overheating. After the thermal shock, heat is dissipated from the PCM to the ambient, the PCM solidifies and prepares for next thermal shock. PCM cooling technique is suitable for power devices which generate heat intermittently, such as portable electronics (Setoh et al. 2010) and power battery pack (Alipanah and Li 2016; Wang and Huang 2016).

Conventionally, organic PCMs (typically paraffin) are widely used for thermal management of power devices. The main drawback of paraffin PCMs lies in their low thermal conductivity, which seriously hinders the heat conduction inside the PCMs and

© Springer Nature Switzerland AG 2019
T. Fakhfakh et al. (Eds.): ICAV 2018, ACM 13, pp. 129–135, 2019.
https://doi.org/10.1007/978-3-319-94616-0_13

thus decreases the heat transfer efficiency. There are generally two methods to improve this situation: (1) increasing the thermal conductivity of the PCM via modification or nano-particle inclusion (He et al. 2012, Abdollahzadeh and Esmaeilpour 2015); (2) providing high conductive paths into the PCM to enhance the heat transfer inside, such as internal fin (Krishnan et al. 2005).

This paper is dedicated to develop a primary PCM heat sink used for coping with 1 W/cm^2 thermal shock. Firstly, an approximate theoretical analysis is conducted to find out the PCM which has the best cooling capability and is most suitable for high thermal shock situation from the view point of thermal performance. Then, based on this PCM, a plate fin structure is investigated, and the effect of fin number, heating power level on a PCM and Melting front position at various times are parametrically studied.

2 Mathematical Model

The heat transfer and fluid flow analysis in the cooling system are assumed to be two-dimensional. The flow of the molten PCM in heat sink assumed to be laminar, incompressed. The PCM is supposed to be pure, homogenous and with isotropic physical properties. The governing equations used here for the PCM are:

Momentum equations:

$$\rho_1\left(\frac{\partial u}{\partial t} + U\frac{\partial u}{\partial x} + V\frac{\partial u}{\partial y}\right) = -\frac{\partial P}{\partial x} + \mu_1\left(\frac{\partial^2 u}{\partial x^2} + \frac{\partial^2 u}{\partial y^2}\right) + Bu \tag{1}$$

$$\rho_1\left(\frac{\partial v}{\partial t} + U\frac{\partial v}{\partial x} + V\frac{\partial v}{\partial y}\right) = -\frac{\partial P}{\partial x} + \mu_1\left(\frac{\partial^2 v}{\partial x^2} + \frac{\partial^2 v}{\partial y^2}\right) + \rho_0 g\beta(T - T_0) + Bv \tag{2}$$

The thermal expansion coefficient β is introduced into the momentum Eq. (2) to include the buoyancy force term, according to the Boussinesq's approximation:

$$\rho = \rho_0 g\beta(T - T_0) \tag{3}$$

H_l is the liquid fraction during the phase change which is defined by the following relations:

$$H_l = 0 \quad \text{if } T \prec T_m \tag{4}$$

$$H_l = 1 \quad \text{if } T \succ T_m \tag{5}$$

The momentum source terms Bu and Bv (Eqs. (1) and (2)) were used for cancel velocities in solid region where $B(H_l)$ is the porosity function:

$$B = \frac{-C(1 - H_l^2)}{(H_l^3 + b)} \tag{6}$$

Where b is a small computational constant used to avoid division by zero, and c is a constant reflecting the morphology of the melting front.

Energy equation

$$Cp_{eq}\frac{\partial T}{\partial t} = \lambda_{eq}\left(\frac{\partial^2 T}{\partial x^2} + \frac{\partial^2 T}{\partial y^2}\right) - Cp_{eq}\left(u\frac{\partial T}{\partial x} + v\frac{\partial T}{\partial y}\right) \tag{7}$$

Where Cp_{eq} and λ_{eq} represent respectively the equivalent volume heat capacity and equivalent thermal conductivity.

$$Cp_{eq} = H_l(\rho Cp)_l + (1 - H_l)(\rho Cp)_s \tag{8}$$

$$\lambda_{eq} = H_l\lambda_l + (1 - H_l)\lambda_s \tag{9}$$

The governing equations previously described are solved using COMSOL MULTIPHYSICS. This code uses discretization and a formulation based on the finite element method.

3 Physical Model

Here, a specific heating condition is investigated first, and more general conclusion for arbitrary operation condition will be discussed later. The heat source works intermittently, the power of which is Q = 1 W. The plate fin structure is investigated as shown in Fig. 1. The plate fin heat sink is evenly divided into n basic units. In the following subsections, the effects of those parameters on the thermal performance of the heat sink will be discussed in detail. Among which, fin height b will be kept constant (b = 9 mm) and c = 1 mm. Boundary conditions include power value of 1 W applied at bottom of PCM. All other walls are treated as adiabatic. The initial condition set for entire domain is 45 °C.

The fin material of aluminium, phase change material used is sodium hydrate. The properties of the PCM, used in the present study, are summarised in Table 1.

4 Numerical Study

4.1 Effect of Heating Power Level on a PCM

Figure 2 shows the effects of different input powers on the set up with heat sink in the vertical position.

We can see at first the same temperature profile for three sources. At the beginning, the temperature increases rapidly and this is similarly observed for all sources. This phase corresponds to the conduction regime. Then we can notice a plateau for all sources which can be explained by the melting onset and conductive regime establishment in each liquid area. After-wards, we can remark a re-increase of temperature for the three cases. The PCM starts to melt earlier as the power is increased. The

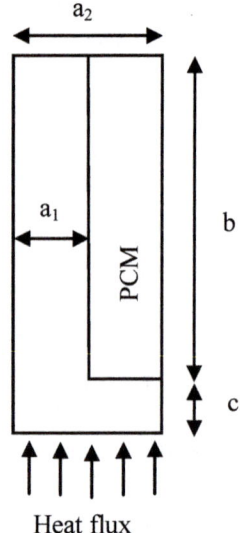

Fig. 1. Plate fin structure

Table 1. Properties of a commercial sodium hydrate-based PCM (Levin et al. 2013)

Melting temperature (°C)	Density (kg/m³) Solid = liquid	Thermal conductivity (W/(mK))	Specific heat (KJ/(kg.K))		Latent heat (KJ/kg)
			Solid	Liquid	
50	1360	0.6	2.7	2.4	113

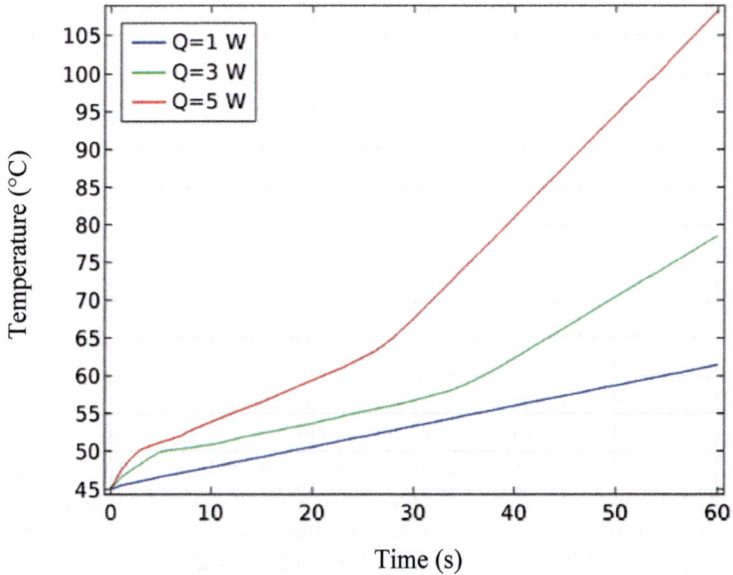

Fig. 2. Effect of heating power level on a PCM

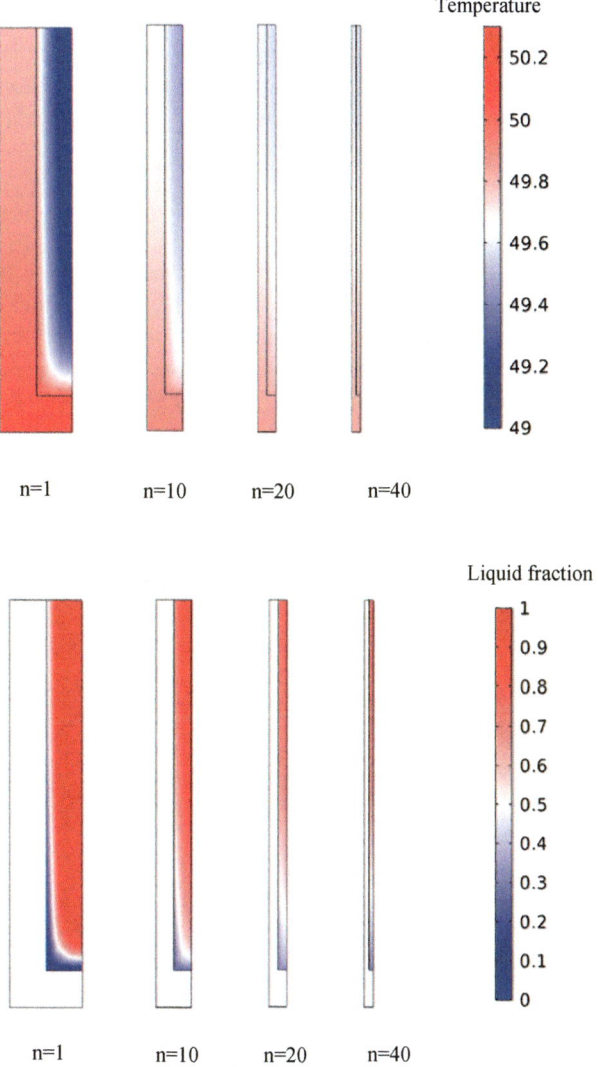

Fig. 3. Effect of the fin number. (a) Temperature on the heat sink bottom; (b) Liquid fraction contour after a thermal shock

maximum temperature attained at 1 W was 61 °C and the PCM was still undergoing phase change after 19 s. At 3 and 5 W the phase changes were completed after about 7 s and their maximum temperatures attained after 60 s were 79 and 104 °C respectively. At 1 W, the heat sink is at a much lower temperature compared to the powers at 3 and 5 W. The temperature of the electronic system device at 5 W is much higher after 60 s when the PCM in the heat sink has completely melted.

4.2   Effect of the Fin Number

Figure 3(a) and (b) intuitively show the temperature contour and the liquid fraction contour of the plate fin heat sink under different fin number conditions in the instant after a thermal shock. With the fin number increases, the width becomes smaller and smaller, bottom temperature Tmax decrease rapidly.

4.3   Melting Front Position at Various Times

The knowledge of the heat transfer mechanism during melting is essential to the understanding of phenomenon. Therefore, in Fig. 4 the melting front position at various times of 18, 20 and 22 s are depicted. In these photographs, the blue and red colors observed represent respectively the liquid and solid phases. At early time (t = 18 s), it can be noticed the formation of thin layer of liquid PCM near the respective heat source. At time = 20 s, convection establishes resulting the interface distortion at the high level of each liquid area. As times progresses, the liquid zone is rapidly merged. After t = 22 s, the majority of PCM is melted.

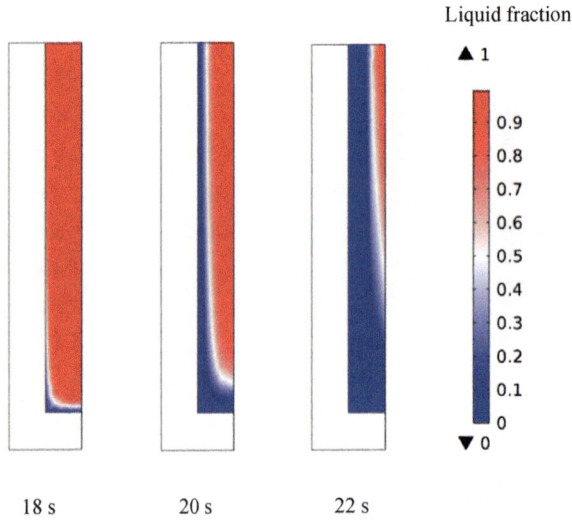

Fig. 4. Melting front position at various times

5   Conclusion

As a new kind of phase change material, low melting point metals own high cooling capability due to their high thermal conductivity and high volumetric latent heat, in which sodium hydrate possesses the best thermal performance. Parametric sodium hydrate based PCM heat sink with internal aluminum fin can cope with 1 W/cm^2. Geometric dimension of the fin structure has great influence on the thermal performance of the heat sink. Thermal performance of the heat sink improves with the number of fin increases.

Acknowledgements. This work is financed by the Project CRIOS and the European Union.

References

Setoh, G., Tan, F., Fok, S.: Experimental studies on the use of a phase change material for cooling mobile phones. Int. Commun. Heat Mass Transf. **37**, 1403–1410 (2010)

Alipanah, M., Li, X.: Numerical studies of lithium-ion battery thermal management systems using phase change materials and metal foams. Int. J. Heat Mass Transf. **102**, 1159–1168 (2016)

Wang, Q., Huang, C.: Investigation of the thermal performance of phase change material/mini-channel coupled battery thermal management system. Appl. Energy **164**, 659–669 (2016)

He, Q., Wang, S., Tong, M., Liu, Y.: Experimental study on thermophysical properties of nanofluids as phase-change material (PCM) in low temperature cool storage. Energy Convers. Manag. **64**, 199–205 (2012)

Abdollahzadeh, M., Esmaeilpour, M.: Enhancement of phase change material (PCM) based latent heat storage system with nano fluid and wavy surface. Int. J. Heat Mass Transf. **80**, 376–385 (2015)

Krishnan, S., Garimella, S.V., Kang, S.S.: A novel hybrid heat sink using phase change materials for transient thermal management of electronics. IEEE Trans. Comp. Packag. Technol. **28**, 281–289 (2005)

Levin, P.P., Shitzer, A., Hetsroni, G.: Numerical optimization of a PCM-based heat sink with internal fins. Int. J. Heat Mass Transf. **61**, 638–645 (2013)

Sensitivity Analysis of Frequency Response Functions for Load Resistance of Piezoelectric Energy Harvesters

Rabie Aloui[1](✉), Walid Larbi[2], and Mnaouar Chouchane[1]

[1] National Engineering School of Monastir (ENIM),
Avenue Ibn Jazzar, 5019 Monastir, Tunisia
{rabie.aloui,mnaouar.chouchane}@enim.rnu.tn
[2] Structural Mechanics and Coupled Systems Laboratory (LMSSC),
Conservatoire National des Arts et Metiers (CNAM),
292, rue Saint-Martin, 75141 Paris Cedex 03, France
walid.larbi@cnam.fr

Abstract. Piezoelectric energy harvesting from ambient energy sources, particularly vibrations, has attracted considerable interest throughout the last decade. Sensitivity analysis is a promising method used for many engineering problems to assess input-output systems based on vibration. In this paper, the formulation of first order sensitivity (FOS) of complex Frequency Response Functions (FRFs) is developed to evaluate the output responses of piezoelectric energy harvesters. The adapted approach for the FOS is the finite difference method, which consists in computing an approximation of the first derivation. Furthermore, the main goal is to study the influence of the variation of the load resistance from the short circuit (load resistance tends to zero) to open circuit (load resistance tends to the infinity) conditions on the tip displacement and the voltage FRFs of a Bimorph Piezoelectric Energy Harvester (BPEH). The determination of FRFs of the harvester are derived using Finite Element Modelling for a bimorph piezoelectric cantilever beam based on Euler-Bernoulli theory, which is composed of an aluminum substrate covered by two PZT-5A layers. The results show a high sensitivity of the FRFs of the BPEH to the load resistance at the natural frequencies. For each excitation frequency, the sensitivity near the resonance frequencies decreases from the short circuit conditions to the open circuit conditions.

Keywords: Sensitivity analysis · Vibration · Energy harvesting
Piezoelectric materials · Finite element method

1 Introduction

Sensitivity analysis of dynamic structures and mechatronic systems is very helpful in solving many engineering problems, such as: parametric identification problems, structural optimization, model updating problems and others (Lasecka-Plura and Lewandowski 2014), especially, for vibration energy harvesting devices using piezoelectric materials, which has been extensively studied over the past decade (Li et al.

© Springer Nature Switzerland AG 2019
T. Fakhfakh et al. (Eds.): ICAV 2018, ACM 13, pp. 136–148, 2019.
https://doi.org/10.1007/978-3-319-94616-0_14

2014). Several studies focused on modeling a multilayer cantilever beam with one, two or multi-piezoelectric layers used for vibration energy harvesting (Erturk and Inman 2011; Paknejad et al. 2016).

Two main approaches have been used by researchers for modeling piezoelectric energy harvesters are: (i) The analytical distributed parameter model (Erturk and Inman 2008, 2011) in which, the beam is modeled by a second-order partial differential equation in terms of beam tip displacement. (ii) The finite element model derived by De Marqui Junior et al. (2009) for an unimorph energy harvester plates, and a bimorph energy harvester cantilever beam (Amini et al. 2015). This formulation uses a standard discretization of beam layers, providing models with less restrictive assumptions, and takes into account the global electrical variables.

Since the first approach is limited to basic models, the finite element modeling is applied in this paper to determine the sensitivity of the frequency response functions (de Lima et al. 2010; Lasecka-Plura and Lewandowski 2014) of the bimorph piezo-electric energy harvester for the load resistance. The finite element equations of electromechanical problems are first presented. Then, the variational formulation of a laminated piezoelectric beam is developed for a Bimorph Piezoelectric Energy Harvester (BPEH) to determine the mechanical and electrical output FRFs. For the first order sensitivity analysis, the finite difference approach is applied to study the influence of the load resistance on the voltage and tip displacement FRFs.

2 Finite Element Modeling of the Energy Harvester

The finite element formulation of elastic structure with bonded piezoelectric patches proposed in (Thomas et al. 2009; Larbi et al. 2014) is used. The governing finite element equations of the dumped electromechanical problem can be expressed as:

$$M_m \ddot{U}(t) + C_m \dot{U}(t) + K_m U(t) + K_c V(t) = F(t) \tag{1}$$

$$K_e V(t) - K_c^T U(t) = Q(t) \tag{2}$$

where M_m is the global $(N \times N)$ mass matrix, K_m is the global $(N \times N)$ stiffness matrix, C_m is the global $(N \times N)$ damping matrix and K_c is the global electrome-chanical coupling matrix $(N \times P)$, K_e is the diagonal global $(P \times P)$ capacitance matrix, $F(t) = Fe^{j\omega t}$ is the global $(N \times 1)$ vector of mechanical forces, $Q(t) = Qe^{j\omega t}$ is the global $(P \times 1)$ vector of electric charge outputs, $U(t) = Ue^{j\omega t}$ is the global $(N \times 1)$ vector of mechanical coordinates and $V(t) = Ve^{j\omega t}$ is the global $(P \times 1)$ vector of voltage outputs. Here, N and P respectively, are the number of mechanical degrees of freedom and the number of piezoelectric elements. The global mechanical damping matrix C_m is assumed a linear combination of the mass and stiffness matrices:

$$C_m = \alpha M_m + \beta K_m \tag{3}$$

where α and β are the proportionality constants which are typically determined experimentally using at least two modal damping associated to two different natural frequencies.

Equation (1) corresponds to the mechanical equation of motion with electrical coupling, and a forcing vector $F(t)$. Whereas Eq. (2) corresponds to the electrical circuit equation with a mechanical coupling term. In this paper, the harvested energy is dissipated through a resistive load R. Using Ohm's law, the following additional equation relates the voltage vector V and the charge vector Q:

$$V(t) = -R\dot{Q}(t) \tag{4}$$

Considering in particular the finite element formulation of a laminated beam with a total of K layers including P piezoelectric layers, which is excited under sinusoidal base motion. Three mechanical degrees of freedom per node are used $\left(u, w, \theta \approx -\frac{\partial w_{rel}}{\partial x}\right)$. The piezoelectric layers of the cantilever beam are poled in the thickness direction with an electrical field applied parallel to this polarization. Such a configuration is characterized in particular by the electromechanical coupling between the axial strain ε_1 and the transverse electrical field E_3 (Thomas et al. 2009). Furthermore, the reduced law behavior of a thin piezoelectric layer is written as follows:

$$\sigma_1 = \bar{c}_{11}\varepsilon_1 - \bar{e}_{31}E_3 \tag{5}$$

$$D_3 = \bar{e}_{31}\varepsilon_1 + \bar{\epsilon}_{33}E_3 \tag{6}$$

where $\sigma_1, \varepsilon_1, E_3$ and D_3 are respective the normal stress, normal strain, electric field and electric displacement, $\bar{c}_{11}$ is the elastic modulus, $\bar{e}_{31}$ is the piezoelectric coupling coefficient and $\bar{\epsilon}_{33}$ is the permittivity at constant strain. The variational formulation, in this case, is defined as follows:

$$\sum_{k=1}^{K}\int_{\Omega^k}\rho^k(\ddot{u}_x\delta u_x + \ddot{u}_z\delta u_z)d\Omega + \sum_{k=1}^{K}\int_{\Omega^k}\bar{c}_{11}^k\varepsilon_1\delta\varepsilon_1 d\Omega + \sum_{p=1}^{P}\frac{V^{(p)}}{h^{(p)}}\int_{\Omega^{(p)}}\bar{e}_{31}^{(p)}\delta\varepsilon_1 d\Omega = 0 \tag{7}$$

$$-\sum_{p=1}^{P}\frac{\delta V^{(p)}}{h^{(p)}}\int_{\Omega^{(p)}}\bar{e}_{31}^{(p)}\varepsilon_1 d\Omega + \sum_{p=1}^{P}\delta V^{(p)}C^{(p)}V^{(p)} = \sum_{p=1}^{P}\delta V^{(p)}Q^{(p)} \tag{8}$$

where ρ^k and Ω^k are the mass density and the domain occupied by the kth layer, $C^{(p)} = \frac{S^{(p)}}{h^{(p)}}\bar{\epsilon}_{33}^{(p)}$ is the capacity of the pth piezoceramic layer, where $S^{(p)}$ and $h^{(p)}$ are respectively the active surface and the thickness of the pth piezoceramic layer. The mechanical displacements u_x and u_z are defined as follow:

$$u_x(x, z, t) = u(x, t) - z\theta(x, t) \tag{9}$$

$$u_z(x, z, t) = w(x, t) = w_b(t) + w_{rel}(x, t) \tag{10}$$

where $w_b(t) = W_b e^{j\omega t}$ is the base displacement and $w_{rel}(x, t)$ is the relative displacement (for clamped-free beam).

The various terms appearing in the variational formulation in Eqs. (7) and (8) are now successively discussed.

- The kinetic energy variation is:

$$\sum_{k=1}^{K} \int_{\Omega^k} \rho^k (\ddot{u}_x \delta u_x + \ddot{u}_z \delta u_z) d\Omega \Rightarrow \delta U^T M_m \ddot{U} - \delta U^T F \qquad (11)$$

where F is the inertial forcing vector due to base excitation which can be expressed as an effective mass vector m^* multiplied by the base acceleration (De Marqui et al. 2009) as follows:

$$\sum_{k=1}^{K} \int_{\Omega^k} \rho^k \delta w_{rel} \ddot{w}_b d\Omega \Rightarrow -\delta U^T m^* \ddot{w}_b = -\delta U^T F \qquad (12)$$

- The mechanical contribution to the internal energy variation is:

$$\sum_{k=1}^{K} \int_{\Omega^k} \bar{c}_{11}^k \varepsilon_1 \delta \varepsilon_1 d\Omega \Rightarrow \delta U^T K_m U \qquad (13)$$

- The piezoelectric contributions to the internal energy variation, related to the direct and inverse effect, are given in the following equations.

$$\sum_{p=1}^{P} \frac{V^{(p)}}{h^{(p)}} \int_{\Omega^{(p)}} \bar{e}_{31}^{(p)} \delta \varepsilon_1 d\Omega \Rightarrow \delta U^T K_c V \qquad (14)$$

$$\sum_{k=1}^{P} \frac{\delta V^{(p)}}{h^{(p)}} \int_{\Omega^{(p)}} \bar{e}_{31}^{(p)} \varepsilon_1 d\Omega \Rightarrow \delta V^T K_c U \qquad (15)$$

- The electrical contribution to the internal energy variation is:

$$\sum_{p=1}^{P} \delta V^{(p)} C^{(p)} V^{(p)} \Rightarrow \delta V^T K_e V. \qquad (16)$$

3 Finite Element Modeling of a BPEH

In this section, the system matrices used in Eqs. (1) and (2) are derived using the finite element formulation of a bimorph piezoelectric vibration energy harvester excited by base motion. The harvester consists in an Euler Bernoulli beam composed of two layers of PZT-5A (piezoelectric material) bonded to an aluminum substrate (elastic material) as shown in Fig. 1. Thus, the total number of layers is equal to 3 ($K = 3$) the number of piezoelectric elements is equal to 2 ($P = 2$).

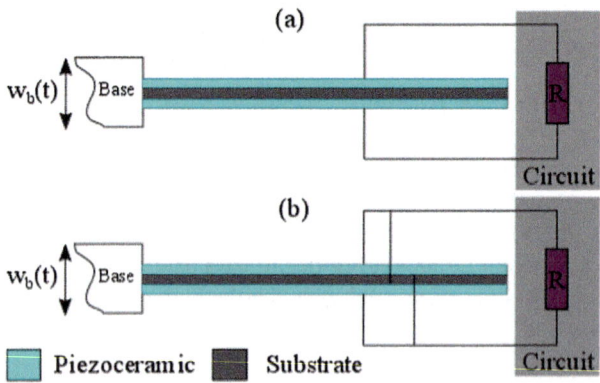

Fig. 1. Cantilever piezoelectric energy harvester configurations under base excitation: (a) bimorph (series connection) and (b) bimorph (parallel connection)

The electrical degrees of freedom associated to the two piezoelectric layers are the voltage vector V and charge vector Q defined as follows:

$$V = \begin{pmatrix} V^{(1)} \\ V^{(2)} \end{pmatrix}; \ Q = \begin{pmatrix} Q^{(1)} \\ Q^{(2)} \end{pmatrix} \tag{17}$$

Each piezoelectric layer is characterized by its capacity $C^{(p)}$ and the electromechanical coupling vector $K_c^{(p)}$, $p = 1, 2$.

$$K_e = diag\left(C^{(1)}, C^{(2)}\right); \ K_c = \begin{pmatrix} K_c^{(1)} & K_c^{(2)} \end{pmatrix} \tag{18}$$

3.1 Equivalent Representation of the Series and the Parallel Connection Cases of a BPEH

The equivalent representation of the finite element electromechanical equations of a BPEH for the series and the parallel connections is very useful to predict the electrical output responses across the resistor (in the circuit). For this purpose, the equivalent terms of the equivalent representation are obtained first.

The two-piezoceramic layers are assumed to be identical (same material, same dimensions). It is therefore reasonable to assume that both of them have the same capacity $\left(C^{(1)} = C^{(2)} = C\right)$ and generate the same output voltage so that $\left(V^{(1)} = V^{(2)} = V\right)$ and $\left(Q^{(1)} = Q^{(2)} = Q\right)$ (De Marqui et al. Junior 2009). Therefore, the nodal forces related to the converse piezoelectric effect ($K_c V$) when a voltage V is applied to the electrodes are given by the following term:

$$K_c V = K_c \begin{pmatrix} 1 \\ 1 \end{pmatrix} V = \tilde{K}_c V \tag{19}$$

Where $\tilde{\boldsymbol{K}}_c = \boldsymbol{K}_c^{(1)} + \boldsymbol{K}_c^{(2)}$ is the apparent electromechanical coupling vector $(N \times 1)$. Furthermore, the resulting charge and voltage in the circuit of the BPEH are given in Table 1 for the series and parallel connections (Erturk and Inman 2011).

Table 1. The charge and voltage in the electrical circuit for series and parallel connection of the two piezoelectric layers with a resistance load

	Series connection	Parallel connection
Charge in the circuit	Q	$2Q$
Voltage in the circuit	2 V	V

After modifying Eqs. (1) and (2) and transforming then to the frequency domain using Laplace transform, the equivalent electromechanical equations of a BPEH become:

$$\left[-\omega^2 \boldsymbol{M}_m + j\omega \boldsymbol{C}_m + \boldsymbol{K}_m\right] \boldsymbol{U} + \boldsymbol{K}_c^{eq} V = \boldsymbol{F} \tag{20}$$

$$\left(j\omega C^{eq} + \frac{1}{R}\right) V - j\omega \boldsymbol{K}_c^{eqT} \boldsymbol{U} = 0 \tag{21}$$

where $\boldsymbol{K}_c^{eq}$ and C^{eq} are respectively the equivalent electromechanical coupling vector and the equivalent capacity of a BPEH, which are given in Table 2, V is the voltage across the load resistance (in the circuit).

Table 2. Equivalent electromechanical coupling and capacitance of a bimorph energy harvester for the series and the parallel connections of the piezoceramic layers

Terms	Series connection	Parallel connection
$\boldsymbol{K}_c^{eq}$	$\tilde{\boldsymbol{K}}_c/2$	$\tilde{\boldsymbol{K}}_c$
C^{eq}	$C/2$	$2C$

3.2   Frequency Response Functions

The FRFs are defined here as the response outputs of the BPEH (displacement, voltage, current, power) per base acceleration (in terms of the gravitational acceleration, $g = 9.81$ m/s^2). The equivalent expression for nodal displacements FRFs relative to the base excitation problem of the BPEH is:

$$\frac{\boldsymbol{U}}{-\omega^2 W_b} = \left(-\omega^2 \boldsymbol{M}_m + j\omega \boldsymbol{C}_m + \boldsymbol{K}_m + \frac{j\omega \, \boldsymbol{K}_c^{eq} \boldsymbol{K}_c^{eqT}}{\left(\frac{1}{R} + j\omega C^{eq}\right)}\right)^{-1} \boldsymbol{m}^* \tag{22}$$

For the mechanical response (vibration), only the transverse tip displacement FRF $(w_n/-\omega^2 W_b)$ is considered in this study (n is the total node number of standard discretization with linear elements).

The voltage FRF is obtained as a function of the nodal displacements FRFs.

$$\frac{V}{-\omega^2 W_b} = \frac{j\omega K_c^{eqT}}{\left(\frac{1}{R}+j\omega C_p^{eq}\right)} \left(-\omega^2 M_m + j\omega C_m + K_m + \frac{j\omega K_c^{eq} K_c^{eqT}}{\left(\frac{1}{R}+j\omega C^{eq}\right)}\right)^{-1} m^* \quad (23)$$

The current FRF and the power FRF are obtained from the voltage FRF as follows:

$$\frac{I}{-\omega^2 W_b} = \frac{1}{R}\left(\frac{V}{-\omega^2 W_b}\right); \quad \frac{P}{-\omega^2 W_b} = \frac{1}{R}\left(\frac{V}{-\omega^2 W_b}\right)^2 \quad (24)$$

The four frequency response functions may be collected into a single vector defined as follows.

$$H = \frac{1}{-\omega^2 W_b}\begin{bmatrix} w_n & V & I & P \end{bmatrix} \quad (25)$$

The global finite element matrices appearing in the FRFs establish the dependence of the response of the system on a set of parameters, and can be expressed in the following form.

$$H = H(\omega, p) \quad (26)$$

Where H is the frequency response functions vector, p is a vector of parameters of the BPEH.

4   Finite Difference Approach to Sensitivity Analysis of FRFs

The finite difference method originates from a Taylor series expansion to approximate the first order sensitivity (FOS) and is undoubtedly the simplest method to implement. The FOS of the responses with respect to a given design parameter p_i, evaluated for a given set of values of the design parameters p^0 is defined as the following partial forward derivative:

$$\left.\frac{\partial H}{\partial p_i}\right|_{p_i^0} = \lim_{\Delta p_i \to 0} \frac{H\left(\omega, p_i^0 + \Delta p_i\right) - H\left(\omega, p_i^0\right)}{\Delta p_i} \quad (27)$$

where Δp_i is the parameter increment in the finite difference scheme, applied to the current value of the parameter p_i^0, while all other parameters are kept unchanged. The sensitivity of the response with respect to p_i can be numerically estimated by finite differences by successively computing the responses corresponding to $p_i = p_i^0$ and $p_i = p_i^0 + \Delta p_i$, and then calculating:

$$\left.\frac{\partial \boldsymbol{H}}{\partial p_i}\right|_{p_i^0} \approx \frac{\boldsymbol{H}\left(\omega, p_i^0 + \Delta p_i\right) - \boldsymbol{H}\left(\omega, p_i^0\right)}{\Delta p_i} \tag{28}$$

The accuracy of the sensitivity estimates depends on the choice of the value of the parameter increment Δp_i, which has to be small compared to the corresponding parameters p_i^0 but there are limitations due to numerical truncation. The choice of Δp_i is critical in the precision of the calculated derivatives. Therefore, Δp_i are chosen following rule proposed by Arruda and Santos (1993).

$$\Delta p_i = \min\{\|H(p_i)\|, \delta_i\} \tag{29}$$

where $\|.\|$ is the Euclidian norm of the output FRFs vector and δ_i is defined as:

$$\delta_i = \begin{cases} 10^{-1} & \text{if } |p_i^0| < 10^{-6} \\ 10^{-3}|p_i^0| & \text{if } |p_i^0| \geq 10^{-6} \end{cases} \tag{30}$$

In order to check the accuracy of the calculated first order sensitivity, an approximation of the FRF for the parameter $(p_i + \Delta p_i)$ is computed using the following formula:

$$\hat{\boldsymbol{H}}(\omega, p_i^0 + \Delta p_i) \approx \boldsymbol{H}(\omega, p_i^0) + \frac{\partial \boldsymbol{H}}{\partial p_i} \Delta p_i \tag{31}$$

where $\hat{\boldsymbol{H}}$ is the first order approximation of the output FRFs of the harvester.

5 Case Study

This section presents an example of a BPEH computed using the previous finite element model. The material properties and geometrical characteristics of the harvester used in this study are given in Table 3 (Erturk and Inman 2011). For the purpose of simulation, the coefficient for the first two modes are chosen to be $\zeta_1 = 0.010$ and $\zeta_2 = 0.012$, the constants α and β are computed using these coefficients. The computing of the first derivation using the finite difference approach consists in varying the nominal value of the parameter by 0.25% ($\Delta p_i = 0.25\%$ of p_i). The load resistance is mounted in series with the piezoelectric layers. The sensitivity analysis of the tip displacement and voltage FRFs of the BPEH for a load resistance R are presented here.

The analysis is carried out for the frequency range from 0 to 5000 Hz. The first three resonance frequencies of the BPEH for short-circuit ($R \rightarrow 0$) and open circuit ($R \rightarrow \infty$) conditions are given in Table 4. The effective electromechanical modal coupling factor $k_{eff,r}$ characterizes the energy exchange between the mechanical structure and the piezoelectric layers. It is usually defined, for the system rth mode, by:

Table 3. Material properties and geometrical characteristics of the reference PEH

Parameters		PZT-5A	Aluminum
L	Beam length (mm)	30	30
b	Beam width (mm)	5	5
h_p, h_s	Layers thickness	0.15	0.05
E_p, E_s	Young's modulus (GPa)	61	70
ρ_s, ρ_p	Mass density (kg/m^3)	7750	2700
$\bar{e}_{31}$	Piezoelectric constant (C/m^2)	−10.4	–
$\bar{\epsilon}_{33}$	Permittivity constant (nF/m)	13.3	–

$$k_{eff,r}^2 = \frac{\left(f_r^{oc}\right)^2 - \left(f_r^{sc}\right)^2}{\left(f_r^{sc}\right)^2} \qquad (32)$$

where f_r^{sc} and f_r^{oc} are, respectively, the short-circuit and open-circuit rth system natural frequencies.

Table 4. First three short-circuit and open-circuit natural frequencies of the BPEH, and the effective electromechanical coupling factor

Mode (r)	f_r^{sc} (Hz)	f_r^{oc} (Hz)	$k_{eff,r}$
1	181.1	191.1	0.0656
2	1159.8	1171.7	0.0206
3	3246.7	3258.0	0.007

The tip displacement and the voltage FRFs of the BPEH are given respectively in μm/g and V/g, and the resistance is expressed in Ω. Hence, the sensitivities relative to the load resistance are given respectively in μm/(g Ω) and V/(g Ω).

Figure 2 shows the first order sensitivity of the tip displacement FRF of the harvester with respect to the electrical load resistance R. Sensitivity analysis is applied for three resistances 100 Ω, 10 kΩ and 100 kΩ. It can be observed that the peaks of the FOS curves occur at the natural frequencies and has low values, around 10^{-7} over a wide range of frequencies.

The enlarged views of the FOS of the vibration to the load resistance given in Fig. 2 are presented as response surfaces using load resistance as an additional axis, Fig. 3. The absolute value of the sensitivity in the vicinity of resonance frequencies decreases from the short circuit conditions ($R \to 0$) to the open circuit conditions ($R \to \infty$). The sensitivity of the tip displacement (vibration) to the load resistance has lower values in the vicinity of the second mode compared to that of the first mode.

The variation of the sensitivity of the tip displacement of the BPEH for load resistance at the fundamental short-circuit resonance frequency and at the fundamental open-circuit resonance frequency are shown in Fig. 4. It is worth to note that the sensitivity curves are not completely monotonic. It should also be noted that the sensitivity of the vibration to the load resistance at the short circuit excitation frequency always remains greater in absolute value than that at the open circuit excitation frequency.

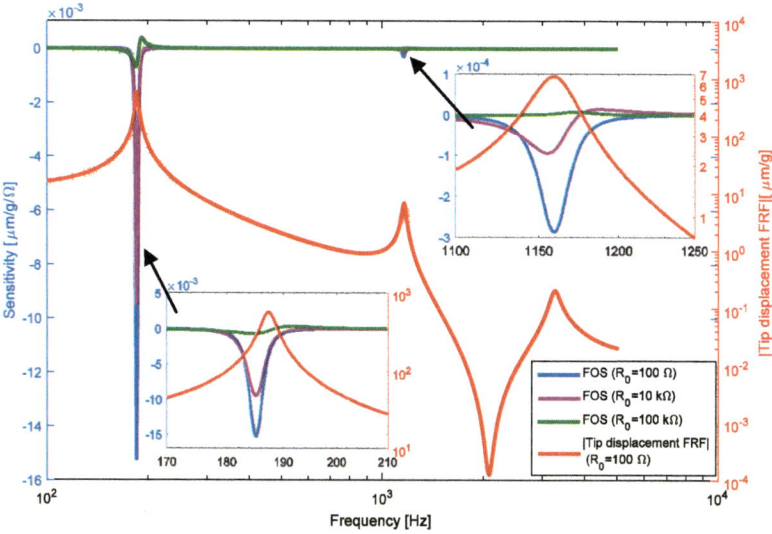

Fig. 2. First order sensitivity of tip displacement FRF modulus versus excitation frequency for three load resistances

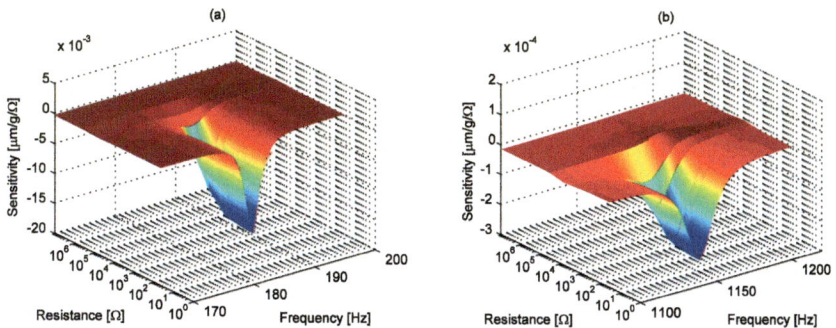

Fig. 3. First order sensitivity surface response of voltage FRF modulus versus excitation frequency and load resistance (a) in the vicinity of mode 1 (b) in the vicinity of mode 2

Figure 5 shows the first order sensitivity of the voltage FRF modulus versus the excitation frequency for three load resistances. We notice that the sensitivity for the load resistance is important in the vicinity of natural frequencies. The resistance value of 100 Ω has the largest sensitivities in the vicinity of the resonance frequencies, it is followed by the sensitivities of 10 kΩ, and finally the 100 kΩ sensitivity for each vibration mode.

Figure 6 shows that the sensitivity of the voltage output decreases when the resistance varies from the short circuit conditions to the open circuit conditions for all excitation frequencies. Furthermore, for each occurs of value of the resistance, the maximum value of sensitivity of voltage output matches the resonance frequency.

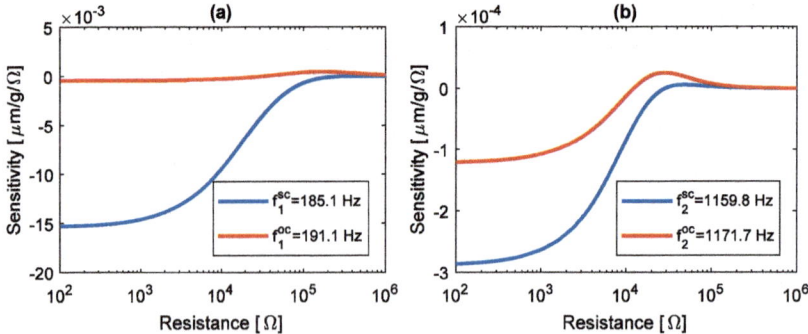

Fig. 4. Variation of the FOS of the tip displacement FRF to the load resistance versus load resistance for excitations at the short-circuit and the open-circuit resonance frequencies of: (a) mode 1 (b) mode 2

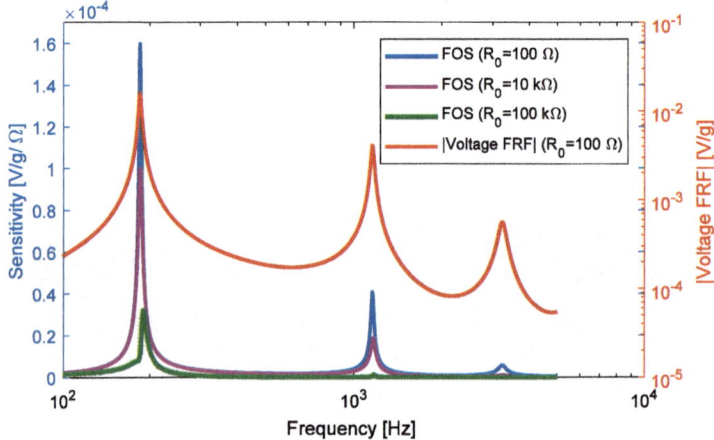

Fig. 5. First order sensitivity of voltage FRF modulus versus excitation frequency for three load resistance values

Therefore, the FOS of voltage FRF is significant at the natural frequencies and at low electrical load resistances (short circuit conditions).

Figure 7 shows the first order sensitivity of the voltage FRF modulus as a function of load resistance for excitations at the fundamental short-circuit and open-circuit resonance frequencies. For the first two modes, as the load resistance increases from the short-circuit to the open-circuit conditions, the sensitivity of the voltage FRF decreases monotonically. One can see clearly that the sensitivity of voltage FRF for the load resistance is more important for the short-circuit frequency excitation then the open-circuit frequency excitation for low resistance (short-circuit conditions). Both the sensitivities of the voltage FRF to the load resistance at the two fundamental resonance frequencies (short-circuit and open-circuit frequencies) have a very low value at the open-circuit condition.

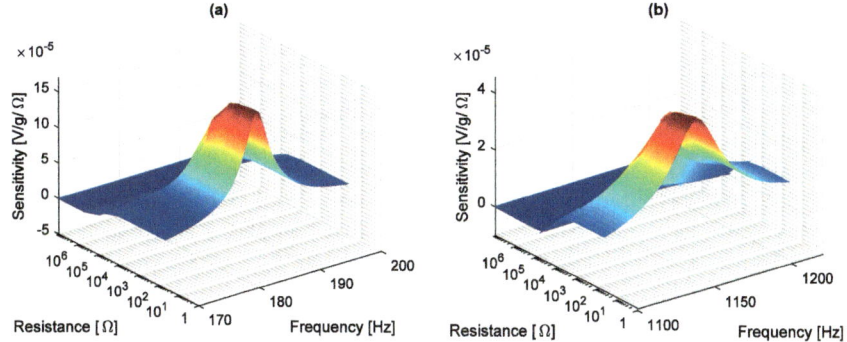

Fig. 6. First order sensitivity of modulus of Voltage FRF versus excitation frequency and load resistance (a) in the vicinity of mode 1 (b) in the vicinity of mode 2

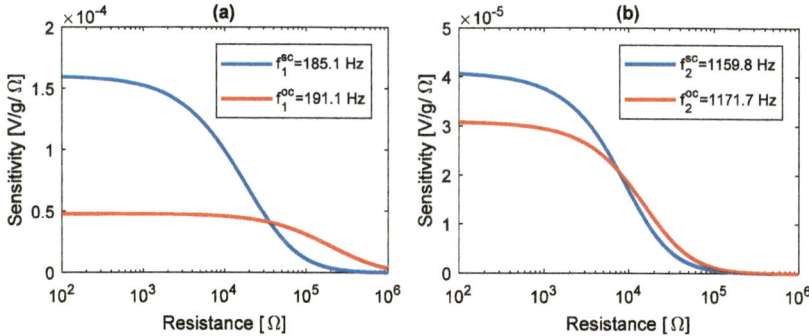

Fig. 7. Variation of the FOS of the voltage FRF for the load resistance versus load resistance for excitations at the short-circuit and the open-circuit resonance frequencies of: (a) mode 1 (b) mode 2

6   Conclusion

In this paper, the sensitivity analysis of frequency response functions has been considered. The finite difference approach has been used to approximate the first order sensitivity of tip displacement (vibration) and the voltage FRFs to a variation of the electrical load resistance of the harvester. The first order sensitivity analysis of the tip displacement and voltage FRFs of the BPEH have shown that the sensitivity to the load resistance is significant at the natural frequencies. Furthermore, the influence of the electrical load resistance variation for the vibration and voltage outputs is more important at the short circuit conditions than at the open circuit conditions. These results are very helpful to determine the optimal load resistance for an optimization study using the load resistance as a parameter.

References

Amini, Y., Emdad, H., Farid, M.: Finite element modeling of functionally graded piezoelectric harvesters. Compos. Struct. **129**, 165–176 (2015). https://doi.org/10.1016/j.compstruct.2015.04.011

Arruda, J.R.F., Santos, J.M.C.: Mechanical joint parameter estimation using frequency response functions and component mode synthesis. Mech. Syst. Signal Process. **7**, 493–508 (1993)

de Lima, A.M.G., Faria, A.W., Rade, D.A.: Sensitivity analysis of frequency response functions of composite sandwich plates containing viscoelastic layers. Compos. Struct. **92**, 364–376 (2010). https://doi.org/10.1016/j.compstruct.2009.08.017

De Marqui Jr., C., Erturk, A., Inman, D.J.: An electromechanical finite element model for piezoelectric energy harvester plates. J. Sound Vib. **327**, 9–25 (2009). https://doi.org/10.1016/j.jsv.2009.05.015

Erturk, A., Inman, D.J.: A distributed parameter electromechanical model for cantilevered piezoelectric energy harvesters. J. Vib. Acoust. **130**, 041002 (2008). https://doi.org/10.1115/1.2890402

Erturk, A., Inman, D.J.: Piezoelectric Energy Harvesting. Wiley, Chichester (2011)

Larbi, W., Deü, J.-F., Ohayon, R., Sampaio, R.: Coupled FEM/BEM for control of noise radiation and sound transmission using piezoelectric shunt damping. Appl. Acoust. **86**, 146–153 (2014). https://doi.org/10.1016/j.apacoust.2014.02.003

Lasecka-Plura, M., Lewandowski, R.: Design sensitivity analysis of frequency response functions and steady-state response for structures with viscoelastic dampers. Vib. Phys. Syst. **26** (2014)

Li, H., Tian, C., Deng, Z.D.: Energy harvesting from low frequency applications using piezoelectric materials. Appl. Phys. Rev. **1**, 041301 (2014). https://doi.org/10.1063/1.4900845

Paknejad, A., Rahimi, G., Farrokhabadi, A., Khatibi, M.M.: Analytical solution of piezoelectric energy harvester patch for various thin multilayer composite beams. Compos. Struct. **154**, 694–706 (2016). https://doi.org/10.1016/j.compstruct.2016.06.074

Thomas, O., Deü, J.-F., Ducarne, J.: Vibrations of an elastic structure with shunted piezoelectric patches: efficient finite element formulation and electromechanical coupling coefficients. Int. J. Numer. Methods Eng. **80**, 235–268 (2009). https://doi.org/10.1002/nme.2632

Effect of Harmonic Excitation on PCB and Component Assembly

Ayda Halouani[1,2](✉), Mariem Miladi Chaabane[1](✉),
Mohamed Haddar[1](✉), and Abel Cherouat[2](✉)

[1] Laboratory of Mechanics, Modeling and Production (LA2MP),
National School of Engineers of Sfax, BP 1173, 3038 Sfax, Tunisia
`mariam.mi@hotmail.fr`, `mohamed.haddar@enis.rnu.tn`
[2] Laboratory of Automatic Generation of Meshing and Advanced Methods
(GAMMA3), University of Technology of Troyes, 12 rue Marie Curie,
10000 Troyes, France
{`ayda.halouani,abel.cherouat`}`@utt.fr`

Abstract. The plastic ball grid array (PBGA) package has become a major packaging type in recent years, due to its high capacity for the input/output counts. However, vibration loading is encountered during the service life of PBGA. This study investigates the effect of vibration loading on the solder ball response. A two-dimensional finite element model of the printed circuit board (PCB) and PBGA component assembly is released using COMSOL Multiphysics software. The natural frequencies and modes were calculated. Forced vibration analysis was performed around the first natural frequency to determine the solder joints having highest stress and strain concentration under harmonic excitation. It showed that the interface between solder ball and the PCB is the most vulnerable part. Displacement and Von Mises stress variation were calculated in the most critical point. It was found that the height amplitude of displacement and Von Mises stress may conduct to decrease the solder interconnects lifetime. Moreover, resonance may conduct to the failure of the solder joints.

Keywords: PBGA · Finite element · Vibration loading · Harmonic excitation

1 Introduction

Printed Circuit Board (PCB) are used in most electronic products to mechanically support and electrically connect chips, capacitors, or other electronic components via soldered joints. During utilisation, these products will experience loading environments that include vibration loading that is why it is necessary to determine the structural integrity of the PCB and its components due to this load. PCB is exposed to vibration loading and it is well known that solder is the most critical part of the assembly as it assure the interconnection between the card and the component. Several work has been established in the literature in order to study the dynamic behavior of an electronic component under vibration load. Grieu et al. (2008) presented a methodology to calculate the damage of solder joint under random vibration. In addition, he used FEM to

© Springer Nature Switzerland AG 2019
T. Fakhfakh et al. (Eds.): ICAV 2018, ACM 13, pp. 149–154, 2019.
https://doi.org/10.1007/978-3-319-94616-0_15

calculate the lifetime of electronic components under random loadings. Mattila et al. (2008) carry out experimental tests of temperature and harmonic vibration to study the effect of different temperatures on solder joint reliability. The maximum strain on the PCB was increased with temperature rise. Cinar et al. (2011) investigated the failure mechanism of finite ball grid array (FBGA) solder joints due to harmonic excitation. They used experiments and finite element method to determine the failure of solder joint. Delmonte et al. (2013) developed a thermomechanical model to predict the fatigue life of the solder joints. He presented an approach to determine strain using Coffin-Masson equation. Zhang et al. (2015) conducted tests in order to investigate the effect of temperature on PCB frequency and strain; moreover, he studied the reliability of solder joints under vibrating loading.

In this work, a finite element model of PCB and component assembly is presented in order to investigate the effect of vibration loading. Forced vibration study was realized to determine the location of the solder joints that has the highest stress concentration under harmonic excitation.

2   Finite Element Model

COMSOL Multiphysics finite element model is used to study the stress-displacement and thermal analysis of the solder balls on the Plastic Ball Grid Array (PBGA) component (Fig. 1a). The PCB is made of two layers: FR4 of 12 mm length, 1.23 mm thickness and Cooper with the same length and 0.07 mm thickness. The Component of 2 mm × 0.5 mm is mounted with 0.76 mm diameter solder balls under 0.48 mm pitch and it is situated in a distance of 5 mm from each face. The boundary conditions for one of the opposite faces of PCB are set as clamped and the other face is free. The material properties of the FR4, Cooper, Solder ball and component are listed in Table 1. In this numerical simulation, the density of mesh will have an impact on the predicted results. For this reason, mesh is refined in the solder ball as represented in Fig. 1b (Fig. 2).

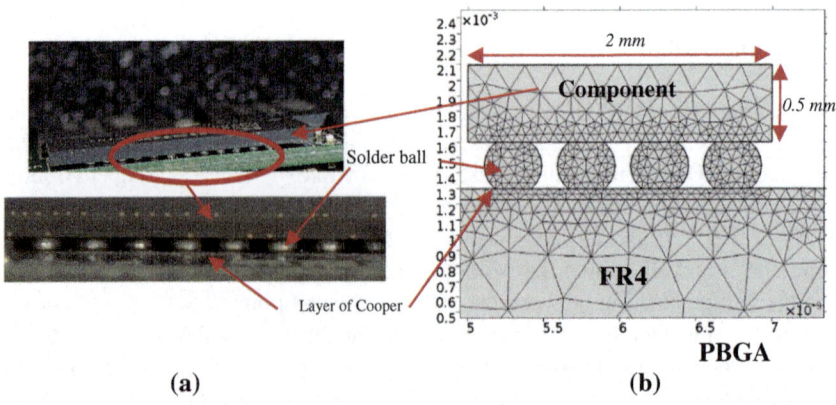

Fig. 1. (a) PCB and component assembly (b) FEM mesh of the PBGA

Table 1. Material elastic properties

Material	E (GPa)	ρ (kg/m^3)	v
FR4	22	1900	0.28
Copper	110	8960	0.35
Solder (60Sn-40Pb)	10	9000	0.4
Silicon	170	2329	0.28

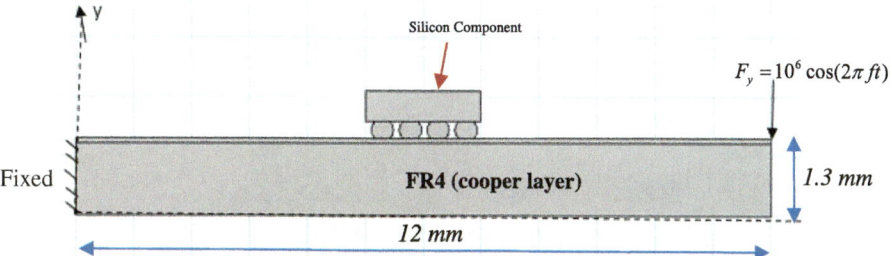

Fig. 2. Geometry loading conditions of the PBGA

3   Numerical Simulation

The aim of this study is to investigate the dynamic response of the PCB. The harmonic excitation force (Eq. 1) is applied in the bending direction (Y) on the free edge. In a similar work (Cinar et al. 2011, 2013), it is shown that the first natural frequency is the most destructive mode. Thus, the amplitude of displacement at the first mode is in the range of 0.1 mm. However, it is very small and under 0.01 mm for mode 2 and 3. Forced vibration analysis was performed around the first natural frequency and the bending harmonic excitation is given by:

$$F_y = F_0 \cos(2\pi ft) \quad \text{with} \quad F_0 = 10^6 \text{ N/m} \tag{1}$$

In this study, the natural frequencies and their modes are calculated using the COMSOL FEM model described in Sect. 2. Table 2 shows the First-four modes shape of vibration, the maximum values of Von-Mises Stress.

The predicted results of the von Mises stress and strain distribution in the solder balls and the component on the first natural frequency are given in Figs. 3 and 4. Nothing that, the maximum stress is located on the corner solder ball in the surface between the solder and the PCB. Therefore, the solder ball at this location under the most stressed condition. The maximum strain is localization in both surface located between the solder-PCB and solder-component. The peak of stress (or strain) at a point A between the solder joint and the component in the frequency (f) is range of 5.4–5.5 kHz. Figure 5 shows the y displacement at the most critical point A (Fig. 4) during the frequency range, the peak of y displacement is 0.026 mm. Figure 6 shows the Von

Table 2. Natural frequency, the natural modes and maximum stress

Mode	Frequency (kHz)	Deformed mode	PCB	Maximum Von-Mises Stress (N/m^2)
1	5.502			5.15 10^5
2	29.878			5.96 10^5
3	67.826			1.15 10^7
4	82.023			1.32 10^7

Mises stress at the same point, the stress amplitude increases to the largest (43 MPa) when the excited frequency is equal to the eigen-frequency of the board.

We can concluded that, the harmonic excitation affects the loading intensity of solder interconnects. If the excitation frequency is equal to 5.502 kHz, the solder stress and displacement increase. This may conduct to decrease the solder interconnects lifetime. Moreover, resonance may cause the failure of the solder joints.

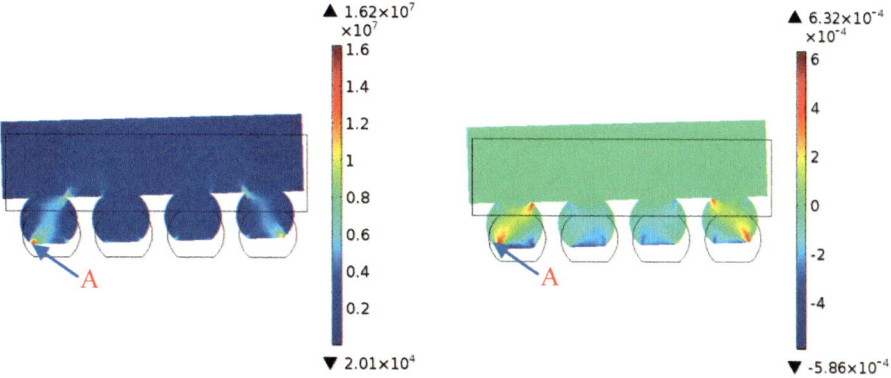

Fig. 3. Von-Mises stress distribution in the solder ball (N/m^2)

Fig. 4. Strain distribution in the solder ball (m/m)

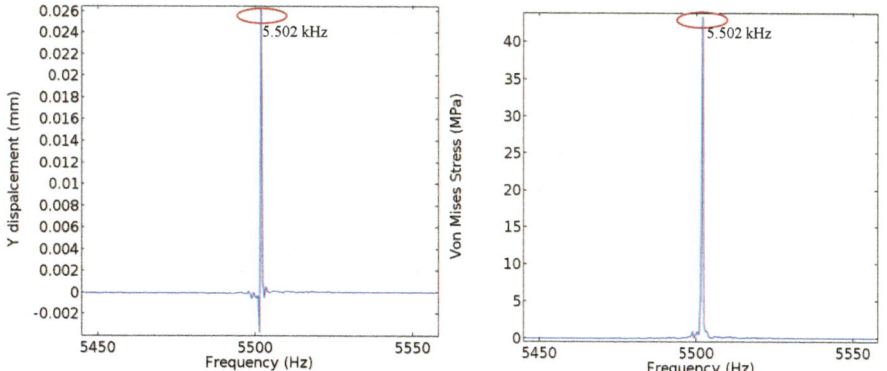

Fig. 5. Bending displacement (Point A)

Fig. 6. Von-Mises stress (Point A)

4   Conclusion

A finite element model is developed to study the effect of harmonic vibration in the behavior of solder ball in 60Sn-40Pb (displacement, stress, strain). The forced vibration study implemented in COMSOL Multiphysics in 2D case showed that the solder joints are the most sensitive part of the PCB and component assembly. More specifically, the weakest point is the point between the PCB ant the solder ball near to the fixture edge. The numerical model will be applied to develop parametric PCB and component model in order to study the reliability of solder joints.

References

Grieu, M., Massiot, G., Maire, O., Chaillot, A., Munier, C., Bienvenu, Y., Renard, J.: Durability modelling of a BGA component under random vibration. In: 9th International Conference on Thermal, Mechanical and Multiphysics Simulation and Experiments in Micro-electronics and Micro-systems (2008)

Cinar, Y., Jang, J., Jang, G., Kim, S., Jang, J., Chang, J., Jun, Y.: Failure mechanism of FBGA solder joints in memory module subjected to harmonic excitation. Microelectron. Reliab. **52** (4), 735–743 (2011)

Cinar, Y., Jang, J., Jang, G., Kim, S., Jang, J., Chang, J., Jun, Y.: Effect of solder pads on the fatigue life of FBGA memory modules under harmonic excitation by using a global-local modeling technique. Microelectron. Reliab. **53**, 2043–2051 (2013)

Mattila, T., Suotula, L., Kivilahti, J.K.: Replacement of the drop test with the vibration test – the effect of test temperature on reliability. In: 58th Electronic Components and Technology Conference, ECTC 2008, pp. 629–637. IEEE (2008)

Delmonte, N., Giuliani, F., Cova, P.: Finite element modeling and characterization of lead-free solder joint fatigue life during power cycling of surface mounting power devices. Microelectron. Reliab. **53**, 1611–1616 (2013)

Zhang, H.W., Liu, Y., Wang, J., Sun, F.L.: Effect of elevated temperature on PCB responses and solder interconnect reliability under vibration loading. Microelectron. Reliab. **55**, 2391–2395 (2015)

Structures Dynamics
and Fluid-Structure Interaction

Alternative Design Strategy for Water-Hammer Control in Pressurized-Pipe Flow

Mohamed Fersi[1,2] and Ali Triki[1(✉)]

[1] Research Unit: Mechanics, Modelling Energy and Materials M²EM,
National Engineering School of Gabès, University of Gabès, Gabès, Tunisia
`mohamedfersi@yahoo.com`, `ali.triki@enis.rnu.tn`
[2] Department of Mechanics, National Engineering School of Sfax,
University of Sfax, B.P. 1173, 3038 Sfax, Tunisia

Abstract. This paper proposed a design technique to dampen water-hammer surges into an existing steel piping system based on replacing a short-section of the transient sensitive region of the main piping system by another one made of polymeric material. The flow behavior was described using a one dimensional unconventional water hammer model based on the Ramos formulation to account for pipe-wall deformation and unsteady friction losses. The numerical solver was performed using the fixed gird Method of Characteristics. The effectiveness of the proposed design technique was assessed with regard to water-hammer up-surge scenario, using a high- or low-density polyethylene (**HDPE** or **LDPE**) for the replaced short-section. Results demonstrated that the utilized technique provided a useful tool to soften severe water-hammer surges. Additionally, the pressure surge softening was slightly more important for the case of a short-section made of **LDPE** polymeric material than that using an **HDPE** polymeric material. However, it was observed that the proposed technique induced an amplification of the radial-strain magnitude and spread-out of the period of wave oscillations. It was also found that the amortization of pressure amplitude, and reciprocally the radial strain magnitude, was strongly dependent upon the short-section size and material.

Keywords: Protective device · Water-hammer control · Polymeric material
LDPE-HDPE-pipe-wall material · Ramos model
Fixed-grid method of characteristics

1 Introduction

Pressurized-pipe systems are subject to water-hammer surge, or flow shocks, whether induced by setting or accidental maneuvers. Incidentally, these maneuvers may trigger a series of positive and negative surges of sharp magnitude large enough to induce undesirable effects such as excessive noise, fatigue and stretch of the pipe wall and disruption of normal control.

Accordingly, water-hammer control constitutes a major concern for hydraulic researchers and designers in order to ensure a global economic efficiency and safety operations of pressurized-pipe systems. Although water-hammer surge cannot be

© Springer Nature Switzerland AG 2019
T. Fakhfakh et al. (Eds.): ICAV 2018, ACM 13, pp. 157–165, 2019.
https://doi.org/10.1007/978-3-319-94616-0_16

avoided completely, certain design measures are commonly taken to mitigate effectively the severe impact of these waves to a desirable extent.

On the other hand, recent researches on pipe-wall materials have shown that polymeric materials, such as high- or low-density polyethylene (**HDPE** or **LDPE**), provide a significant damping of transient pressure fluctuations during high and low pressure surge loading (Pezzinga 2002; Covas et al. 2004a, b, 2005; Brinson and Brinson 2008; Triki 2016, 2017a, b, c). Thereby, the rheological behavior of viscoelastic materials brings about a great damping of the fluid pressure fluctuations, in contrast to elastic pipes where there is no delay between the pressure rise/drop and the pipe wall expansion/contraction (Covas et al. 2004a, b, 2005).

Considering the foregoing behavior of polymeric material, Massouh and Comolet (1984) examined experimentally the efficiency of adding a short rubber pipe in series to a main pipeline as an up-surge suppressor. The authors showed that the over pressure was significantly damped with gradually varied oscillations and a relatively long period. Concurrently, Triki (2016, 2017a, b, c) investigated the efficiency of the inline/branching design strategy using (**HDPE** or **LDPE**) short section. Specifically, the author (Triki 2016) used the Ramos formulation based one-dimensional water-hammer model for numerical simulation. Results addressed only pressure-head evolutions.

In order to deliver more desirable design estimates of supplement parameters such as the circumferential-stress and the radial-strain evolutions, numerical investigations are extended in this paper to illustrate the two latter parameters which are importantly embedded in the design stage of hydraulic systems.

This paper is outlined into four parts: following this introduction, the one-dimensional (**1-D**) pressurized-pipe flow model using the Ramos formulation, to describe both pipe-wall viscoelasticity and unsteady friction effects, is briefly presented. The transient flow computation is based on the Fixed Grid Method of Characteristics (**FG-MOC**), with specified time step. Thereafter, typical water-hammer up-surge scenarios are analyzed and discussed. Finally, summary and conclusions are drawn in Sect. 4.

2   Materials and Methods

One of the simplest (**1-D**) pressurized-pipe flow models, characterizing unsteady frictions and pipe-wall viscoelastic behavior, is the one proposed by Ramos et al. (2004):

$$\frac{\partial H}{\partial t} + \frac{a_0^2}{gA}\frac{\partial Q}{\partial x} = 0 \tag{1}$$

$$\frac{1}{A}\frac{\partial Q}{\partial t} + g\frac{\partial H}{\partial x} + g\left(h_{f_s} + \frac{1}{gA}\left(k_{r1}\frac{\partial Q}{\partial t} + k_{r2}a_0\,Sgn(Q)\left|\frac{\partial Q}{\partial x}\right|\right)\right) = 0 \tag{2}$$

where H is the piezometric head; Q is the flow discharge; A is the cross sectional area of the pipe; g is the gravity acceleration; $a_0 = \sqrt{K/\rho/1 + \alpha(D/e)KJ_0}$ is the wave speed; x and t are the longitudinal coordinates along the pipeline axis and the time, respectively; α is a dimensionless parameter that depends on the pipe cross-section and axial constraints ($\alpha = 1$, for thin wall elastic pipes (Wylie and Streeter 1993); D is the pipe inner diameter; e is the pipe wall thickness; K is the bulk elasticity modulus of the fluid; ρ is the fluid density; J_0 is the elastic creep compliance; $k_{r1} = 0.003$ and $k_{r2} = 0.04$ are two decay coefficients (Ramos et al. 2004), affecting the phase shift and the damping of the transient pressure waves, respectively.

The quasi-steady head loss component per unit length, h_{f_s}, is computed for turbulent and laminar flow, respectively, as follows:

$$h_{f_s} = RQ|Q| \quad \text{and} \quad h_{f_s} = \frac{32v'}{gD^2A}Q \qquad (3)$$

where, $R = f/2DA$ is the pipe resistance; v' is the kinematic fluid viscosity and f is the Darcy-Weisbach friction factor.

On notes that the total circumferential stress σ and the total radial strain ε may be expressed as follows (Wylie and Streeter 1993):

$$\sigma = \frac{\alpha' \Delta pD}{2e} \quad \text{and} \quad \varepsilon = \frac{\sigma}{E_0} \qquad (4)$$

where: p is the pressure and $E_0 = 1/J_0$ is the Young modulus.

The numerical solution of the initial boundary value problem governed by the momentum and continuity Eqs. (1) and (2) is typically developed using the (FG-MOC) for handling multi-pipe systems with variable wave speeds. Briefly, the corresponding compatibility equations, solved by the finite difference scheme along the set of characteristic lines, yield (Ramos et al. 2004):

$$\mathbf{C}^{\pm} : \left(H_{i,t}^j - H_{i\pm1,t-\Delta t}^j \right) \pm \frac{a_0^j}{gs^j} \left(Q_{i,t-\Delta t}^j - Q_{i\pm1,t-\Delta t}^j \right) \pm a_0^j \Delta th_{f_{i\pm1,t-\Delta t}^j} = 0 \quad \text{along}$$

$$\frac{\Delta x^j}{\Delta t} = \pm \frac{a_0^j}{C_r^j} \qquad (5)$$

in which, C_r^j is the Courant number used to allow the grid points to coincide with the intersection of the characteristic curves; the upper subscript j refers to the pipe number ($1 \leq j \leq np$) and the lower subscript i refers to the section number of the jth pipe ($1 \leq i \leq n_s^j$); n_s^j is the number of sections of the jth pipe and np is the number of pipes; Δt and Δx are the time and the space step increments, respectively.

For the series junction of multi-pipes, constant flow rates (i.e., no flow storage at the junction) and a common hydraulic grade-line elevation (i.e., continuous) are assumed at the junction, for each time step.

Accordingly, these assumptions yield:

$$Q_{x=L^{j-1},t}^{j-1} = Q_{x=0,t}^j \quad \text{and} \quad H_{x=L^{j-1},t}^{j-1} = H_{x=0,t}^j \qquad (6)$$

where the right hand of Eq. (6) refers to the values of the hydraulic parameters just upstream of the junction, and the left hand refers to the location just downstream of the junction.

3 Application, Results and Discussion

This section aims to apply the protection technique to dampen water-hammer surges. The hydraulic system considered herein (Fig. 1), initially consists of (i.e. without implementing the protection technique) a constant head reservoir ($H_0 = 45$ m) and a main steel pipeline equipped with a free discharge valve at its outlet. The main steel pipeline specifications are illustrated in Table 1. The initial steady state flow rate is $Q_0 = 0.58$ l/s. The water-hammer surge is generated by a fast and full closure of the downstream valve with a constant pressure-head condition maintained at the upstream reservoir. The boundary conditions, associated with such a scenario, may be expressed as follows:

$$Q|_{x=L} = 0 \quad \text{and} \quad H|_{x=0} = H_{OR} \ (t \succ 0) \tag{7}$$

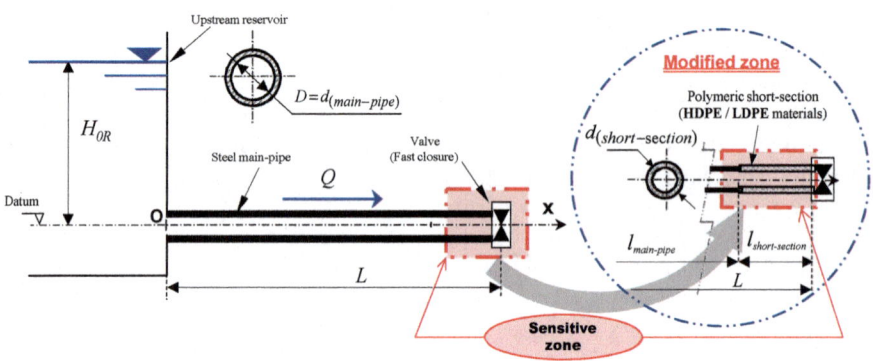

Fig. 1. Definition sketch of the hydraulic system.

Figure 1 presents a schematic layout for the implementation of the protection technique. This technique consists in replacing a downstream short-section (i.e. at the location where the surge disturbance is initiated) of the main steel piping system by another one made of a polymeric pipe-wall material, including **HDPE**- or **LDPE** material. The short-section specifications are listed in Table 1. It is worth noting that the length of the initial steel piping system (i.e. without protection) is $L = 100$ m; however, after modification, this length is reduced to $l_{main\text{-}pipe} = 95$ m.

One notes that the calculations of water-hammer courses were performed using an algorithm based on the **FG-MOC**, using a specified time step $\Delta t = 0.018$ s and Courant numbers $C_r^{main\text{-}pipe} = 0.9709$ and $C_r^{short\text{-}section} = 1$, corresponding to the steel main pipe and the polymeric short section.

Table 1. Characteristics of applied pipelines

Parameters	Steel	HDPE	LDPE
Length L [m]	100.0	5.0	5.0
Diameter D [m]	50.6	50.6	50.6
Young modulus [GPa]	210.0	1.43	0.643

Figure 2a, b and c displays the comparison between the piezometric head, the circumferential stress and the radial strains, respectively, versus time, computed at the downstream end $(x = L)$ predicted from water-hammer calculations into a piping system made of a steel main-pipe (i.e. system without protection), along with the corresponding results computed for the protected system composed of series junctions of a steel main-pie and **HDPE** or **LDPE** short-section.

Figure 2a illustrates the pressure-head amortization effects of the first peak along with the spread-out of the pressure-head oscillations period, in the protected system cases. Results reveal that, for the first cycle of pressure-head oscillations, the larger overpressure is observed for the steel main-pipe case $(H_{Max.}^{steel\,pipe} = 82.719\,\text{m})$, while the corresponding value is attenuated when implementing the protection technique using HDPE and LDPE materials for the short-section $(H_{Max.}^{(steel\,+\textbf{HDPE})\,pipe} = 76.758\,\text{m}$ and $H_{Max.}^{(steel\,+\textbf{LDPE})\,pipe} = 69.263\,\text{m})$. In other words, the up-pressure attenuations obtained using **HDPE** and **LDPE** short-section materials are, respectively: $\Delta H = H_{Max.}^{steel\,pipe} - H_{Max.}^{(steel\,+\textbf{HDPE})\,pipe} = 5.961\,\text{m}$ and $\Delta H = H_{Max.}^{steel\,pipe} - H_{Max.}^{(steel\,+\textbf{LDPE})\,pipe} = 13.456\,\text{m}$. Consequently, the employed technique allows a significant amortization of the first pressure peak compared with that predicted for the same transient event initiated into the steel piping system. More precisely, this amortization is slightly more important for the case using an **LDPE** short-section (51.29%) than the one obtained using an **HDPE** short-section (23.07%).

Similarly, Fig. 2b illustrates that the employed technique also allows a significant amortization of the first circumferential-stress peak compared with the one predicted into the non-protected system. More precisely, these amortizations are slightly more important for the case using an **LDPE** short-section (i.e.: 18.31% of the first circumferential-stress peak) than those obtained using an **HDPE** short-section (i.e.: 58.28% of the first circumferential-stress peak).

Inversely, Fig. 2c shows that the damping effects of pressure-head and circumferential-stress peaks, discussed above, are accompanied with an amplification of the total radial strain peaks. More precisely, for the case using a short-section made of **HDPE**, the magnitude of the first strain peak is $\Delta\varepsilon_{up-surge}^{(steel\,+\textbf{HDPE})\,pipe} = 2.22 \times 10^{-3}\text{m/m}$. A more important amplitude is observed for the case using an **LDPE** short-section, corresponding to: $\Delta\varepsilon_{up-surge}^{(steel\,+\textbf{LDPE})\,pipe} = 3.16 \times 10^{-3}\,\text{m/m}$. This result may be physically explained by the viscoelastic behavior of polymeric pipe wall material which has a retarded strain, in addition to the instantaneous strain observed in elastic pipe wall material. Incidentally, the corresponding amplitude was equal to $\Delta\varepsilon_{up-surge}^{steel\,pipe} = 2.88 \times 10^{-5}\text{m/m}$, for the non-protected system case, which corresponds to the elastic radial deformation component only.

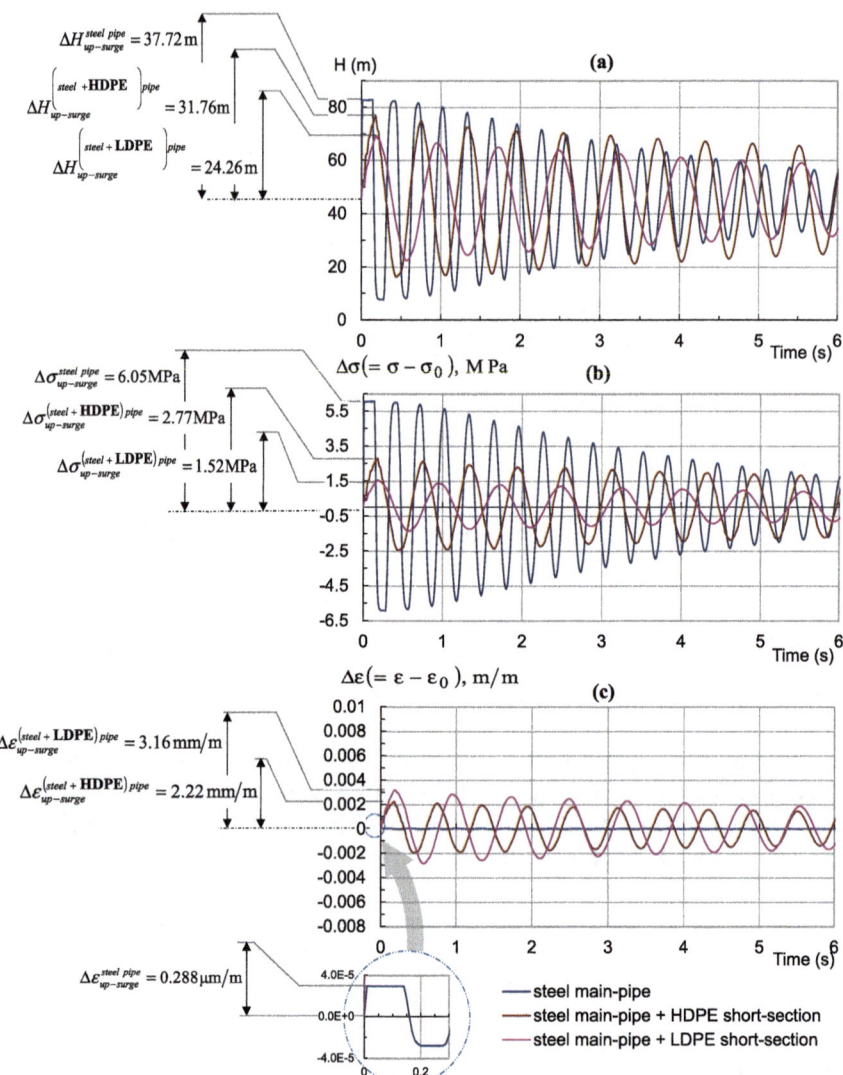

$\Delta H^{steel\ pipe}_{up-surge} = 37.72\,\mathrm{m}$

$\Delta H \left(\substack{steel\ +\mathbf{HDPE} \\ up-surge} \right)^{pipe} = 31.76\,\mathrm{m}$

$\Delta H \left(\substack{steel\ +\mathbf{LDPE} \\ up--surge} \right)^{pipe} = 24.26\,\mathrm{m}$

$\Delta\sigma^{steel\ pipe}_{up-surge} = 6.05\,\mathrm{MPa}$

$\Delta\sigma^{(steel\ +\mathbf{HDPE})pipe}_{up-surge} = 2.77\,\mathrm{MPa}$

$\Delta\sigma^{(steel\ +\mathbf{LDPE})pipe}_{up-surge} = 1.52\,\mathrm{MPa}$

$\Delta\varepsilon^{(steel\ +\mathbf{LDPE})pipe}_{up-surge} = 3.16\,\mathrm{mm/m}$

$\Delta\varepsilon^{(steel\ +\mathbf{HDPE})pipe}_{up-surge} = 2.22\,\mathrm{mm/m}$

$\Delta\varepsilon^{steel\ pipe}_{up-surge} = 0.288\,\mu\mathrm{m/m}$

— steel main-pipe
— steel main-pipe + HDPE short-section
— steel main-pipe + LDPE short-section

Fig. 2. Comparison of (a) piezometric heads, (b) circumferential stresses and (c) radial strains at the downstream valve section versus time for the hydraulic system with and without implementation of the protection procedure.

In addition, based on Fig. 2a, b and c, it is remarkable to observe that the periods of the first cycle of pressure-head oscillations, predicted for the protected system, are: $T^{(steel\ +\mathbf{HDPE})pipe}_{1^{st}\ Cycle} = 1.3\,\mathrm{s}$ and $T^{(steel\ +\mathbf{LDPE})pipe}_{1^{st}\ Cycle} = 3.73\,\mathrm{s}$ for the cases of short-sections made of **HDPE** and **LDPE** polymeric materials, respectively, while the corresponding period, for the piping system without protection (i.e. steel main pipeline), is equal to $T^{steel\ pipe}_{1^{st}\ Cycle} = 0.4\,\mathrm{s}$. Thus, the use of polymeric short-sections induces the spread-out of the period of pressure-head oscillations. Consequently, the final subsequent steady state regime takes more time to be reached in the case of the protected system than in the case of the system without protection.

The first phase of test experiments has shown the ability of the proposed technique to soften water-hammer surge. It will be interesting to study the magnitude sensitivity of the first maximum pressure peak to the size of the replaced polymeric short-section.

So as to accurately depict this sensitivity, the maximum pressure-head peak traces at the downstream end versus time for the protected system using **HDPE** and **LDPE** polymeric materials, with the short-section length and diameter being the controlling variables, are shown in Fig. 3a and b, respectively. Specifically, the following set of diameters and lengths are performed: $d_{(short-section)} = \{0.025; 0.0506; 0.075 \text{ and } 0.1 \text{ m}\}$ and $l_{(short-section)} = \{1; 2.5; 5; 7.5 \text{ and } 10 \text{ m}\}$.

As expected, these graphs reveal that the variation of the short-section size affects the magnitude of the maximum peak of transient pressure oscillations. In other words, as the replaced short-section volume increases, the associated damping effect of the maximum pressure head increases. More precisely, Fig. 3a clearly illustrates that, for the length

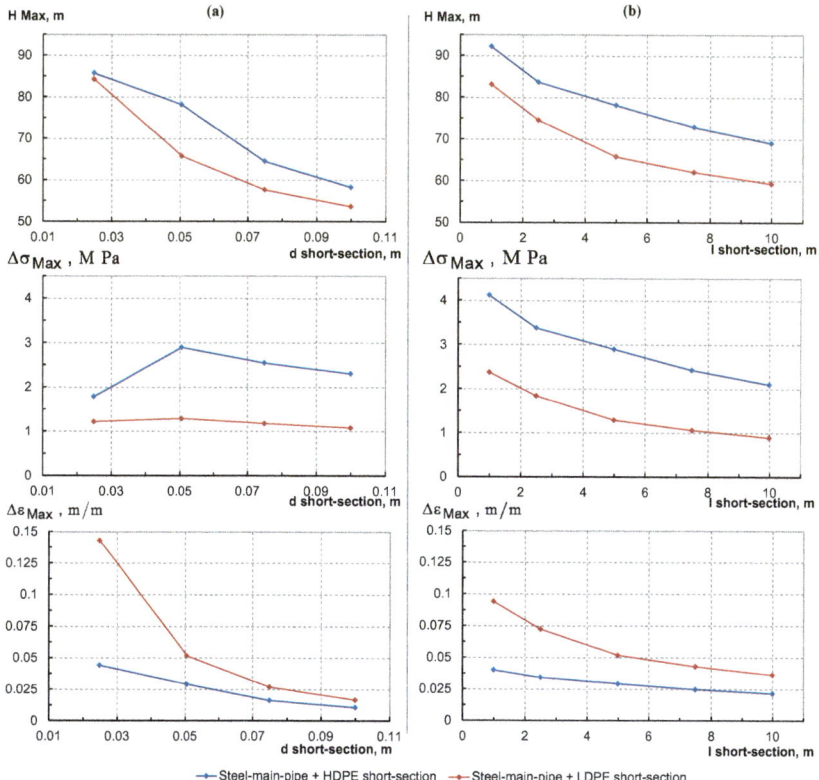

Fig. 3. Variation of maximum piezometric heads, stresses and strains, at the downstream valve section, for the protected system with a polymeric (**HDPE/LDPE**) short-section: variation depending on the short-section (a) diameter and (b) length.

values $l_{(short-sec\,tion)} = 1$ m and 2 m, the maximum peak decreases significantly. However, for the length values beyond $l_{(short-sec\,tion)} \geq 5$ m, the variation of the maximum transient pressure peak is slightly affected. Similarly, analysis of Fig. 3b indicates that as the diameter of the polymeric short section increases from $d_{(short-sec\,tion)} = 0.025$ m to 0.0506 m, the maximum pressure peak is significantly damped. However, this amortization is not pronounced for the diameter values beyond $d_{(short-sec\,tion)} \geq 0.075$ m. Thereby, $l_{(short-sec\,tion)} = 5$ m and $d_{(short-sec\,tion)} = 0.075$ m may be considered as the optimal values of the polymeric short section diameter and length.

4 Conclusion

In summary, the present study has illustrated that the proposed protection technique is effective in softening severe water-hammer surge. It is remarkable to observe that the employed technique provides a large damping of the first pressure peak associated with a transient initiating event. However, the foregoing behavior is accompanied with the amplification of radial strain peaks and the spread-out of the period of wave oscillations. In addition, the pressure damping (and reciprocally, the radial-strain amplification) is observed to be more pronounced when using an **LDPE** polymeric material for the replaced short sections than an **HDPE** material. It is also shown that other factors contributing to the damping rate of pressure head and the radial-strain amplification depend upon the short-section size (i.e. length and diameter). On the other hand, examination of the sensitivity of the pressure peak magnitude, with the short-section length and diameter being the controlling variables, verifies that significant volumes of the short section provide important pressure surge damping and radial-strain amplification. However, this correlation is not significant beyond optimum diameter and length values.

Overall, such a technique may greatly enhance the reliability and improve the cost-effectiveness of pressurized-pipe utilities, while safeguarding operators. It is estimated that these findings are of practical importance in the design measure side for the mitigation of severe water-hammer surges.

References

Brinson, H.F., Brinson, L.C.: Polymer Engineering Science and Viscoelasticity: An Introduction. Springer, Berlin (2008)

Covas, D., Stoianov, I., Ramos, H., Graham, N., Maksimovic, C., Butler, D.: Water hammer in pressurized polyethylene pipes: conceptual model and experimental analysis. Urb. Water J. **1** (2), 177–197 (2004a)

Covas, D., Stoianov, I., Ramos, H., Graham, N., Maksimovic, C.: The dynamic effect of pipe-wall viscoelasticity in hydraulic transients. Part I-experimental analysis and creep characterization. J. Hydraul. Res. **42**(5), 517–532 (2004b)

Covas, D., Stoianov, I., Mano, J.F., Ramos, H., Graham, N., Maksimovic, C.: The dynamic effect of pipe-wall viscoelasticity in hydraulic transients. Part II-model development, calibration and verification. J. Hydraul. Res. **43**(1), 56–70 (2005). https://doi.org/10.1080/0022168050950 0111

Güney, M.S.: Waterhammer in viscoelastic pipes where cross-section parameters are time dependent. In: Proceedings of 4th International Conference on Pressure Surges, BHRA, Bath, UK, pp. 189–209 (1983)

Massouh, F., Comolet, R.: Étude d'un système anti-bélier en ligne-Study of a water-hammer protection system in line. La Houille Blanche **5**, 355–362 (1984)

Pezzinga, G.: Unsteady flow in hydraulic networks with polymeric additional pipe. J. Hydraul. Eng. **128**(2), 238–244 (2002)

Triki, A.: Water-hammer control in pressurized-pipe flow using an in-line polymeric short-section. Acta Mech. (2016). https://doi.org/10.1007/s00707-015-1493-13

Triki, A.: Water-hammer control in pressurized-pipe flow using a branched polymeric penstock. J. Pipeline Syst. Eng. Pract. ASCE. **8**(4), 04017024 (2017a). https://doi.org/10.1061/(asce)ps.1949-1204.0000277

Triki, A.: Further investigation on water-hammer control inline strategy in water-supply systems. J. Water Suppl. Res. Technol. AQUA (2017b). https://doi.org/10.1061/(asce)ps.1949-1204.0000277

Triki, A.: Dual-technique based inline design strategy for water-hammer control in pressurized-pipe flow. Acta Mech. (2017c). https://doi.org/10.1007/s00707-017-2085-z

Wylie, E.B., Streeter, V.L.: Fluid Transients in Systems. Prentice Hall, Englewood Cliffs (1993)

Parametric Study on the Efficiency of an Inverse Energetic Approach to Identify the Boundary Acoustic Sources

Ahmed Samet[1,2(✉)], Mohamed Amine Ben Souf[1,2], Olivier Bareille[2], Tahar Fakhfakh[1], Mohamed Ichchou[2], and Mohamed Haddar[1]

[1] Laboratoire de Mécanique, Modélisation et Productique (LA2MP), École Nationaled' Ingénieurs de Sfax, Université de Sfax, 3038 Sfax, Tunisia
samet.ahmed14@hotmail.com,
bensouf.mohamedamine@gmail.com,
tahar.fakhfakh@gmail.com, mohamed.haddar@enis.rnu.tn
[2] Laboratoire de Tribologie et Dynamique des Systèmes (LTDS), École Centrale de Lyon, 36 venue Guy de Collongue, 69134 Écully Cedex, France
{Olivier.Bareille,mohamed.ichchou}@ec-lyon.fr

Abstract. This paper presents a parametric study of the microphones distribution effect in the identification of the boundary acoustic sources acting in the acoustic cavities through the knowledge of the acoustic energy densities measurement. An energetic approach, also called the simplified energy method (MES) was developed to predict the energy densities distribution for the acoustic applications. MES can also be applied to structures to determine energy densities. This energy method can solve inverse problems in order to localize and quantify the structural and the acoustic boundary sources at medium and high frequency ranges, thanks the inverse formulation of this energetic approach (IMES). The main novelty of this paper is to study the performances of this inverse energetic approach in the quantification and localization of the boundary acoustic sources acting in the acoustic cavities. Numerical investigation concerning 3D acoustic cavity was performed to test the validity of the presented technique using different number of acoustic sources and distance between the microphones repartition and the cavity walls. The numerical results show that the inverse simplified energy method (IMES) has an excellent performance in identifying and detecting the boundary acoustic sources at medium and high frequency ranges from the knowledge of the acoustic energy densities measurement.

Keywords: Acoustic sources identification · Inverse energetic approach Microphones reparation · Medium and high frequency

1 Introduction

The identification of acoustic sources from operating measurement has been a current topic and constitute a particular attention in the academic and industrial projects. The choice of the used tool or method depends on the frequency band of study since there are appropriate approaches for each frequency domain.

© Springer Nature Switzerland AG 2019
T. Fakhfakh et al. (Eds.): ICAV 2018, ACM 13, pp. 166–175, 2019.
https://doi.org/10.1007/978-3-319-94616-0_17

In the low-frequency ranges, several methods have been used such as the inverse FEM, BEM or FRF methods, since these techniques are well adapted to this frequency range (Weber et al. 2008; Drenckhan et al. 2004; Schuhmacher et al. 2003; Djamaa et al. 2007). Indeed, the low-frequency domain is a frequency range where the modal overlap is low and modal information clearly appears; therefore, in this frequency band, cavities do not require fine meshing and FEM or BEM calculation costs are not too high.

For higher frequency ranges, several approaches can be used. Some techniques are based on low-frequency methods such as BEM or FEM. However, these techniques require very fine meshes and very high calculation costs. For example, an FEM or BEM calculation on complex cavity can take several hours or days whereas energy methods such as the statistical energy analysis (SEA) or the simplified energy method (MES) would take seconds or minutes, since this kind of method does not require fine meshes. For this reason, the energy methods based on energy quantities are often used. Among these methods the simplified energy method. The direct theory formulation has been applied in various domains including beam (Ichchou et al. 2001), membrane and plates (Ichchou and Jezequel 1996) and acoustic applications (Besset et al. 2010).

The IMES has been developed by Chabchoub et al. (2011) to estimate the structural force applied in a plate through the measurement of the structural density field. This inverse method was also developed for complex structures modeled with many coupled plates to identify structural loads (Samet et al. 2017a; Samet et al. 2018a). Samet et al. (2017b) developed this inverse method for structure-acoustic interaction to identify vibration sources from acoustic measurements. Recently, the IMES method was extended to the field of damage detection to localize the geometrical or material discontinuities presented in the structure through the knowledge of the energy density field (Samet et al. 2018b, c).

The main novelty of this paper is to study the performances of this inverse energetic approach in the identification of acoustic sources using different numbers of sources and repartitions of microphones.

This paper is structured as follow. First, the direct and the inverse formulation of the simplified energy method are presented in Sect. 2. Next, the influence of number of boundary sources and the microphones repartition are presented to studies the efficiency of this predictive tool to identify the acoustic sources in Sect. 3.

2   Inverse Energetic Approach: IMES

2.1   Direct Formulation

The inverse simplified energy method (IMES) is a vibro-acoustic approach developed to identify the structural and acoustic sources at medium and high frequencies (Chabchoub et al. 2011, Samet et al. 2017a, Samet et al. 2017b, Samet et al. 2018a). This approach based on the description of two local energy quantities: the first is the total energy density W defined as a sum of the potential energy density and the kinetic energy density, the second energy quantity $\overrightarrow{I}$ presents the energy flow inside the system.

The inverse formulation of MES method used the non-correlation of the propagating waves assumption. The derivation this method is done considering the n-dimensional (*nD*) space. So, for the case of acoustic cavity, the following equation will be considered with 3-dimensional (*3D*) space. Considering symmetrical waves correspond to the propagating fields from a point source *s* in (*3D*) space. These fields depend only on the distance *r* between the source *s* and the measurement point *m*. The energy balance can then be written as follows:

$$\frac{1}{r^{n-1}}\frac{\partial}{\partial r}(r^{n-1}\vec{I}) + \eta\omega W\vec{n} = 0 \tag{1}$$

Considering only *W*, this equation can be as follows:

$$\frac{1}{r^{n-1}}\frac{\partial}{\partial r}(r^{n-1}W) + \eta^2\omega^2 W\vec{n} = 0 \tag{2}$$

The solutions of this equation in terms of energy density and active intensity are expressed by *G* and $\vec{H}$ respectively:

$$G(r) = \frac{1}{\gamma_0 c}\frac{e^{-\eta\omega r}}{r^{n-1}}; \quad \vec{H}(r) = \frac{1}{\gamma_0}\frac{e^{-\eta\omega r}}{r^{n-1}}\vec{u_r} \tag{3}$$

where γ_0 is the solid angle of the considered space (4π for *3D* space), *r* is the distance between the source *s* and the measurement point *m* and $\vec{u_r}$ is the unit vector from *m* to *s*.

The total acoustic energy field is constructed by the superposition of a direct field P_{inj} (primary source) coming from the input power in the system surface Ω and a reverberant field σ (secondary sources) coming from the fictitious sources localized in the system boundaries $\partial\Omega$, as shown in Fig. 1.

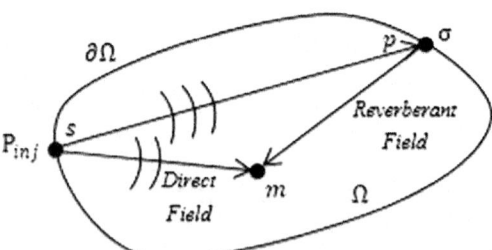

Fig. 1. Source description.

$$W(m) = \int_{\Omega} P_{inj}(s)G(s,m)d\Omega + \int_{\partial\Omega} \sigma(p)G(p,m)\vec{u_r}.\vec{n}\, d\Omega \tag{4}$$

Finally, W can be expressed thanks an operator K.

$$(W) = K(P_{inj}, \sigma) \tag{5}$$

Thus, the matrix formulation of Eq. (4) can be written as follows:

$$\{W\} = [S].\{P_{inj}\} \tag{6}$$

where $\{W\}$ is the acoustic energy density field and $[S]$ is a (Ne, Nm) matrix sensitivity operators. Let us recall that Ne is the number of sources and Nm is the number of microphones.

2.2   Inverse Formulation

The inverse simplified energy method will be presented in this section. The boundary acoustic sources will be detected thanks to measurements made on the system. Thus, the MES Eq. (6) is discretized to provide the following matrix:

$$\left\{ \begin{array}{c} W_1 \\ \vdots \\ W_{N_m} \end{array} \right\} = \left[\begin{array}{ccc} S_{11} & \cdots & S_{1N_e} \\ \vdots & \ddots & \vdots \\ S_{N_m 1} & \cdots & S_{N_m N_e} \end{array} \right] . \left\{ \begin{array}{c} P_1^{inj} \\ \vdots \\ P_{N_e}^{inj} \end{array} \right\} \tag{7}$$

The IMES aim to invert Eq. (7). The boundary acoustic sources can then be identified through the knowledge of a set energy densities within the cavity. Then, the IMES formulation is expressed as follows:

$$\{P_{inj}\} = [S]^+ . \{W\} \tag{8}$$

where + is the pseudo-inverse. Let us recall that we only deal with boundary acoustic sources because our parametric studies focus on the effectiveness of this inverse method to detect the external excitations in cavities like aircraft cabinsor other industries, such as the automotive, etc.

3   Results and Discussion

This section deals with numerical tests for different reparations of microphones in order to study the performance of the IMES technique to identify the boundary acoustic sources.

3.1   Geometry System and Boundary Conditions

In this section, we consider an acoustic cavity with absorbent boundary. Figure 2 present the geometry of the cavity. Two boundary acoustic sources located in the wall of cavity with coordinates: source 1 ($X = -0.3082$ m, $Y = 1.225$ m, $Z = 0.8815$ m) and sources 2 ($X = 0.9082$ m, $Y = 0.5$ m, $Z = 0.5733$ m), and respectively injecting an input power of 3 and 2 W/m^{-2}. The cavity absorption is assumed to be uniform of $\alpha = 0.075$. The displayed results were computed in an octave band with a frequency center $f_c = 2000$ Hz, where the modal overlap is quite high, and the use of energy quantities is essential.

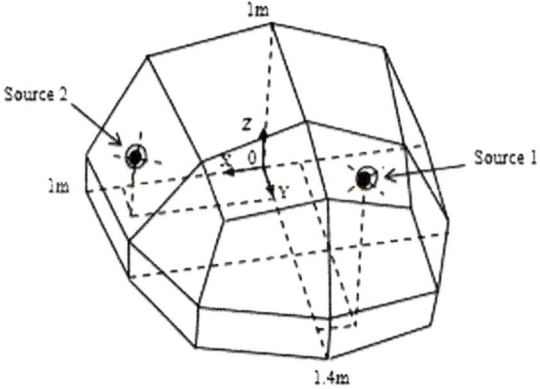

Fig. 2. Geometry of acoustic cavity.

The numerical methodology presented in Fig. 3 consists on discretizing the walls of cavity on *Ne* number of facets and we consider each facet as an acoustic source. After that, a *Nm* number of microphones are distributed in the cavity to predict the acoustic density energy. Finally, the inverse approach is applied to identify acoustic sources.

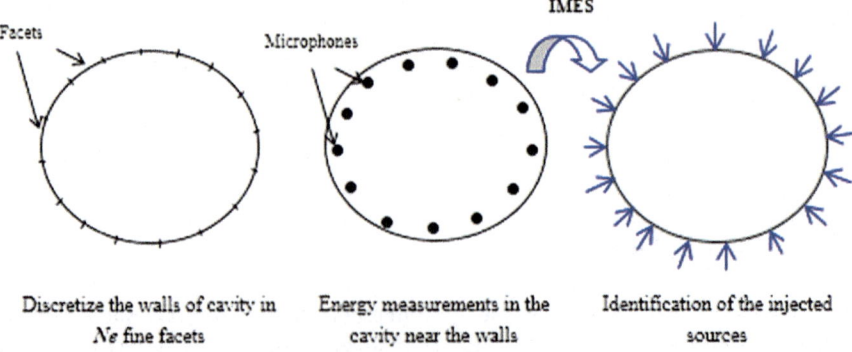

Fig. 3. Flow chart of numerical methodology.

3.2   Influence of the Number of Facets

In this section, the effect of the number of facets *Ne* is presented in order to study the performance of the IMES approach in the identification of boundary sound sources. The measurement grid is composed in *Nm* = 143 microphones located at a distance

$E = 0.05$ m from the walls of cavity, which implies that 572 energy quantities are measured. Figure 4 shows the repartition of microphones on cavity walls.

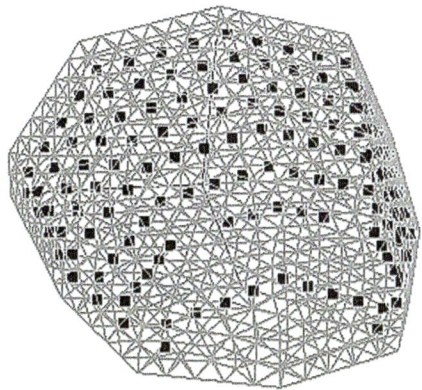

Fig. 4. Microphones repartition (■ measurement grid).

After that, the walls of cavity are discretized in Ne different number of facets in order to study the influence of this parameter in the identification of sound sources. The cavity walls are discretized in the first example into coarse facets $Ne = 300$ and in the second example into fine facets $Ne = 1350$. Figure 5 presents the distribution of the estimated power in the walls cavity. It is clear that two boundary sound sources are well quantified and in addition when the number of facets increase the acoustic sources are well located.

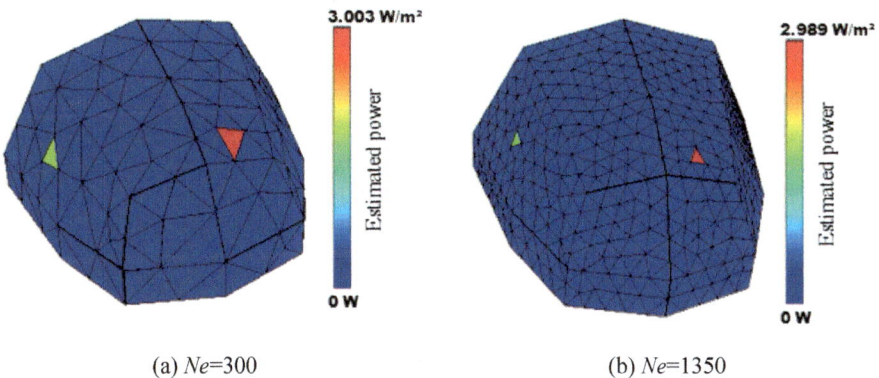

(a) $Ne=300$ (b) $Ne=1350$

Fig. 5. Identification of boundary acoustic sources for different number of facets Ne.

Figure 6 present the estimated power recalculated in each facet. For the first example, the injected powers are well identified, on the other hand two other powers (parasites) are detected which perturb the estimation of the injected powers.

Contrariwise, in the second example when the facets number increase the parasites are decreased. Therefore, it is preferable that the facets are fines for well localize the injected sound sources. Then, this test shows that more the number of facets is important more the parasites and the calculation errors are reduced.

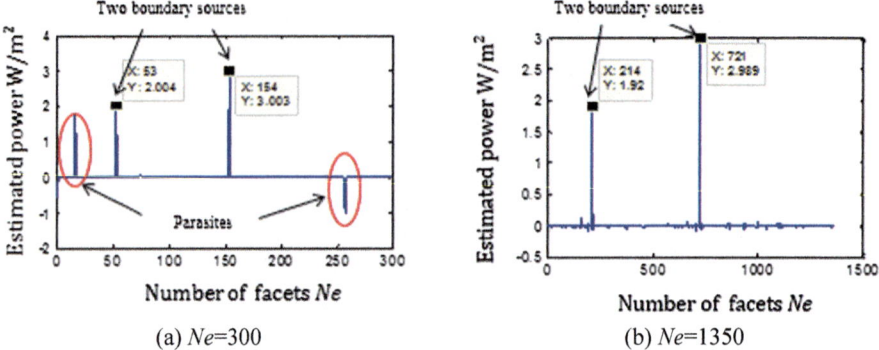

(a) Ne=300   (b) Ne=1350

Fig. 6. Injected power recalculated for different number of facets Ne.

In the next section, the distance between the cavity walls and microphones repartition will be changed in order to study the influence of this parameter in the identification of acoustic sources.

3.3   Influence of Distance Between the Microphones Repartition and Cavity Walls

For this test, the results found by the president test are used to study the influence of the distance E between the microphones and the cavity walls on the performance of the IMES method. As before, the cavity walls are discretized into $Ne = 1350$ facets, and the microphones are simulated in $Nm = 143$.

Two simulation tests were performed; the first one with distance $E_1 = 0.05$ m and the second one with distance $E_2 = 0.1$ m, as shown in Fig. 7.

Figure 8 shows the distribution of the estimated power in the cavity walls. It is clear that two boundary sound sources are well quantified and located for $E_1 = 0.05$ m. On the other hand, for $E_2 = 0.1$ m parasites are detected around the first acoustic source.

For more clarity, Fig. 9 present the estimated power recalculated in each facet. It can be observed that for $E_2 = 0.1$ m the parasites are clearly appearing around the source 1 more than for $E_1 = 0.05$ m.

The errors of the estimated power are summarized in Table 1. It clear that the amount of source information decreases as the microphones are more distant from the cavity walls. Therefore, it is preferable that the distances between cavity walls and measurement points are as short as possible. Then, this test shows that more the distance is short more the parasites and the calculation errors are reduced.

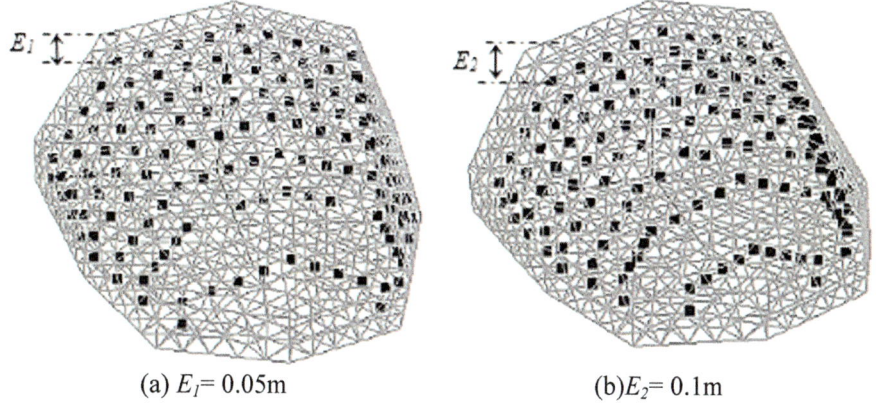

(a) E_1= 0.05m   (b)E_2= 0.1m

Fig. 7. Microphones repartition for the two distances.

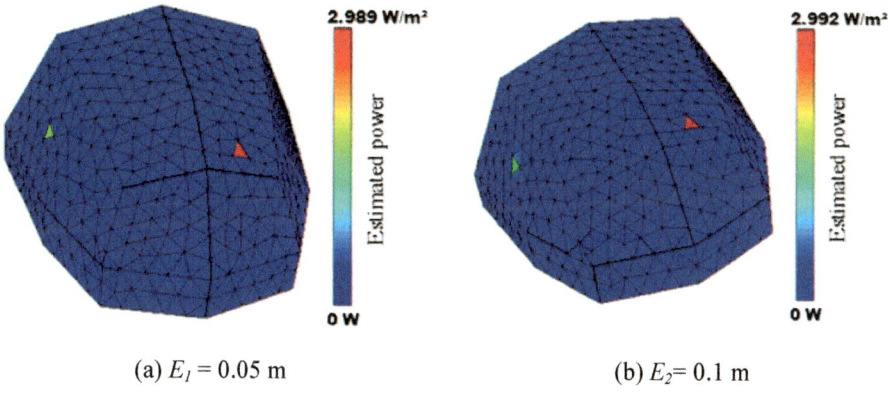

(a) E_1 = 0.05 m   (b) E_2= 0.1 m

Fig. 8. Identification of boundary acoustic sources.

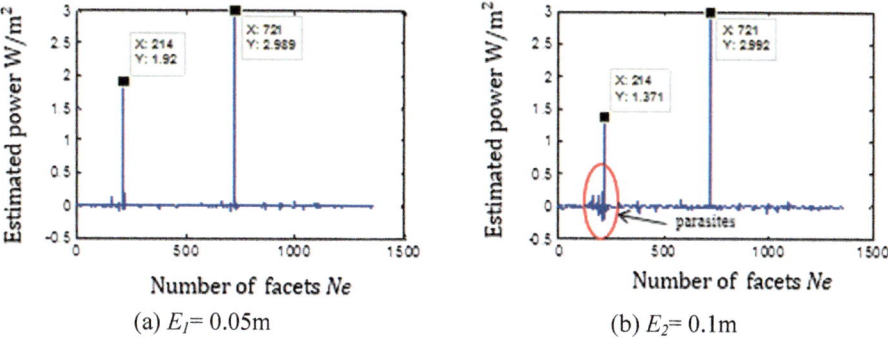

(a) E_1= 0.05m   (b) E_2= 0.1m

Fig. 9. Injected power recalculated for different distance E.

Table 1. Estimated power error.

Distance	Exact injected power (W.m^{-2})		Estimated power (W.m^{-2})		Error (%)	
	Source 1	Source 2	Source 1	Source 2	Source 1	Source 2
$E_1 = 0.05$ m	2	3	1.92	2.989	4	0.3
$E_2 = 0.1$ m	2	3	1.371	2.992	31	0.26

4 Conclusion

The objective of this paper was to study the performance of the inverse simplified energy method to detect the boundary acoustic sources acting in acoustic cavities. Several numerical test cases involving different numbers of sources and position of microphones were considered. This parametric study confirms the efficiency of this inverse method to identify the boundary acoustic sources.

In the future works, an approach should be developed to optimize the distribution of vibro-acoustic absorbers and to treat the effect of acoustic sources already identified by the IMES. In other words, the efficiency of the absorber distribution is currently being investigated.

References

Weber, M., Kletschloski, T., Sachau, D.: Identification of noise sources by means of inverse finite element method using measured data. J. Acoust. Soc. Am. **123**(5), 3064 (2008)

Drenckhan, J., Sachau, D.: Identification of sound sources using inverse FEM. In: 7th International Symposium on Transport Noise and Vibration, St. Petersburg (2004)

Schuhmacher, A., Hald, J., Rasmussen, K.B., Hansen, P.C.: Sound source reconstruction using inverse boundary element calculations. J. Acoust. Soc. Am. **113**(1), 114–127 (2003)

Djamaa, M.C., Ouelaa, N., Pezerat, C., Guyader, J.L.: Reconstruction of a distributed force applied on a thin cylindrical shell by an inverse method and spatial filtering. J. Sound Vib. **301**, 560–575 (2007)

Ichchou, M.N., Le Bot, A., Jezequel, L.: A transient local energy approach as an alternative to transient sea: wave and telegraph equations. J. Sound Vib. **246**(5), 829–840 (2001)

Ichchou, M.N., Jezequel, L.: Letter to the editor: comments on simple models of the energy flow in vibrating membranes and on simple models of the energetic of transversely vibrating plates. J. Sound Vib. **195**(4), 679–685 (1996)

Besset, S., Ichchou, M.N., Jezequel, L.: A coupled BEM and energy flow method for mid-high frequency internal acoustic. J. Comput. Accoust. **18**(1), 69–85 (2010)

Chabchoub, M.A., Besset, S., Ichchou, M.N.: Structural sources identification through an inverse mid-high frequency energy method. Mech. Syst. Signal Process. **25**(8), 2948–2961 (2011)

Samet, A., Ben Souf, M.A., Bareille, O., Ichchou, M.N., Fakhfakh, T., Haddar, M.: Vibration sources identification in coupled thin plates through an inverse energy method. Appl. Acoust. **128**, 83–93 (2017a)

Samet, A., Ben Souf, M.A., Bareille, O., Ichchou, M.N., Fakhfakh, T., Haddar, M.: Structural sources localization in 2D plate using an energetic approach. In: International Conference Design and Modeling of Mechanical Systems, pp. 449–458. Springer, Cham (2018a)

Samet, A., Ben Souf, M.A., Bareille, O., Ichchou, M.N., Fakhfakh, T., Haddar, M.: Structural source identification from acoustic measurements using an energetic approach. J. Mech. **34**, 1–11 (2017b)

Samet, A., Ben Souf, M.A., Bareille, O., Ichchou, M.N., Fakhfakh, T., Haddar, M.: Structural damage localization from energy density measurements using an energetic approach. Arch. Appl. Mech. **88**, 1–13 (2018b)

Samet, A., Hui, Y., Ben Souf, M., Bareille, O., Ichchou, M., Fakhfakh, T., Haddar, M.: Experimental investigation of damage detection in plate-like structure using combined energetic approaches. Proc. Inst. Mech. Eng., Part C: J. Mech. Eng. Sci. 0954406218771102

Statistical Investigations of Uncertainty Impact on Experiment-Based Identification of a Honeycomb Sandwich Beam

Ramzi Lajili[1,2,3(✉)], Khaoula Chikhaoui[1],
and Mohamed Lamjed Bouazizi[1,4]

[1] Research Unit of Structural Dynamics, Modelling and Engineering
of Multi-Physics, Preparatory Engineering Institute of Nabeul (IPEIN),
8000 Mrezgua, Nabeul, Tunisia
ajiliramsis@gmail.com, chikhaoui2013@gmail.com,
mohamedlamjed@gmail.com
[2] Laboratory of Tribology and Dynamics of Systems (LTDS),
Ecole Centrale de Lyon, 36 Avenue Guy de Collongues, 69130 Ecully, France
[3] National School of Engineers of Tunis (ENIT), University of Tunis el Manar,
BP 37 Le Belvedere, 1002 Tunis, Tunisia
[4] Mechanical Department, College of Engineering,
Prince Sattam Bin Abdulaziz University, Al-Kharj, Kingdom of Saudi Arabia

Abstract. Experimentally, errors on measurement points' coordinates, among others, could affect identification results. These errors can be committed by engineer or result from measuring tools and conditions. Resulting coordinates' variability is modeled in this work by uncertainties and is included into an experiment-based identification process to identify, in a wave propagation framework, the wavenumber and the wave attenuation of a honeycomb sandwich beam. The proposed process combines a Variant of the Inhomogeneous Wave Correlation (V-IWC) method and a sample-based uncertainty propagation method: the Latin Hypercube Sampling. Vibratory fields, which are used as inputs of the identification process, are computed experimentally. Both deterministic and statistical investigations of identified wavenumber and damping are performed. Results prove the efficiency of the proposed V-IWC method on wide frequency ranges and the robustness of identification against uncertainties. Moreover, if some measured vibratory fields do not match associated measurement points' coordinates, no damping sensitivity to such uncertainty is detected.

Keywords: Damping · Wavenumber · Identification
Inhomogeneous Wave Correlation · Honeycomb sandwich beam
Uncertainties

1 Introduction

Damping modeling and identification is obviously necessary when designing structures. Such phenomenon, on which vibration problems are directly dependent, forms an ever growing emphasis in vibroacoustic applications. Complexity of damping identification is amplified if more complex structural properties' extraction, such as that of

© Springer Nature Switzerland AG 2019
T. Fakhfakh et al. (Eds.): ICAV 2018, ACM 13, pp. 176–185, 2019.
https://doi.org/10.1007/978-3-319-94616-0_18

composite materials, occurs. Honeycomb sandwich structures have been increasingly used in several engineering fields. Their integration is due to high strength-to-weight ratios, interesting mechanical and material properties and high energy dissipation.

In the literature, modal-based identification approaches are frequently used at low frequencies. Nevertheless, their use becomes of limited interest in mid and high frequencies where great modal density exists. Wave propagation offers an interesting alternative framework which is based on the wavenumber space (k-space). K-space-based approaches allow efficient identification on wide frequency ranges. The McDaniel method (McDaniel et al. 2000) is one of the most frequently used approaches in the literature. It consists on adjusting iteratively the wavenumber and the damping, for each frequency, considering those of neighboring frequencies as initial estimates. The applications of the method in the literature include both 1D and 2D identification problems. It was used, for instance, in (McDaniel and Shepard 2000) to identify the damping of a freely suspended beam which was excited by an arbitrary transient load. An accurate estimation of the damping loss factor was allowed at any frequency. However modal-based methods, such as half-power point method, permitted to estimate the damping loss factor only near resonance frequencies. The McDaniel method was later extended by Ferguson et al. (2002) to 2D identification problems. A combination of Continuous Fourier Transform and least square minimization allowed identifying a single dominant homogeneous wave when using a windowed field far away from the near-field sources which would otherwise create disturbances.

A second interesting k-space-based approach is the Inhomogeneous Wave Correlation (IWC). It consists on correlating the vibratory field with an inhomogeneous wave. A frequency and direction-dependent dispersion equation is hence obtained from a space vibratory field. The IWC method applications in the literature include both isotropic, anisotropic, 1D and 2D problems (Berthaut et al. 2005; Ichchou et al. 2008b). With special emphasis on composite structures, the IWC method efficiency was illustrated by Ichchou et al. (2008a) and Inquiété (2008) in mid and high frequency ranges. An experiment-based IWC method was used by Chronopoulos et al. (2013) for composite panel identification. Vibratory data was measured experimentally and results were compared to the Wave Finite Element Method estimations. Moreover, both experiment-based and numerical-based IWC methods were used by Cherif et al. (2015) to identify orthotropic honeycomb panels' damping. Hence, either experimental or numerical vibratory data were used, respectively. A numerical-based IWC method was recently applied by Lajili et al. (2017) to identify propagation parameters of a honeycomb sandwich beam and compared to experiment-based estimations. The above cited works, among others, illustrated the efficiency of the IWC method on mid-high frequencies and highlighted its limits at low frequencies, when low modal overlaps occur, especially for damping estimates. To overcome inaccurate identification at low frequencies, several improved forms of the IWC method have been, recently, proposed. Van Belle et al. (2017) proposed an extended form of the IWC method which takes into account the experimental excitation location when expressing the correlated inhomogeneous wave. Roozen et al. (2017) proposed to use only half of the measurement data in the IWC, either to the left or to the right of the excitation position. Disturbing influence of the measurement data occurs on the left of the excitation point when fitting the right running waves, and vice versa. The proposed method was compared to the

Prony method and the spatial Fourier approach. In the same context, the main purpose of this work is to propose a Variant of the IWC method (V-IWC). Its principle is to correlate experimentally the measured vibratory fields with a sum of inhomogeneous forward and backward propagating waves. The constructed experiment-based identification process is used to identify the wavenumber and the wave attenuation (spatial damping), compared to the McDaniel method, which is considered as reference.

Experimentally, identification complexity does not depend only on considered frequency band and structural properties but also on experiments reliability. Errors which could be committed by engineer or could result from measuring tools and conditions should have great effects on identification accuracy. One of the influential error types is that affecting measurement points' coordinates. Indeed, if some vibratory fields do not match associated measurement points' coordinates, what effect could one obtain on estimates? For more realistic modeling, errors on measurement points' coordinates are here supposed to vary randomly and are thus modeled by parametric uncertainties. To investigate the impact of uncertainties on identification, uncertainty propagation is performed. Statistical investigations could be allowed by sample-based methods. The Monte Carlo Simulations (MCS) (Fishman 1996; Rubinstein and Kroese, 2008) and the Latin Hypercube Sampling (LHS) method (McKay et al. 1979; Helton and Davis 2003) are the most frequently used. Both methods are based on a succession of deterministic evaluations corresponding to a set of realizations of random variables and allow accurate results through simple implementations. The LHS method permits to reduce the prohibitive computational time required by the MCS without a significant loss of accuracy, by partitioning the variability space into regions of equal probability and selecting one sampling point in each region. In the context of structural identification, the LHS method has been recently combined by Lajili et al. (2018) with the standard IWC method to identify propagation parameters of an isotropic beam through an analytical-numerical model. The main purpose of the present paper is to construct an identification process which combines the LHS method with the V-IWC method in order to identify the wavenumber and the damping of a honeycomb sandwich beam. The proposed identification process is then compared to other processes combining the LHS method with either McDaniel method or standard IWC method to evaluate the efficiency of its estimates and their robustness against uncertainties.

2 Theoretical Backgrounds

2.1 McDaniel Method

The McDaniel method (McDaniel et al. 2000; McDaniel and Shepard 2000) consists on iteratively adjusting, for each frequency, the wavenumber to approximate accurately the response. Wavenumbers of neighboring frequencies are considered as initial estimations. It considers a harmonic displacement field which depends on space coordinates $u = \Re\{Ue^{-i\omega t}\}$, where $\Re\{.\}$ refers to the real part and U to the displacement amplitude. The McDaniel method consists on solving the linear differential equation of motion of the neutral axe or surface of the structure which takes the form:

$$-\omega^2 U + \mathcal{L}\{U\} = 0, \tag{1}$$

where $\mathcal{L}\{U\}$ is a linear operator containing the displacement derivatives with respect to the space coordinate x.

Accounting for boundary conditions $\mathcal{L}_b\{u\}|_{(x=x_b)} = \Re\{Be^{-i\omega t}\}$, where x_b represents boundaries and B is complex valued, the solution of Eq. (1) is expressed as:

$$U(x) = \sum_{n=1}^{N} \left\{ F_n e^{ik_n(1+i\gamma)x} + B_n e^{ik_n(1+i\gamma)(L-x)} \right\}, \tag{2}$$

where γ is the wave attenuation and N the number of different waves. Each wave n is characterized by a wavenumber k_n of complex value, containing positive real and imaginary parts, and an amplitude F_n or B_n according to forward or backward propagation, respectively, computed using boundary conditions.

To verify the validity of the initially supposed wavenumber, it is compared to the obtained wavenumber through the error function defined as:

$$\varepsilon^2(k) = \sqrt{\frac{\sum_{m=1}^{M} \rho(x_m)|U^{mes}(x_m, \omega) - U(x_m, \omega)|^2}{\sum_{m=1}^{M} \rho(x_m)|U(x_m, \omega)|^2}}, \tag{3}$$

where M is the number of measurement points, $\rho(x_m)$ the coherence function and $U^{mes}(x_m, \omega)$ and $U(x_m, \omega)$ are the measured and real wave fields, respectively. To minimize this error, an optimization algorithm varying the wavenumber is applied.

2.2 Proposed V-IWC Method

Correlating the vibratory field with inhomogeneous waves is the principle of the IWC method. Mathematically, the harmonic field $\hat{u}(x, y)$ is calculated either from a harmonic excitation or from temporal Fourier transforms:

$$u(x, y, t) = \int_0^{+\infty} \hat{u}(x, y)e^{i\omega t} d\omega. \tag{4}$$

The ω-dependence of the field $\hat{u}(x, y)$ is comprised in the hat $\hat{}$.

Whereas the McDaniel method accounts for both incident (forward propagating: e^{-ik} and backward propagating: e^{ik}) waves and evanescent waves (e^{-k} and e^{k}), the standard form of the IWC method considers only one term corresponding to forward propagating wave and neglects three terms corresponding to backward propagating wave and evanescent waves:

$$u_{IWC} = e^{-ik(\theta)(1+i\gamma(\theta))(x\cos(\theta) + y\sin(\theta))}, \tag{5}$$

where θ is the wave direction.

To improve the IWC method identification and overcome the lack of terms in its standard form, the proposed V-IWC method accounts for both forward and backward propagating waves (e^{-ik} and e^{ik}) through a correlation with a sum of inhomogeneous waves of the form:

$$u_{V\text{-}IWC} = e^{-ik(\theta)(1+i\gamma(\theta))(x\cos(\theta)+y\sin(\theta))} + e^{ik(\theta)(1+i\gamma(\theta))(x\cos(\theta)+y\sin(\theta))}. \qquad (6)$$

The correlation between the vibratory field and the proposed waves is performed through an *IWC* criterion which is defined as a Modal Assurance Criterion (MAC) (Ewins 1984) and must be maximized to optimize the wavenumber and damping (wave attenuation) identification:

$$IWC(k, \gamma, \theta) = \frac{\left| \int_S \hat{u}.u^*_{V\text{-}IWC} dxdy \right|}{\sqrt{\int_S \hat{u}.\hat{u}^*.dxdy \times \int_S u_{V\text{-}IWC}.u^*_{V\text{-}IWC}.dxdy}}, \qquad (7)$$

where $u^*_{V\text{-}IWC}$ is the complex conjugate of the wave $u_{V\text{-}IWC}$. This criterion represents the wave contribution in the field $\hat{u}(x, y)$ or also the ratio of the energy carried by the wave and the total energy contained in the field.

3   Experiment-Based Identification of a Honeycomb Sandwich Beam

Structural identification of a sandwich beam with aluminum honeycomb core, Fig. 1, is illustrated in this section. The length and width of the beam are, respectively, 1 m and 0.029 m. Materials' properties of the beam components are listed in Table 1. The beam is freely suspended and a flexural loading is applied.

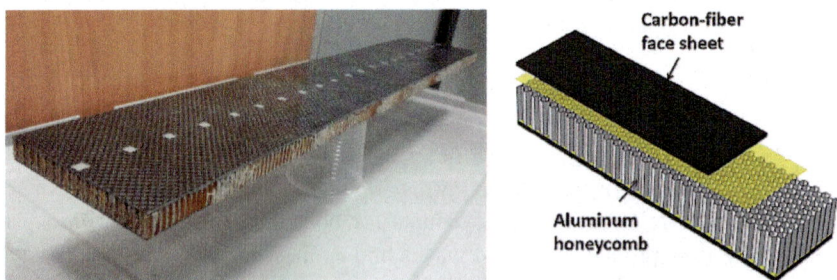

Fig. 1. Honeycomb-core sandwich beam

Experiment-based identification processes are applied here to estimate propagation parameters of the sandwich beam. The vibratory fields which are used as inputs of identification processes are computed experimentally at each measurement point, using a Scanning Laser Vibrometer (Ometron VPI+), Fig. 2. An electrodynamic shaker Bruel & Kjær 4809 is used to excite mechanically the freely suspended beam. The phase

Table 1. Material properties of the sandwich beam components.

Material		Properties										
Type	Reference	Thickness (m)	Young modulus (GPa)			Poisson's ratio			Shear modulus (GPa)			Density (Kg m^{-1})
			E_1	E_2	E_3	v_{12}	v_{13}	v_{23}	G_{12}	G_{13}	G_{23}	ρ
Face plates	Vicotex G803/914	0.002	60.27	60.27	5	0.029	0.35	0.35	5	5	5	1594
Core plate	5056 3.1 3/16.001	0.011	0.415	0.267	0.668	0.29	0.3	0.3	0.131	0.310	0.137	49.65

reference is obtained by a force transducer Bruel & Kjær 8001. Both signals are sampled with a Hewlett Packard Paragon 35654A.

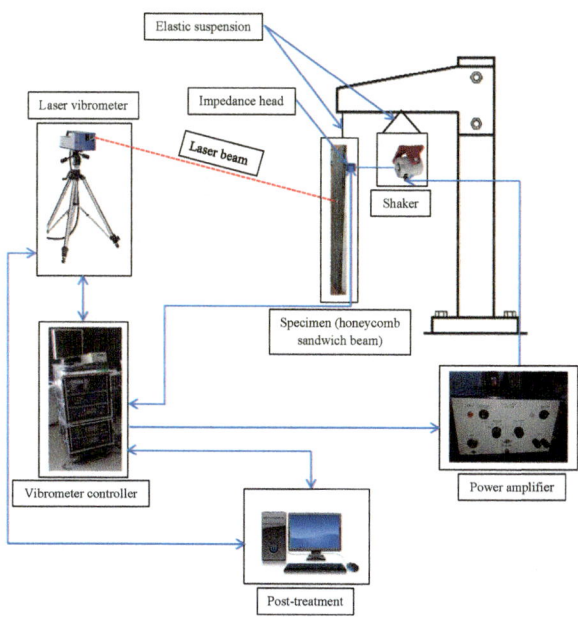

Fig. 2. Experimental set-up for measuring vibratory fields.

Two main types of experiment-based identification processes are applied in this section, depending on if uncertainties are taken into account or not. In deterministic case, which do not account for uncertainties, the main purpose is to illustrate the efficiency of the proposed V-IWC identification compared to the standard IWC method, the McDaniel method being considered as reference. Statistical investigations are then performed to evaluate the robustness of the V-IWC estimates against uncertainties.

Several types of uncertainties could be considered in this study. We focus only on parametric uncertainties quantifying the errors on measurement points' coordinates. Actually, if some measurement points do not match the associated displacement fields, how would this affect identified parameters?

In a probabilistic framework, the variability of the measurement points' coordinates is quantified as $x = x_0(1 + \delta_x\xi)$, where x_0 is the vector containing the mean coordinates of the measurement points, δ_x is the dispersion value and ξ is a Gaussian random variable.

Combined with the McDaniel method, the standard IWC method or the V-IWC method, uncertainty propagation is performed using 1000-samples LHS method. 1000 successive deterministic simulations are thus generated and statistical post-processing evaluations are then performed to quantify each identified parameter's variability. A dispersion level $\delta_x = 3\%$ of the measurement points' coordinates is considered here.

The wavenumber and the wave attenuation are at first identified without taking into account uncertainties. Comparisons of the V-IWC estimates with those obtained using the McDaniel method and the standard IWC method are illustrated in Figs. 3 and 4.

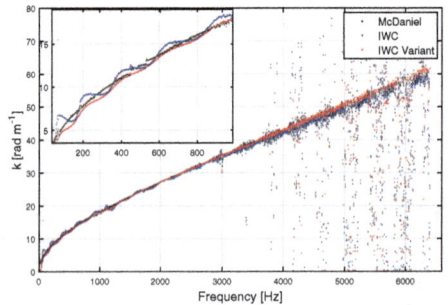

Fig. 3. Variation of the wavenumber according to frequency.

Note, at first, that results show the identification sensitivity to measurement disturbance and noise, especially at high frequencies, due to experimental conditions.

Small oscillations affect the curves of the V-IWC estimates compared to those obtained by the standard IWC method, especially for higher frequencies where the convergence of the IWC-V results is faster. The standard IWC method estimates are inaccurate, especially for wave attenuation. Good agreement is obtained between the McDaniel and the V-IWC estimates. The curves follow the same trend and have nearly the same order of magnitude throughout the whole frequency band. Moreover, the wavenumber estimates are more accurate than those of the wave attenuation.

In theory, maximal *IWC* criterion gives more accurate identification. Computing the maximal *IWC* criterion for each applied method at each frequency step is thus interesting: it consists on the objective function illustrated by Fig. 5. The maximal *IWC* criterion of the V-IWC method is greater than that of the standard IWC method and

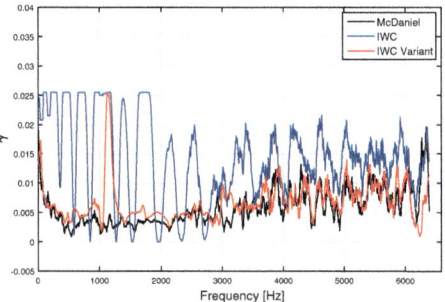

Fig. 4. Variation of the wave attenuation according to frequency.

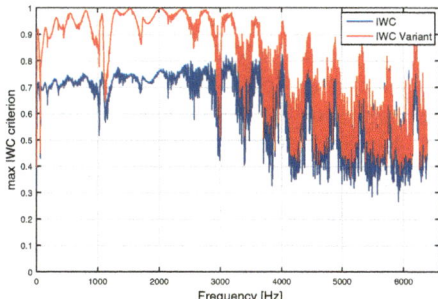

Fig. 5. Objective functions variation according to frequency.

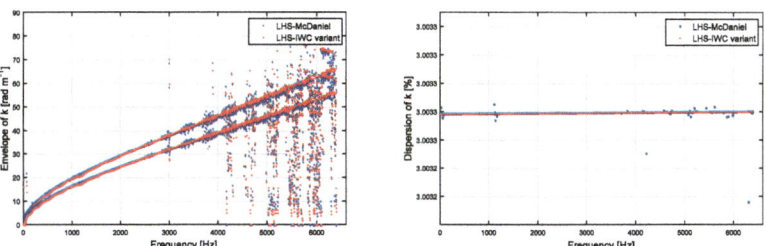

Fig. 6. Envelope and dispersion of the wavenumber.

tends to the theoretical limit which is 1. Greatest difference between objective functions is obtained at lowest frequencies.

Extending the study to presence of uncertainties, the randomness of identified wavenumber and wave attenuation is statistically investigated as illustrated by Figs. 6 and 7.

Larger variability is obtained on the wavenumber at higher frequencies. This is illustrated by larger envelopes, which represent extreme statistics, Fig. 6. The wavenumber is dispersed by 3%, if a similar dispersion (3%) is imposed on

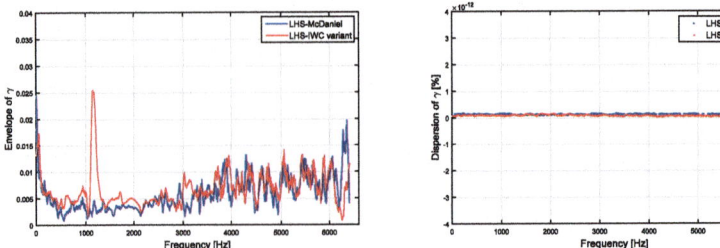

Fig. 7. Envelope and dispersion of the wave attenuation.

measurement points' coordinates. Dispersion is here computed by the ratio between standard deviation and mean. Errors on measurement points' coordinates are hence very influential on identified wavenumber. These results show the importance of accurate experiments for identification. Such errors must be reduced as possible. Here, one must address each measurement point to its vibratory field.

Regarding the wave attenuation's variability, Fig. 7, extreme statistics are perfectly superposed and a null dispersion is found. Subsequently, the randomness of the measurement points' coordinates does not affect the spatial damping of the beam.

At last, results illustrate the accurate identification allowed by the proposed V-IWC method over large frequency range. It can thus be considered as an alternative to the standard IWC method, especially at low frequencies. The V-IWC identification remains robust in spite of the large variability of identified parameters due to uncertainty impact. Furthermore, no damping sensitivity to such uncertainty is detected.

4   Conclusion

Correlating the experimentally-measured vibratory fields with a sum of inhomogeneous forward and backward propagating waves is the principle of the structural identification method proposed in this paper. The method is a variant which allow overcoming the limits of the standard Inhomogeneous Wave Correlation method, especially at low frequencies. Identification was performed in particular to estimate the wavenumber and the wave attenuation of a honeycomb sandwich beam.

If, experimentally, some errors affect the measurement points' coordinates, this could be investigated statistically by according to these coordinates a variability which can be modeled by parametric uncertainties. The sample-based Latin Hypercube Sampling method was used to propagate uncertainties. Statistical investigations of the variability of the identified wavenumber and wave attenuation illustrated the impact of uncertainties. Uncertainties are very influential on identified wavenumber, but do not affect the wave attenuation. No sensitivity of spatial damping is thus obtained to such type of uncertainty.

Besides, identification could be sensitive to other parameters' variability which could be the purpose of either sensitivity or uncertainty analysis. As perspective, it is interesting to evaluate identified damping sensitivity, in particular.

References

Berthaut, J., Ichchou, M.N., Jezequel, L.: K-space identification of apparent structural behavior. J. Sound Vib. **280**, 1125–1131 (2005)

Cherif, R., Chazot, J.-D., Atalla, N.: Damping loss factor estimation of two-dimensional orthotropic structures from a displacement field measurement. J. Sound Vib. **356**, 61–71 (2015)

Chronopoulos, D., Troclet, B., Bareille, O., Ichchou, M.N.: Modeling the response of composite panels by a dynamic stiffness approach. Compos. Struct. **96**, 111–120 (2013)

Ewins, D.J.: Modal Testing: Theory and Practice. Research Studies Press, London (1984)

Ferguson, N.S., Halkyard, C.R., Mace, B.G., Heron, K.H.: The estimation of wavenumbers in two dimensional structures. Proc. ISMA **II**, 799–806 (2002)

Fishman, G.S.: Monte Carlo: Concepts, Algorithms, and Applications. Springer, New York (1996)

Helton, J.C., Davis, F.J.: Latin hypercube sampling and the propagation of uncertainty in analyses of complex systems. Reliab. Eng. Syst. Saf. **81**, 23–69 (2003)

Ichchou, M.N., Bareille, O., Berthaut, J.: Identification of effective sandwich structural properties via an inverse wave approach. Eng. Struct. **30**, 2591–2604 (2008a)

Ichchou, M.N., Berthaut, J., Collet, M.: Multi-mode wave propagation in ribbed plates: Part I, wavenumber-space characteristics. Int. J. Solids Struct. **45**, 1179–1195 (2008b)

Inquiété, G.: Numerical simulation of wave propagation in laminated composite plates. Ph.D. thesis, Ecole Centrale de Lyon, France (2008)

Lajili, R., Bareille, O., Bouazizi, M.L., Ichchou, M.N., Bouhaddi, N.: Parameter identification of a sandwich beam using numerical-based inhomogeneous wave correlation method. Adv. Acoust. Vib. Appl. Cond. Monit. **5**, 65–75 (2017)

Lajili, R., Bareille, O., Bouazizi, M.L., Ichchou, M.N., Bouhaddi, N.: Inhomogeneous wave correlation for propagation parameters identification in presence of uncertainties. In: CMSM 2017. Design and Modeling of Mechanical Systems-III, pp. 823–833 (2018)

McDaniel, J.G., Dupont, P., Salvino, L.: A wave approach to estimating frequency-dependent damping under transient loading. J. Sound Vib. **231**, 433–449 (2000)

McDaniel, J.G., Shepard, W.S.: Estimation of structural wave numbers from spatially sparse response measurements. J. Acoust. Soc. Am. **108**, 1674–1682 (2000)

McKay, M.D., Beckman, R.J., Conover, W.J.: A comparison of three methods for selecting values of input variables in the analysis of output from a computer code. Technometrics **2**, 239–245 (1979)

Rubinstein, R.Y., Kroese, D.P.: Simulation and the Monte Carlo Method, 2nd edn. Wiley, Hoboken (2008)

Roozen, N.B., Labelle, L., Leclère, Q., Ege, K., Alvarado, S.: Non-contact experimental assessment of apparent dynamic stiffness of constrained-layer damping sandwich plates in a broad frequency range using a Nd:YAG pump laser and a laser Doppler vibrometer. J. Sound Vib. **395**, 90–101 (2017)

Van Belle, L., Claeys, C., Deckers, E., Desmet, W.: On the impact of damping on the dispersion curves of a locally resonant metamaterial: Modelling and experimental validation. J. Sound Vib. **409**, 1–23 (2017)

Characterization of the Mechanical and Vibration Behavior of Flax Composites with an Interleaved Natural Viscoelastic Layer

Daoud Hajer[1,2(✉)], El Mahi Abderrahim[1], Rebiere Jean-Luc[1],
Taktak Mohamed[2], and Haddar Mohamed[2]

[1] Laboratoire d'Acoustique de l'Université du Maine (LAUM) UMR CNRS
6613, Université du Maine, Av. O. Messiaen, 72085 Le Mans Cedex 9, France
daoud.hajer@yahoo.fr
[2] Laboratoire de Mécanique, Modélisation et Production (LA2MP),
Ecole Nationale d'Ingénieurs de Sfax (ENIS), Université de Sfax,
Route de Soukra, 3038 Sfax, Tunisia

Abstract. This study presents an analysis of the mechanical and vibration behavior of a flax fibre reinforced composites with and without an interleaved natural viscoelastic layer. Two types of elastic and viscoelastic cross ply laminates $[0_2/90_2]_s$ and $[0_2/90_2/NR]_s$ have been characterized experimentally using different mechanical and vibrational tests. The elastic laminate is composed of natural long flax fibre and greenpoxy resin while the viscoelastic laminate is composed of a natural viscoelastic layer and two elastic composites. First, both types of specimen composites were studied using uni-axial tensile tests under the same conditions. A comparison between the two composites behaviors has been realized. Then, Acoustic Emission (AE) has been often used for the identification and characterization of micro failure mechanisms and damage in laminates. Finally, experimental vibration analyses were carried out on the composites with and without an interleaved natural viscoelastic layer. Throughout a series of resonance vibration tests, the evolution of the Young modulus and the modal damping were evaluated. The effect of the viscoelastic layer on the mechanical and vibration behavior of the elastic composite has been investigated and analyzed. It has been shown that the viscoelastic layer improves with a significant way the modal properties of the flax fibre reinforced composite.

1 Introduction

Composite materials are increasingly used in various fields of application. Nowadays, the interest of environmental and ecological concerns in recent years has led to the development of biosourced composites. These types of materials have good mechanical performances sometimes better than those of synthetic fibre reinforced composites [1]. Several researchers of natural fibres have been studied [1–4]. Their results showed that these materials have many advantages such as their low density, their biodegradability and their relatively high specific mechanical properties. Hemp and Flax composites are the most studied in the two last decades [5]. Knowing that France is the first producer

© Springer Nature Switzerland AG 2019
T. Fakhfakh et al. (Eds.): ICAV 2018, ACM 13, pp. 186–194, 2019.
https://doi.org/10.1007/978-3-319-94616-0_19

of flax, this type of natural fibre has been chosen to be investigated in this present work. Many researchers have studied the vibration behavior of flax composites [1, 3, 6]. They demonstrate that they possess very high modal properties.

However, the ability of viscoelastic materials to dissipate energy makes this type of material a good element in many areas of the industry. Thus, several works are interested in increasing the damping of composite structures by the insertion of viscoelastic layers on their materials [6, 7].

The aim of this study is to analyse the mechanical and vibration behavior of a flax fibre reinforced composites with and without an interleaved viscoelastic layer made of natural rubber (NR). Two types of elastic and viscoelastic cross ply laminates $[0_2/90_2]_s$ and $[0_2/90_2/NR]_s$) were studied.

Quasi static tensile loading was realized on the two types of composites. Moreover, the tests were monitored by acoustic Emission analysis (AE). This study has often been used for the identification and characterization of damage modes on classical composites [8] as well as on natural fibre composites [4]. The data obtained was processed and classified with NOESIS software [9]. The 'K-means' algorithm was used for unsupervised model recognition. This study make it possible to identify and characterize the various mechanisms of damage that occurred during the tests. Finally, the modal properties of the composites were also investigated and the effect of the insertion of the viscoelastic layer on the laminate was studied.

2 Implementation of the Composite

2.1 Materials Used

The materials considered in this study are elastic and viscoelastic laminated materials. The elastic composite laminates are composed of natural long flax fibres trated and manufactured by LINEO [2] and greenpoxy resin 'SR GreenPoxy 56' produced by Sicomin, while the viscoelastic composite laminates are composed of a viscoelastic layer of natural rubber confined between two elastic composites. The obtained laminates $[0_2/90_2]_s$ and $[0_2/90_2/NR]_s$ were tested with quasi static tensile loading and vibrational tests.

2.2 Method of Implementing the Composite

The laminate materials used in this study were made by vacuum molding process. The procedure consists in producing laminates by alternately laying layers of long flax fibres and layers of liquid resin. Different demolding fabrics are interposed on both sides between the molds and the composite in order to facilitate the demolding, and to ensure the homogeneity of the absorption of the resin. Then, the obtained laminates were introduced into a vacuum bag of 0.6 bar for at least 6 h until the total polymerization of the matrix. Hence, they could be demolded after 24 h.

The laminated sheets were manufactured with 8 fibre's layers so as to obtain a total nominal thickness of 4 mm for the elastic composites and 5 mm for the viscoelastic one. The volume fraction was estimated between 38% and 45% for all the specimens.

The porosity in the material was calculated by comparing the measured density of the composite and its theoretical one calculated from the fibre volume fraction and that of the matrix. It has been estimated between 3 and 10%.

3   Study of the Tensile Behavior of the Composite

3.1   Experimental Setup

Axial tensile tests were performed on cross ply laminates $[0_2/90_2]_s$ and $[0_2/90_2/NR]_s$. The tests were carried out on specimens with the dimensions $250 \times 25 \times 4$ mm^3 for the elastic composites and $250 \times 25 \times 5$ mm^3 for the vicoelastic ones according to ASTM standard test standard D3039/D3039M. The specimens have been tested using a standard MTS hydraulic traction machine with a capacity of ± 100 KN.

Figure 1 presents the results of the stress/strain curves for the elastic and vis-coelastic laminates. The obtained results show that the evolution of the stress as a function of the strain present two phases separated by a bend. The first one is a short linear elastic domain and the second one presents a non linear part until break. The point of inflection is very visible on the curves; it occurs for a very low level of deformation (0.1%) and causes a significant loss of rigidity. This non linear behavior is very different from the one of synthetic fibres reinforced composites, It was also observed in the study conducted in the laboratory by Monti et al. for Lin/Helium composites [4]. Thus, the obtained non-linearity could be attributed to the intrinsic behavior of natural flax fibre and more specifically to the behavior of lignin and amorphous cellulose fibre due to shear stresses in the cell walls.

The comparison between the two types of laminated composites show that the first part of the stress/strain curves is identical for both types. Then, the breakage of the viscoelastic composites is obtained at very low strain level comparing with the breakage of the elastic specimens. In fact, the characteristics at break of the elastic

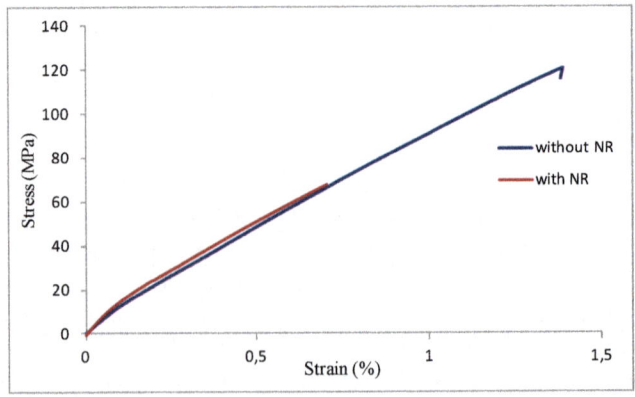

Fig. 1. Typical stress/strain curves of elastic and viscoelastic laminates

laminate is almost two times superior of those corresponding to the viscoelastic composites. The value of the tensile stress for the viscoelastic composite decreased by about 40% compared to that of the laminate composite, while the breaking strain increased by 50%. This decrease is due to the viscoelastic behavior of the natural rubber.

3.2 Damage Analysis by Acoustic Emission

In this part, and in order to analyze the damage mechanisms that appear during tensile tests, the various tests described above were repeated, and the monitoring of the damage was performed by acoustic emission (AE) using the Noesis software. The classification of the AE data was performed using an unsupervised model. Hence, a K-mean algorithm was used to classify and separate the different events into a k clusters based on five acoustic signal parameters: amplitude, rise time, number of counts to peak, duration and energy.

The results obtained from this classification methodology for the cross ply laminates $[0_2/90_2]_s$ and $[0_2/90_2/NR]_s$ are presented in Fig. 2. The results have been described by analyzing the superposition of the amplitudes of the AE signals as a function of time with the evolution of the applied stress (Fig. 2i). Then, the chronology and the appearance of the events of the different classes were studied by the analyze of the counts to peak as a function of time (Fig. 2ii). Finally, and to highlight the separation or overlap between the different classes, the acoustic signals were projected in the plane of two principal components (PCA) (Fig. 2iii). Four classes were obtained for the elastic composite (Fig. 2a). The signals of class A appear at very low stress level. Their appearance coincides with the end of the elastic part of the stress/time curve. This class possesses the lower values of amplitudes which vary between 45 and 50 dB. Classes B and C appear simultaneously during the test. Their appearance was detected at the beginning of the non linear phase of the stress/time curve. The amplitudes of the class B signals vary between 50 and 60 dB while those of the class C vary between 55 and 70 dB. According to the analyzes and the results founded in literature [4, 8], class A corresponds to matrix cracking, class B corresponds to the debonding fibre/matrix and class C corresponds to the delamination. Finally, the events of the Class D appear at the end of the test, just before the breackage of the specimens. The amplitudes of the signals of this class have the highest values, which vary between 70 and 95 dB. So this class is attributed to the fibre breakage.

Figure 2(b) present the results of the classification obtained for the viscoelastic laminate $[0_2/90_2/NR]_s$. The results present five classes. The comparison between the results of the elastic and viscoelastic laminates show that an additional class 'E' is obtained for the viscoelastic laminate, which could concern a domage mechanism related to the viscoelastic layer. The amplitudes of these events' classes vary between 55 and 85 dB. This class could be attributed to the debonding composite/viscoelastic layer.

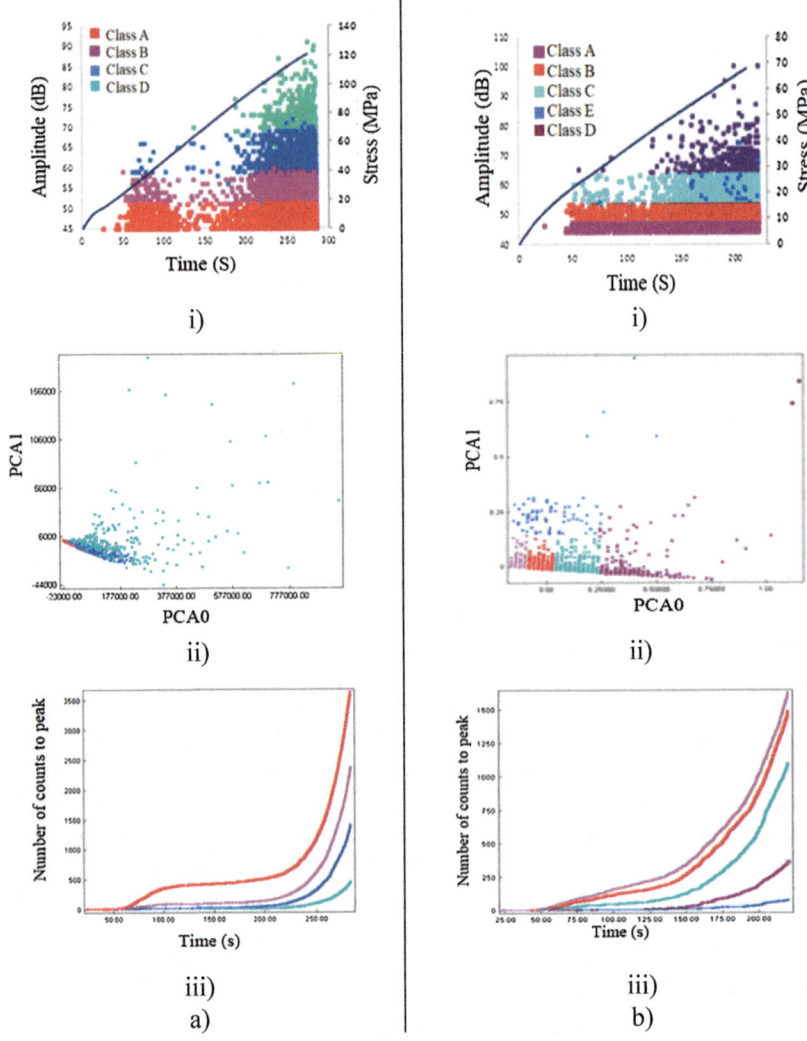

Fig. 2. Analysis of acoustic emission data: (i) amplitude/ time, (ii) principal component analysis and (iii) chronology of occurrence of the different classes of composites: (a) $[0_2/90_2]_s$ and (b) $[0_2/90_2/NR]_s$.

4   Study of the Vibration Behavior of the Composite

The second part of this work is to investigate the dynamic properties of flax fibre reinforced composites with and without an interleaved natural viscoelastic layer. The composite specimens are supported horizontally in a clamping block. The excitation of the flexural vibrations of the beams was induced using an impact hammer and the beam response was detected by an accelerometer. The signals obtained from the acceleration transducer (PCB 352C23, model SN109866) and the force transducer (PCB 086B03, model SN5909) are analyzed by a dynamic signals analyser. Frequency response analyses were carried out to obtain the natural frequencies. Then, the damping factor was calculated using the Half Power Bandwidth (HPB) method used by Hammamiet al. [10] and Daoud et al. [6] Moreover, the young modulus E was calculated for every resonance mode by:

$$E = \frac{12\rho L^4 \omega_n^2}{h^2 \alpha_n^4}$$

The elastic and viscelastic cross ply laminates $[0_2/90_2]_s$ and $[0_2/90_2/NR]_s$ were carried out. The composite specimens were cut in different directions so as to obtain composites with stacking sequences of the type $[\theta_2/(90 + \theta)_2]$ with different orientations of the fibres (Fig. 3). These laminates are labeled Cr-θ.

The modal analysis was carried out on the first four bending modes, according to the experimental analysis presented previously, on specimens of 25 mm width, a nominal thickness of 4 mm for elastic laminates and 5 mm for viscoelastic laminates and for three lengths (150, 200 and 250 mm). Different orientations of the fibres were studied: 0°, 15°, 30°, 45°, 60°, 75° and 90°.

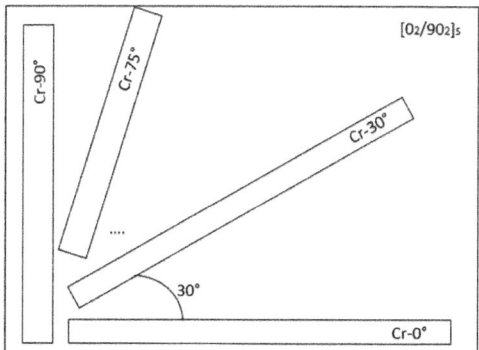

Fig. 3. Cross ply laminated specimens with different orientations

The evolution of Young modulus as a function of the frequencies for the different orientations is presented in Fig. 4.

It can be seen that the young modulus of the orientation 60° (Cr-60°) corresponding to a stacking sequence of the type ($[60_2/30_2]_s$) presents the least values and this for the elastic and viscoelastic composites. So, the specimens Cr-60° are the least rigid comparing to the other configurations of the cross laminates.

For a given orientation, the modulus E varies slightly with the increase in frequency for the elastic composite, while it decrease with a significant way for the viscoelastic laminates. This decrease is attributed by the viscoelastic behaviour of the natural rubber.

Hence, the insertion of the viscoelastic layer in the laminate increases its thickness and mass but decreases its rigidity in relative flexion. So, the modulus decreases drastically with the increase of the vibration frequency.

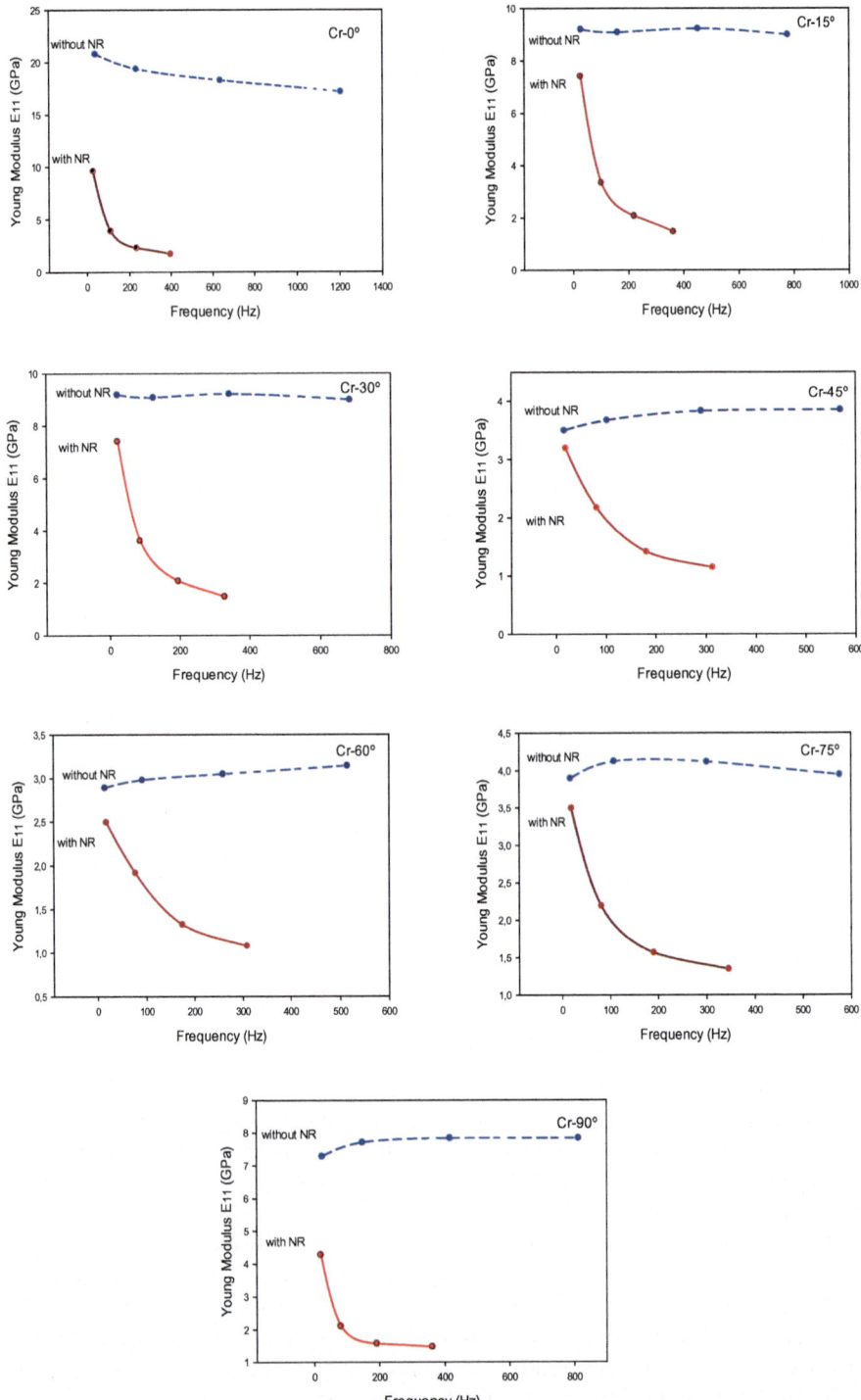

Fig. 4. Variation of the longitudinal Young's modulus E as a function of the vibration frequency for different orientations of the fibres

The effect of the insertion of the viscoelastic layer on the damping factor of flax compsoites was evaluated. The variations of this factor as a function of the frequency for the different stacking sequences and for the two studied materials (elastic and viscoelastic) are shown in Fig. 5. The figure show that, for a given stacking sequence, the damping increases as the frequency increases for the two laminates. This increase is much higher in the case of viscoelastic laminates.

The values of the damping factor for the elastic composite is between 1.5 and 2% for low frequencies and 3% for high frequencies (2000 Hz), while Ithe values corresponding to he viscoelastic laminate are in the order of 3% for low frequencies and could reach the 8% for high frequencies (2000 Hz). The integration of a viscoelastic layer in the cross ply laminates improves significantly its damping and thus its vibration behavior.

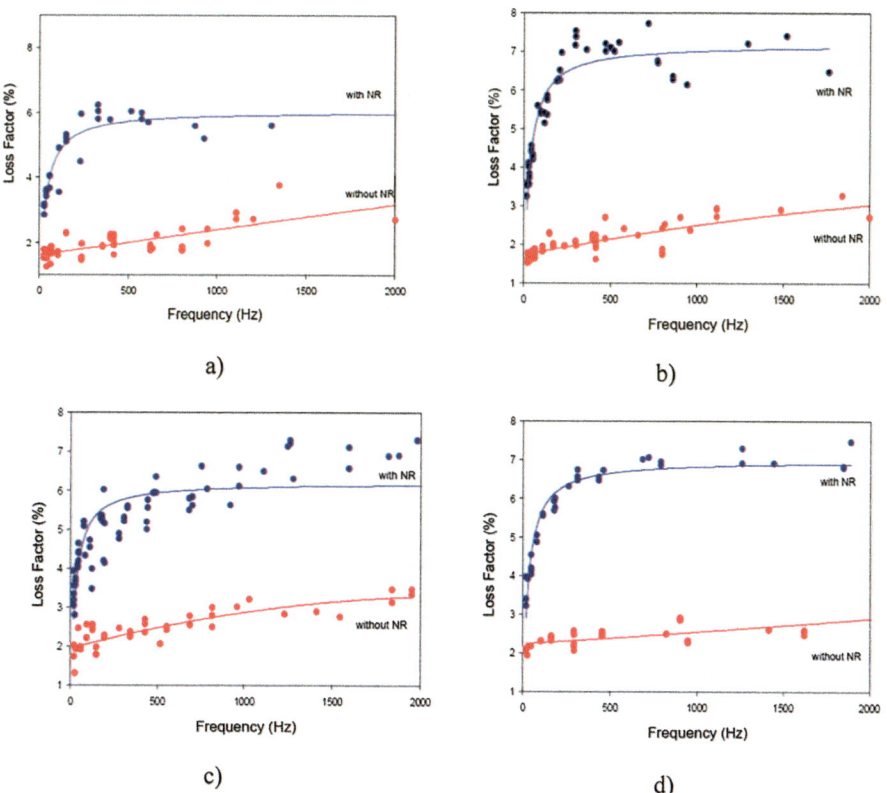

Fig. 5. Variation of damping factor versus frequency of elastic and viscoelastic cross laminates for different orientations: (a) Cr-0° and Cr-90°, (b) Cr-15° and Cr75°, (c) Cr-30° and Cr-60° and (d) Cr-45°.

5 Conclusion

This work focuses on the study of the mechanical and vibration behavior of a flax fibre reinforced composites with and without an interleaved natural viscoelastic layer. The composite materials have been characterized experimentally using different mechanical and vibrational tests. First, both types of composites were studied using uni-axial tensile tests. Acoustic emission (AE) has been often used for the identification and characterization of micro failure mechanisms in composites. The results showed that these composites have very high specific characteristics. It can be used for applications currently using composites reinforced with synthetic fibres such glass, carbon…. Next, experimental vibration analyses were carried out on the composites. It has been shown that the viscoelastic layer plays a major role in damping because it has a high level of energy dissipation. Therefore, it improves with a significant way the modal properties of the composite.

References

1. Assarar, M., Zouari, W., Sahbi, H., Ayad, R., Berthelot, J.: Evaluation of the damping of hybrid carbon-flax reinforced composites. Compos. Struct. **132**, 148–154 (2015)
2. Faruk, O., Bledzki, A., Fink, H., Sain, M.: Biocomposites reinforced with natural fibres: 2000–2010. Prog. Polym. Sci. **37**, 1552–1596 (2012)
3. Yan, L., Jayaraman, K.: Flax fibres and its composites. Compos. Part B **56**, 296–317 (2014)
4. Monti, A., El Mahi, A., Jendli, Z., Guillaumat, L.: Mechanical behaviour and damage mechanisms analysis of a flax-fibre reinforced composite by acoustic emission. Compos. Part A **90**, 100–110 (2016)
5. Monti, A.: Elaboration et caractérisation mécanique d'une structure sandwich à base de constituant naturels. Thèse de doctorat, Université de Maine, Le Mans, France, Décember-2016
6. Daoud, H., El Mahi, A., Rebiere, J., Taktak, M., Haddar, M.: Characterization of the vibrational behaviour of flax fibre reinforced composites with an interleaved natural viscoelastic layer. Appl. Acoust. (2016)
7. Khalfi, B.: Modélisation analytique pour l'étude du comportement vibratoire en régime transitoire d'une plaque avec tompon amortissant contraint. Thèse de doctorat, Ecole polytechnique de Montéral (2012)
8. Masmoudi, S., El Mahi, A., Turki, S.: Fatigue behaviour and structural health monitoring by acoustic emission of E-glass/epoxy laminates with piezoelectric implant. Appl. Acoust. **108** (50–58), 2016 (2016)
9. NOESIS Software: Advanced acoustic emission data analysis pattern recognition and neural networks software (2004)
10. Hammami, M., El Mahi, A., Karra, C., Haddar, M.: Experimental analysis of the linear and nonlinear behavior of composites with delaminations. Appl. Acoust. **108**, 31–39 (2016)

Experimental and Numerical Analysis of Sound Transmission Loss Through Double Glazing Windows

Chaima Soussi$^{(\boxtimes)}$, Walid Larbi, and Jean-François Deü

Conservatoire National des Arts et Métiers (Cnam),
Laboratoire de Mécanique des Structures et des Systèmes Couplés (LMSSC),
2 rue Conté, 75003 Paris, France
chaima.soussi@lecnam.net

Abstract. The domestic windows in the exterior building facade play a significant role in sound insulation against outdoor airborne noise. The prediction of their acoustic performances is classically carried out in laboratory according to standard ISO 10140. In this work, a 3D elasto-acoustic finite element model (FEM) is proposed to predict the sound reduction index of three different glazing configurations of domestic window follows the ISO recommendations for acoustic measurements, which are compared to laboratory measurements. Two acoustic cavities with rigid-boundaries on both sides of the window are used to simulate respectively the diffuse sound field on the source side and the pressure field on the receiver one. By using a simplified FEM for the double-glazed windows, the sound reduction index is calculated from the difference between the source and receiving sound pressure levels in the one-third octave band from 100 to 500 Hz. Although the comparison between numerical and experimental results shows a relatively good agreement which highlights the interest of this kind of approaches to avoid expensive experiments, many improvements should be taken to ameliorate the model such as the different components of the frame and the design of the two rooms to avoid the problematic of multi-resonant frequency ranges.

Keywords: Window · Sound transmission · Experimental measurements
Numerical simulation

1 Introduction

The windows in the exterior building facade play an important role in sound insulation against outdoor noise sources such as road traffic and aircrafts in the low frequency range. Therefore, the acoustic performances of domestic window have been the subject of numerous studies to ensure compliance with their sound insulation capabilities.

The principal acoustic indicator is the sound reduction index values R measured according to the standards in laboratory practice which takes into consideration two reverberation chambers: a source chamber and a receiving chamber separated by a common wall containing an opening in which the test element is mounted. An important topic in sound insulation measurements is the reproducibility since

© Springer Nature Switzerland AG 2019
T. Fakhfakh et al. (Eds.): ICAV 2018, ACM 13, pp. 195–203, 2019.
https://doi.org/10.1007/978-3-319-94616-0_20

significant differences have been observed in results from different laboratories especially in the low frequency range (Cops et al. 1987; Kihlman and Nilsson 1972). For
this issue, the niche effect due to the specific positioning of the test window in the wall
has been investigated numerically by Sakuma et al. (2011) and validated with experimental results by Dijckmans and Vermeir (2012). Results showed that the position of
the test specimen in the opening and the depth of the niche have significant influences
on the sound transmission loss. Also, research has mainly been focused on evaluation
the sound insulation effect of geometry parameters of the window such as type of glass
and type of connections between window frame and the wall. Miskinis et al. (2015)
compared experimentally different double glass models and results showed that the best
choice is with the combination of one ordinary and one laminated glass. For the same
objective, a combined experimental and analytical approach was developed to predict
the sound transmission loss of homogenous and laminated glazing by Ruggeri et al.
(2015). In a similar context, a finite element model of a sandwich plate with viscoelastic core was developed by Larbi et al. (2016) to evaluate the effects of such
material on the sound transmission through the studied system. Recently, Løvholt et al.
(2017) studied the effect of window connections with a 3D numerical modeling in the
very low frequencies (below 100 Hz). Close agreements with results from laboratory
measurements was obtained and showed that the connections have a large importance
in the structural vibration and sound transmission. They showed also that the low
frequency transmission from 15 to 30 Hz is controlled by the windows, whereas the
walls controlled the transmission from 30 to 100 Hz.

This paper is organized as follows: second section describes (i) the test method used
for the laboratory measurements required by ISO 10410 and (ii) the acoustic indicators
for the evaluation of the sound insulation performances of building elements. The third
section is dedicated to the laboratory conditions in which measurements were carried
out and to the 3D numerical model. Finally, comparison and discussion between
numerical and experimental results of three double glazing windows are proposed.

2   Experimental Measurements of Sound Reduction

The airborne sound insulation of building elements such as walls, doors, windows can
be evaluated in laboratory according to standards norms such as (ISO 10140, 2010)
and (ASTM E90-09, 2009). These standards allow the determination of the acoustic
performances of the structures by the measurement of the sound reduction index R
(called also Sound Transmission Loss STL in the English-speaking countries).

In this section, we present a description of the test method used on the experimental
measurements according to the standard ISO 10140 carried out for this research project,
and the indicators used to evaluate acoustic performances of the window.

2.1   Summary of Laboratory Measurement Method

According to different subparts of the standard ISO 10140, the laboratory test measurement takes into consideration two reverberation rooms, horizontally or vertically
adjacent, separated by a common wall containing an opening in which the test element

is mounted as shown in Fig. 1. The minimum volume of two chambers is 50 m³ with a volume difference of 10%. The rooms are mounted on elastic supports and the only significant sound transmission should be through the test element. The reverberant time should not be greater than 2 s and the total absorption in the receiving chamber should be low to ensure the best possible condition of the diffuse field. During all measurements, the average relative humidity should be at least 30% and temperatures should be in the range 20 ± 3 °C.

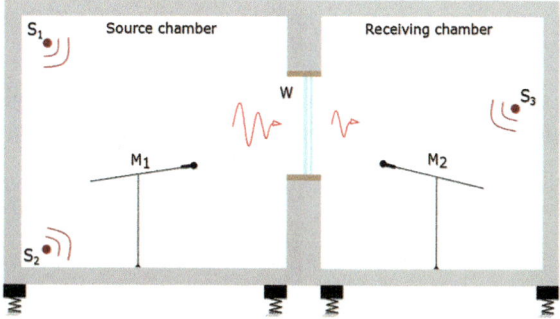

Fig. 1. Experimental set-up: (i) rotating microphones M1 and M2, (ii) loudspeakers S_1, S_2 and S_3 and (iii) tested window W.

In the source room, the sound signals should be random and have a continuous spectrum within each frequency band. For that, one or more loudspeakers can be used, which preferably should be omnidirectional to excite the sound field in the room as uniformly as possible, so the diffuse condition is approximately satisfied. This enables the acoustic pressure to be characterized by a space and frequency averaging which used to calculate the acoustic indicators in one-third octave bands within the frequency range 100 Hz–5000 Hz as explained in the next section.

2.2 Acoustic Indicators

The sound reduction index R is calculated in decibels (dB) as the logarithmic inverse of the sound transmission coefficient τ

$$R = 10 \log_{10} \left(\frac{1}{\tau} \right) \tag{1}$$

where τ is defined as the ratio of the sound power radiated by the test element w_r to the sound power incident w_i.

During the laboratory measurements, R is defined as the difference between the average sound pressure levels in the source room and in the receiving room, respectively L_s and L_r considering the total absorption of the receiving room A (m^2) and the area of the test specimen S (m^2):

$$R = L_s - L_r + 10 \log_{10} \frac{S}{A} \qquad (2)$$

The total absorption A is determined with Sabine's formula in which the reverberation time T_r (s) in the receiving room and its volume V (m^3) are used:

$$T_r = \frac{0.16\ V}{A} \qquad (3)$$

L_s and L_r expressed in dB are ten times the logarithm of the ratio of the square of the sound pressure p (Pa) in the considering chamber to the square of a reference value, p_0 (Pa):

$$L_s = 10 \log_{10} \frac{p^2}{p_0^2} \qquad (4)$$

where the reference value p_0 is 20 μPa.

3   Experimental and Numerical Examples

3.1   Experimental Set-Up

Eight types of double glazing windows have been investigated experimentally, but only the results of three models 6/18/4, 4/20/4 and 10/14/4 are presented in this paper and compared to numerical results. We recall that double glazing units are made up of two pieces of glass with a spacer (i.e. 6/18/4 means that the double glazing unit consists of a 6 mm pane, a 18 mm spacer and a 4 mm pane).

The measurements of the sound insulation performance of windows were done in laboratory considering one-third-octave bands from 100 up to 5000 Hz. The test rooms are composed of a source chamber (volume 73 m^3) and a receiving chamber (volume 65 m^3). Three loudspeakers (RCF-C5215-W), two rotating microphones (Brüel & Kjær Type 3923) and a multi-channel measuring and analyzing device are used (see Fig. 1). The accuracy of the equipment is verified in an accredited metrology described in the standard ISO 10140-5.

The loudspeakers S_1 placed in the top left corner and S_2 in the bottom left corner of the source room generate a pink noise with a total power of 100 dB. A combination of two heights of the loudspeaker S_3 (1.67 and 2.045 m from the floor) and three positions of the microphone M_2 (120°, 240° and 360°) is used to measure the average of the reverberation time T_r.

During the measurements, the test window is fixed with wooden sticks and its perimeter, between the wood frame and the opening, is sealed with silicon in both sides to prevent acoustic leakages. The temperature, pressure and relative humidity are kept constant (respectively 20 °C, 1036 Pa and 46%) and controlled using a sensor.

The frame of the window tested (Fig. 2) is 1.45 m wide by 1.48 m high and the frame thickness is 47 mm. The frame material is Sapele wood with density of 690 kg/m^3 and elastic modulus of 14 GPa. The glass material has density of 2500 kg/m^3.

Fig. 2. The CAD model of test window

3.2 Numerical Model

We describe in this section the numerical model used for the simulations. On usual assumptions, a vibro-acoustic problem couples a structure (elastic material) to an acoustic domain (fluid). In the finite element context, usually such a model is described in terms of structure displacement u and acoustic pressure p. The discretization of the weak formulation of the problem in the frequency domain leads to the following system of equations:

$$\left(\begin{bmatrix} K_s & -C \\ 0 & K_a \end{bmatrix} + i\omega \begin{bmatrix} D_s & 0 \\ 0 & D_a \end{bmatrix} - \omega^2 \begin{bmatrix} M_s & 0 \\ C^T & M_a \end{bmatrix} \right) \begin{bmatrix} u(\omega) \\ p(\omega) \end{bmatrix} = \begin{bmatrix} f_s(\omega) \\ f_a(\omega) \end{bmatrix} \quad (5)$$

where K_s, M_s and D_s are the structural stiffness, mass and damping matrices, K_a, M_a and D_a are the associated acoustic matrices. C is the coupling matrix, f_s and f_a are the structural and acoustic loads, respectively.

In the present study, the vibro-acoustic system is composed of the window and the two rooms containing air. To solve numerically this problem, the commercial software Actran is employed. The 3D finite element model is carried out, as shown in Fig. 3, composed of a source and receiving chambers which are modeled as two rigid acoustic cavities and have the same dimensions of the laboratory configuration. At this stage, a simplified configuration of the window frame was used; a double ordinary glazing (1.45 m wide by 1.48 m high) modeled as clamped thin shells separated by an airtight cavity which is filled with argon whose sound speed is 317 m/s and the density is 1.6 kg/m^3. A diffuse sound field is modeled as the acoustic excitation in the source room.

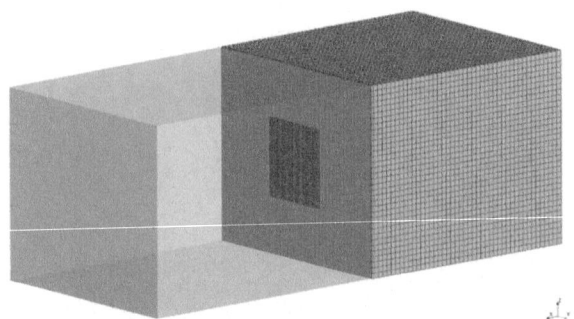

Fig. 3. Numerical model with Actran

A direct frequency analysis is used to compute the response of this model. It covered the range from 0 to 600 Hz with 2 Hz steps and results are presented in the one-third octave band from 100 to 500 Hz. It can be noted that the output noise reduction is given for each frequency and one-third-octave filter was needed to transform results.

Linear finite elements were employed throughout the model. The mesh sizes for the structure and the acoustic domain are controlled by the wavelength λ (m) which depends on the frequency range of interest. For the acoustic domain, the acoustic wavelength is $\lambda_a = c/f$ where c (m/s) is the speed of sound. For the structure, the flexural wavelength is $\lambda_f = \sqrt{2\pi/f}(D/M)^{1/4}$ where D (Pa m^3) is the flexural rigidity and M(kg/m^2) the surface mass density. Since it is recommended to use 6 elements per wavelength, we obtain a model containing around 1.5×10^5 degrees of freedom.

3.3   Results and Discussion

Numerical third-octave band results are compared to experimental results within the frequency range 100 Hz to 500 Hz (8 frequency values). For the three systems studied 10/14/4, 4/20/4 and 6/18/4, we may conclude from Fig. 4 that, overall, experimental and numerical results follow similar pattern but there are some differences especially in the low frequency range. It is important to keep in mind that experimental data are for the global structure of windows while the numerical results are for simplified double glazing plates. In fact, the overall sound insulation of the window is affected by the frame, its various materials, the sealing of the glazing and other parameters. In addition, the reverberant time T_r measured during the laboratory tests is around 1.5 s which leads to a crossover frequency "Schroeder frequency" f_s about 280 Hz. According to (Schroeder 1996), for airborne sound in reverberant room with a volume V, this frequency is given by $f_s = 2000\sqrt{T_r/V}$, and it marks the limit between the well-separated resonances and many overlapping normal modes (diffuse mode). So that, the difference between experimental and numerical results in the range frequency below 300 Hz can be explained by the fact that the sound field is not perfectly diffused.

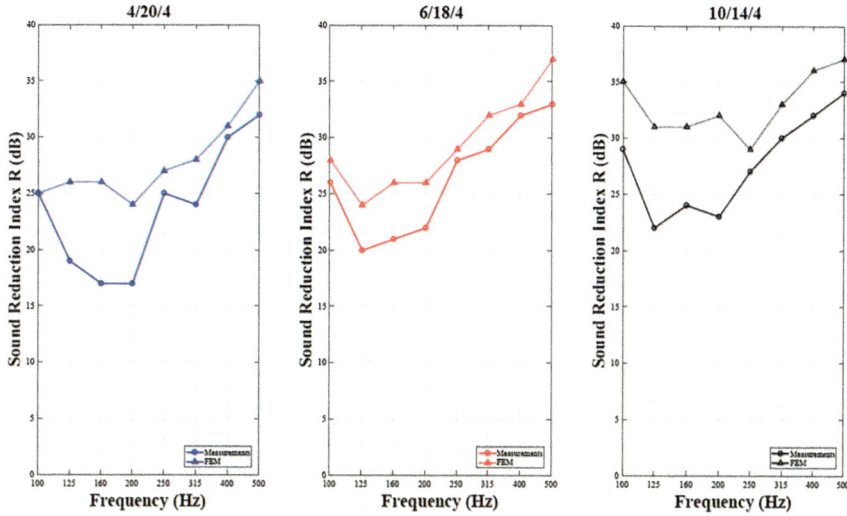

Fig. 4. Numerical and experimental results for the three types of window

Figure 5 shows the effect of the type of glazing systems in the acoustic insulation performance. This figure presents experimental and numerical sound reduction index *R* for the three glazing studied in this project. The sound transmission loss of a double

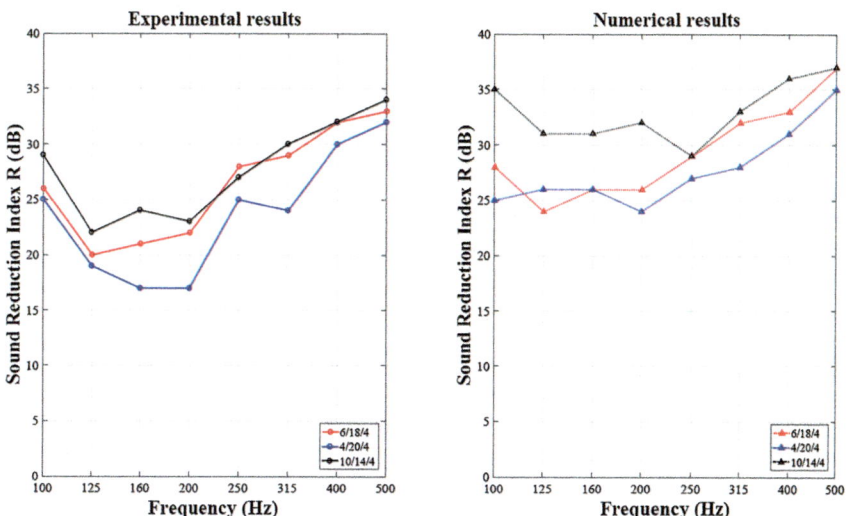

Fig. 5. Effect of the type of glazing systems in the sound reduction index

plate with a fluid gap is classically characterized by four domains. The first one is called the stiffness controlled region which is located at the range of frequencies below the first resonant frequency of the structure. In this domain, the system with higher mass per unit area has the higher R. This behavior is confirmed in Fig. 5. The 10/14/4 which is the heaviest has the best acoustic performances and the 4/20/4 has the lowest R. The second region is the mass-air-mass resonant frequency. In this region, the sound reduction index is reduced due to the coupling of the two plates with the fluid gap. Differences between the experimental and numerical results can be due to the boundary conditions which have an important effect in the resonant frequency of the system. In fact, in the present numerical model the two plates are clamped on their edges, which is not perfectly representative of real boundary conditions.

4   Conclusion

The main objective of this work is to propose an efficient numerical tool based on the finite element method to evaluate the acoustic performances of windows in the exterior building facade. In a first approach, the window is modeled as two glazing plates with an argon gap. Finite elements results for three glazing configurations are obtained and compare to laboratory measurements showing a relatively good agreement between numerical and experimental sound reduction index. Due to the fact that various simplifications are considered in the model, some discrepancies, especially in the low frequency range, are observed. The future works will consist in improving the model by taking into account the elasticity of the window frame, the junction of the glass plates, the damping effect, etc. Improvements should also focus on the development of reduced order models to reduce the computing cost and be able to perform parametric analyses.

Acknowledgements. The authors would like to express their thanks to CODIFAB (Comité professionnel de développement des industries françaises de l'ameublement et du bois) for its financial support.

References

ASTM E90-09: Standard test method for laboratory measurement of airborne sound transmission loss of building paryitions and elements (2009)

Cops, A., Minten, M., Myncke, H.: Influence of the design of transmission rooms on the sound transmission loss of glass-intensity versus conventional method. Noise Control Eng. J. 121–129 (1987)

Dijckmans, A., Vermeir, G.: A wave based model to predict the niche effect on sound transmission loss of single and double walls. Acta Acust. United Acust. **98**, 111–119 (2012)

Kihlman, T., Nilsson, A.: The effects of some laboratory designs and mounting conditions on reduction index measurements. J. Sound Vib. **24**, 349–364 (1972)

Larbi, W., Deü, J.F., Ohayon, R.: Vibroacoustic analysis of double-wall sandwich panels with viscoelastic core. Comput. Struct. **174**, 92–103 (2016)

Løvholt, F., Norèn-Cosgriff, K., Madshus, C., Ellingsen, S.E.: Simulating low frequency sound transmission through walls and windows by a two-way coupled fluid structure interaction model. J. Sound Vib. **396**, 203–216 (2017)

Miskinis, K., Dikavicius, V., Bliudzius, R., Banionis, K.: Comparison of sound insulation of windows with double glass units. Appl. Acoust. **92**, 42–46 (2015)

Ruggeri, P., Peron, F., Granzotto, N., Bonfiglio, P.: A combined experimental and analytical approach for the simulation of the sound transmission loss of multilayer glazing systems. Energy Procedia **78**, 146–151 (2015)

Sakuma, T., Adachi, K., Yasuda, Y.: Numerical investigation of the niche effect in sound insulation measurement. In: 40th International Congress and Exposition on Noise Control Engineering 2011, INTER-NOISE 2011, vol. 3, pp. 2272–2279 (2011)

Schroeder, M.R.: The "'Schroeder frequency'" revisited. J. Acoust. Soc. Am. **99**, 3240–3241 (1996)

Standard ISO 10140: Acoustics - laboratory measurement of sound insulation of building elements: Part 1: Application rules for specific products; Part 2: Measurement of airborne sound insulation; Part 4: Measurement procedures and requirements; Part 5: Requirements for test facilities and equipment (2010)

Efficient Cultural Algorithm for Structural Damage Detection Problem Based on Modal Data

Najeh Ben Guedria[1,3](✉) and Hichem Hassine[2,3]

[1] Laboratory of Mechanic of Sousse (LMS), Sousse, Tunisia
najeh.benguedria@istls.rnu.tn
[2] Laboratory of Mechanic, Modeling and Production (LA2MP), Sfax, Tunisia
hassinehichem@yahoo.fr
[3] Higher Institute of Transportation and Logistics of Sousse,
University of Sousse, Sousse, Tunisia

Abstract. This paper aims to present an efficient cultural algorithm to solve the target optimization problem of vibration-based damage detection. The modal flexibility error residual is employed as objective function to be minimized. Cultural algorithms are inspired from the cultural evolutionary process in nature and use social intelligence to solve optimization problems. A cultural algorithm is composed of a belief space which consists of different knowledge sources, a population space and a set of communication protocols that enables interaction of these two spaces. Cultural algorithm offers powerful tools to solve various optimization problems as a result of its robustness as well as computation effectiveness. In this work, cultural algorithm is applied to generate solutions using three knowledge sources namely situational knowledge, normative knowledge, and domain knowledge. The core idea of using domain knowledge source is to speed up the convergence of the algorithm and thus, reducing its computational cost. The performance of the proposed algorithm is demonstrated through a numerical example, with different damage scenarios and noise levels. Comparison of the proposed algorithm with other basic and state-of-the-art algorithms reveals its superiority in accurately detecting the sites and the extents of structure damages in spite of contaminated vibration data by noise.

Keywords: Structural damage detection · Flexibility matrix
Cultural algorithm · Cantilevered beam

1 Introduction

Damage might result in early aging of structural systems and even prompt dreadful effects. These awful consequences have encouraged researchers and professionals to formulate, implement and employ several non-destructive techniques for damage identification and structural health monitoring. Amid these kinds of techniques, vibration-based damage detection methods are often employed regarding early diagnosis of structural damage in civil, mechanical and aerospace engineering fields. Vibration-based damage detection strategies take into account numerous kinds of data:

© Springer Nature Switzerland AG 2019
T. Fakhfakh et al. (Eds.): ICAV 2018, ACM 13, pp. 204–217, 2019.
https://doi.org/10.1007/978-3-319-94616-0_21

modal parameters (Khoo et al. 2004; Tomaszewska 2010), time series (Cacciola et al. 2011), frequency responses (Santos et al. 2005), and consequently require diverse mathematical formulations and computational algorithms for solving the corresponding inverse problem (Ren and Chen 2010; Na et al. 2011).

The fundamental concept of vibration-based damage detection methods is that damage trigger modifications within structural physical properties (mass, stiffness and damping) in the damaged sites. These local modifications will entirely be mirrored throughout modal parameters i.e. frequencies, mode shapes as well as modal damping or even additional dynamic characteristics of the structural, which are subsequently measured and employed to identify location as well as the extent of the damage. Almost, all these methods take into consideration modal analysis and depend on the general framework of a finite element model (Doebling et al. 1998). Thus, based on this concept, the damage detection problem is transformed into an optimization problem where a set of decision parameters, which describe sites and extend of damage, is sought in order to minimize an objective function established by minimizing the discrepancies between the measured data, obtained from modal testing on the assumed damaged structure, and those obtained from the finite element model. Objective functions are, generally, expressed using natural frequencies (Hasan 1995), natural frequencies and mode shapes (Maeck and de Roeck 2003; Kim et al. 2003), mode shapes curvatures (Pandey et al. 1991; Parloo et al. 2003), modal strain energy (Cornwell et al. 1999; Alvandi and Cremona 2006) or the flexibility matrix (Park et al. 1998; Stutz et al. 2005).

Damage detection problems that belong to class of inverse problems are generally understood to be a bounded nonlinear optimization problem with non-convex and multimodal objective functions (Gonçalves and de Cursi 2001; Lopez et al. 2011). To solve this problem, two types of techniques tend to be primarily used that are the gradient based numerical optimization methods and metaheuristic algorithms. Because gradient based methods usually do not ensure convergence to a unique optimum solution, due to their dependence on the quality of the starting point, and result in local minima, metaheuristic algorithms have lately gained substantial interest. Metaheuristic algorithms are widespread since they are free-gradient techniques, possess a global convergence behavior and they mimic a number of particular strategies taken from nature, social behavior, physical laws and so on. Amongst all the heuristic algorithms, genetic algorithm (GA), neural network (NN) techniques, particle swarm optimization (PSO), ant colony optimization (ACO), evolutionary strategy (ES), differential evolution (DE) algorithms, artificial bee colony (ABC) and harmony search (HS) algorithm have acquired growing interest.

Mares and Surace (1996) developed a procedure, based on GA and the residual force method, to simultaneously locate and quantify structural damage from measured natural frequencies or mode shapes. Friswell et al. (1998) used genetic and eigensensitivity algorithms to identify the location and extent of structural damage, respectively, by minimizing the output error for natural frequencies. Many other authors solved structural damage detection problem using hybrid approach based on GA (Yi and Liu 2001; Chou and Ghaboussi 2001; Guo and Li 2009).

Begambre and Laier (2009) developed a hybrid particle swarm optimization–simplex algorithm (PSOS) for structural damage identification using frequency domain data. The proposed procedure showed high accuracy in noise-polluted scenarios and allowed for the proper identification of damaged elements in the truss and the beam studied. Then, based on PSO algorithm, other methodologies were presented to solve this type of problems (Fallahian and Seyedpoor 2010; Seyedpoor 2011; Kanovic et al. 2011; Kang et al. 2012; Mohan et al. 2013).

Stutz et al. (2015) presented an approach built on the flexibility matrix and DE algorithm for structural damage identification. Ding et al. (2016) proposed an approach for structural damage detection using the artificial bee colony (ABC) algorithm with hybrid search strategy based on modal data. Miguel et al. (2012) presented an approach for structural damage detection in ambient vibration context. This method combines stochastic system identification and the evolutionary harmony search (HS) algorithm.

Despite the fact that all the above-mentioned methods have generally reached satisfactory results in solving the structural damage detection problem, the hunt for acquiring more effective procedures remains on the rise, in order to handle the expanding complexities of engineering optimization problems arising from different fields. Within this context, cultural algorithms (CA), introduced by Reynolds (1994), have recently attracted increasing attention than most other metaheuristic techniques in several fields of science and engineering optimization. CA offers powerful tools to solve various optimization problems as a result of its robustness as well as computation effectiveness (Ali et al. 2016; Reyes et al. 2010). CA can be stated as class of computational models provided by the observation of cultural evolution process in nature. It can be defined as an evolutionary model that composed of a population space, a belief space and a set of communication protocols that enables interaction of these two spaces. A wide bibliographical revision revealed that CA is one of the most promising methods when it was employed to solve a test suite of 25 large-scale benchmark functions as well as a set of real-world problems (Ali et al. 2016). Also, it has been successfully applied for various engineering problems such as job shop scheduling problems (Becerra and Coello 2005a, b; Ho et al. 2004), structural design problems (Ali and Awad 2014), fuzzy-PID control (Wang et al. 2015), power system stabilizer (Khodabakhshian and Hemmati 2013), urban public transportation (Reyes et al. 2010), and so forth. Nevertheless, to the best of our knowledge, CA has been never employed to solve the damage detection problems. Therefore, in this paper we propose pertaining to the first time, to utilize an efficient CA for the structural damage identification problems. In this study, the domain knowledge source is appended to the most used knowledge sources, which are situational knowledge and normative knowledge, to form the belief space. Here, the domain knowledge source is used as a repair operator to avoid detection of elements with low damage level. This will accelerate the convergence of the algorithm and thus, reduce computational cost. In addition, the objective function is expressed based on modal flexibility matrix, which is a very useful as a damage indicator mainly when used data are truncated.

The remainder of this paper is organized as follows. Section 2 presents the mathematical formulation of the inverse damage detection problem built on the flexibility matrix. Section 3 provides an overall description of cultural algorithms. Section 4 details the proposed cultural algorithm used for the structural damage detection problem. Section 5 discusses the results of the numerical simulations, and finally the conclusion of the paper is summarized in Sect. 6.

2 Problem Formulation

2.1 Parameterization of Damage

The basic theory of vibration-based damage detection methods is that damage, as a combination of different failure modes, trigger changes in the structural physical properties (mass, stiffness and damping). These modifications will entirely be mirrored throughout spectral parameters i.e. frequencies, mode shapes as well as modal damping or even additional dynamic characteristics of the structural. A practical technique to represent damage is to consider it as a reduction in the stiffness properties of the structure. Thus, we can assume that structural damage can be quantified as a reduction of Young's modulus in each finite element.

Based on these assumptions, damage is modeled through an elemental stiffness reduction index α_i, which enables the preservation of the original structural connectivity. In the context of discretized finite elements, the global stiffness matrix of the damaged structure is formulated as a summation of damaged and undamaged element stiffness matrices as:

$$[\mathbf{K}] = \sum_{i=1}^{Nd} (1 - \alpha_i)[k_e]_i \tag{1}$$

In this equation, Nd is the total number of elements of the structure, and $[\mathbf{K}]$ is the global stiffness matrix, which is assembled from the element stiffness matrices $[k_e]_i$. The index α_i is always between 0 and 1, where 0 signifies no damage in the element and 1 means that the element loses its stiffness entirely.

2.2 Objective Function Based on the Modal Data

In this paper, the problem of detecting damage is formulated as a bounded nonlinear optimization problem, where a set of damage indices α_i is sought. The objective function, to be minimized, represents the discrepancies between the measured data, obtained from modal testing on the assumed damaged structure, and data obtained from the numerical model. The optimization problem can be formulated as:

$$\begin{aligned} &Find: \boldsymbol{\alpha}^T = \{ \alpha_1 \quad \alpha_2 \quad \dots \quad \alpha_{Nel} \}, \\ &Minimize: f(\boldsymbol{\alpha}), \\ &Subject\,to: \boldsymbol{\alpha}^l \leq \boldsymbol{\alpha} \leq \boldsymbol{\alpha}^u. \end{aligned} \tag{2}$$

$f(\boldsymbol{\alpha})$ is the objective function, $\boldsymbol{\alpha}$ is a damage variable vector containing the locations and extents of Nd unknown damages; $\boldsymbol{\alpha}^l$ and $\boldsymbol{\alpha}^u$ are the lower and upper bounds of the damage vector.

In the specialized literature, different objective functions have been formulated based on the natural frequencies, natural frequencies and mode shapes, mode shapes curvatures, modal strain energy or the modal flexibility matrix. Among all these damage indicators, the flexibility matrix is considered as the most promising indicator. The flexibility matrix $[\mathbf{F}]$ is related to the modal data as follows:

$$[\mathbf{F}] = [\mathbf{\Phi}][\mathbf{\Omega}]^{-1}[\mathbf{\Phi}]^T = \sum_{i=1}^{n} \frac{1}{\omega_i^2} \phi_i \phi_i^T \tag{3}$$

where $[\mathbf{\Phi}] = [\phi_1, \phi_2, \ldots, \phi_n]$ is the mass-normalized mode shape matrix and $[\mathbf{\Omega}] = diag[\omega_1^2, \omega_2^2, \ldots, \omega_n^2]$ represents the diagonal matrix of squared natural frequencies.

As can be mentioned from Eq. (3), the flexibility matrix is highly affected by the lowest natural frequencies, since the modal contribution to the flexibility matrix decreases as frequency increases, and therefore the flexibility promptly converges to a good approximation with only few lower vibration modes. Such characteristics make the flexibility matrix very useful as a damage indicator mainly when data are truncated such as obtained by modal test.

Therefore, in this paper, the objective function of the optimization problem to identify locations and extents of damage is expressed as follows

$$f(\boldsymbol{\alpha}) = \left\| [\mathbf{F}]_E - [\mathbf{F}(\boldsymbol{\alpha})]_D \right\|_{Fro}^2 \tag{4}$$

Where $[\mathbf{F}]_E$ is the modal flexibility matrix from the experimental results, $[\mathbf{F}(\boldsymbol{\alpha})]_D$ is the modal flexibility matrix calculated from the numerical model with the stiffness reduction index and $\| \ \|_{Fro}^2$ indicates the Frobenius norm of the residual matrix.

3   Cultural Algorithm Principle

Cultural algorithm (CA) is a stochastic optimization technique, firstly proposed by Reynolds (1994, 2003), which derived from the cultural evolution process in nature. CA consists of three components: the population space, the belief space and a set of communication protocols. The population space, as in all evolutionary algorithms, is a set of individuals employed as possible candidate solutions to the optimization problem. It can be evolved by any population-based algorithms including genetic algorithms (Michalewicz 1994), evolution strategies (Back 1993), and evolutionary programming (Kim et al. 1995; Yao and Liu 1996). The belief space stores the cultural knowledge sources, acquired during the evolutionary process by selected members of the population space, which are used to influence the search of individuals in the population. In the basic CA, there are mainly five types of knowledge sources (Peng and Reynolds 2004), including normative, topographical, domain, situational and history knowledge sources. The communication protocols provide the means that these knowledge sources can be exchanged between these two spaces. As shows Fig. 1, in CA individuals are evaluated by the performance function to represent the problem-solving experience. And then, some individuals from the population space are chosen, through the acceptance function, to adjust the knowledge in the belief space. The knowledge is used to guide the inheritance of individuals by means of the influence function. The selection function is charged to select the offspring for the next generation.

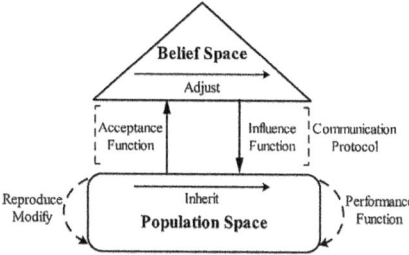

Fig. 1. Overview of cultural algorithm.

4 Proposed Cultural Algorithm for Damage Detection Problems

As described in the previous section, CA model consists of a population space, belief space and a set of communication protocols to adjust knowledge sources and to evolve future generations. Here, the population space is modeled by the evolutionary programming approach. The Belief space includes some knowledge sources that are suitable to the optimization problems under study here. In this paper, situational, normative and domain knowledge sources are considered. The communication protocols include an acceptance function and an influence one. The former is charged to choose a number of individuals, from the population space, that are among the top performers with respect to the optimization criteria. The latter, which guides evolution in the population space, will be in turn controlled through only the situational and the normative knowledge sources residing in the belief space. These two knowledge sources will be dynamically adjusted, during the optimization process, via leaders of the population space. In contrast, there is no updating for the domain knowledge source. In this paper, domain knowledge source is characterized by a single rule incorporating problem specific knowledge. That is, this knowledge source is considered as a repair operator to modify individuals during optimization process. In addition, it is well known that the influence function has a high impact on the success of cultural algorithm. In Reynolds (1997), proposed four influence function to update positions. Here, an efficient function, based on situational and normative knowledge sources, is used to guide the population in better search directions and to enhance the quality of generated solutions. Likewise, diversity of the population will be maintained due to heterogeneous knowledge sources used in the Belief space.

The procedure of the proposed cultural algorithm for a solution of a damage detection optimization problem is described as follows:

Step 1: *Define the optimization problem and initialize the optimization parameters*

- Define the optimization problem as Eq. (2), number of design variables (N_d) and limits of design variables ($\mathbf{L}_b$, $\mathbf{U}_b$).

- Initialize the population size (Np) and the number of maximum iterations $(t_{\max})$.

Step 2: *Initialize the population*

- Generate the initial population according to the population size and number of design variables. For the CA, initial location of the jth particle is given as: $\mathbf{x}_j^0 = \mathbf{L}_b + rand(\mathbf{U}_b - \mathbf{L}_b)$

Step 3: *Create and initialize the belief space*
Step 4: *Repeat until the stopping criteria is met*: $(t = t_{\max})$

- Update iteration counter $(t = t + 1)$,

Step 4.1: *Evaluation*

- Evaluate each individual in population space.
- Select the top $p\%$ of individuals based on the fitness values to adjust belief space.

Step 4.2: *Adjust Belief space*

- Adjust the situational knowledge components using accepted individuals.
- Adjust the normative knowledge components using accepted individuals.

Step 4.3: *Influence population*

- Update the position of each individual by influence function.
- Apply repair operator (with domain knowledge).

5 Numerical Simulation

In order to investigate the effectiveness and robustness of the proposed cultural algorithm for structural damage detection a cantilevered beam, selected from (Guo and Li 2009), is used. The parameters of the beam are as follows: the length 1 m, the height 0.01 m, the width 0.01 m, Young's modulus 210 GPa, and mass density 7860 kg/m³. The beam is discretized into 20 Euler-Bernoulli beam elements. Each element has two nodes and each node has two degrees of freedom: a vertical translation along the axis y and a rotation around the axis z perpendicular to the plan xy, as shown in Fig. 2. The first six normalized frequencies of the beam are listed in Table 1.

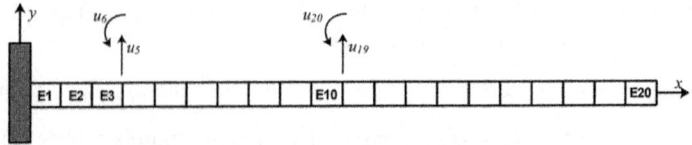

Fig. 2. Cantilevered beam with element numbers.

Table 1. First six frequencies of the cantilevered beam (Hz).

Mode	Undamaged case	Damage case 1	Damage case 2
1	8.3498	8.3498	8.3162
2	52.3277	52.3275	50.3342
3	146.5212	146.5167	140.8189
4	287.1364	287.1055	281.1916
5	474.7077	474.5817	462.2707
6	709.2767	708.8977	696.7403

The performance of proposed CA is compared with four other basic and state-of-the-art algorithms which are the standard PSO, a real-coded ant colony optimization (RCACO) algorithm, a differential evolution (DE) algorithm and a real-coded genetic algorithm (RCGA). The parameters used for PSO are recommended in (Shi and Eberhart 1998). The Inertia Weight w decreases linearly from 0.9 to 0.4 with respect to iterations counter. The maximum velocity V_{max} and minimum velocity V_{min} are set at half value of the upper bound and lower bound, respectively. The acceleration constants c_1 and c_2 are both 2.0. For DE algorithm, the parameters are suggested in (Storn and Price 1997). In this paper, the control parameters are set as F = 0.5 and Cr = 0.9. For RCGA, the uniform crossover and random mutation operators are adopted. The crossover probability is $pc = 0.8$, and the mutation probability is $pm = 0.2$. For RCACO, intensification factor is $q = 0.5$ and the deviation-distance ratio is $\xi = 0.3$, as suggested in (Fu et al. 2013). Meanwhile, for all algorithms, the maximum number of generation is set to 200 and the population size Np is 50. All employed algorithms and the codes of the finite element analysis for the beam and plate are coded in MATLAB programming software and the simulations and numerical solutions are run on an Intel Core i5-3337U 1.8 GHz Personal Computer with 6 GB of Random Access Memory (RAM) under a 64-bits Windows operating system.

Case Study 1: Identification of a Single Damage
First of all, single damage identification is conducted for the beam. A local damage is simulated by a reduction of 25% in Young's modulus in the 20*th* element, that means $\alpha_{20} = 0.25$. The first six natural frequencies of the damaged beam are shown in Table 1. As can be noted, damage on the 20*th* element has very low influence on the first three frequencies. This is leads the detection of location and extent of damage more difficult. To compare the performance of the proposed CA and its rivals, which are PSO, DE, RCGA and RCACO, they are applied to damage detection case 1, using only the first three frequencies and their associated mode shapes. The evolution process of the logarithmic best fitness values of the five techniques are shown in Fig. 3(a). From this figure, it is observed that the objective function value from the present method is much closer to zero than those from the other optimizers, which imply that the proposed CA has converged to the optimal solution. In contrast, DE, RCGA and RCACO have converged to a near optimal solution, as shown in Fig. 3(b). Moreover, it can be noted that PSO algorithm fails to detect neither the location nor the extent of the simulated damage. Hence, it can be concluded, from this example, that the proposed algorithm offers better results than its rivals in terms of accuracy and convergence speed.

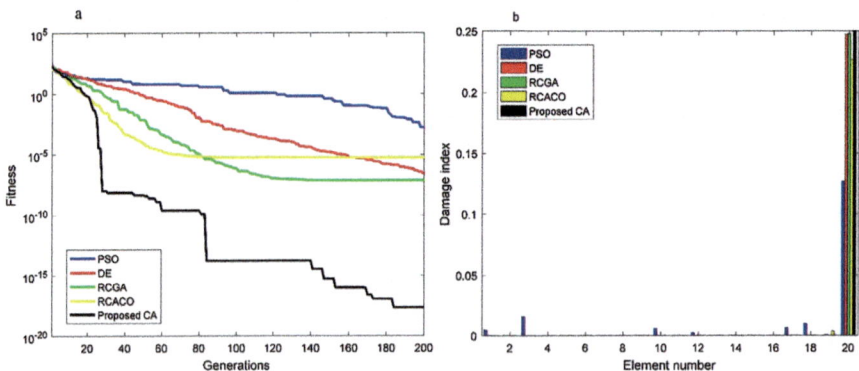

Fig. 3. Evolutionary process in case 1 of the beam: (a) the best fitness; (b) comparison of damage detection results in case 1 of the beam.

Case Study 2: Identification of Multiple Damages

The second damage case assumes that elements 12, 13 and 15 of the beam have 20%, 30% and 30% of stiffness reduction, respectively (Guo and Li 2009). That means $\alpha_{12} = 0.20$ and $\alpha_{13} = \alpha_{15} = 0.30$. The algorithms PSO, DE, RCGA, RCACO and the proposed CA are used to optimize this damage detection problem instance. Figure 4(a) illustrates the evolution processes of logarithmic best fitness values for the five employed techniques. The evolution process of elements damage indices using the proposed algorithm is depicted in Fig. 4(b). Note that the proposed CA have a good converging behavior and quickly reaches the true solution thanks to the perfect balance between exploration and exploitation during search process. Damage detection results of different algorithms are shown in Fig. 5(a). Results indicate that proposed CA is the best compared to others used optimizers in terms of the best solution. In fact, as can be observed, PSO, DE, RCGA and RCACO has detected approximately location and extent of damage. Moreover, for PSO, some false alarm elements but they are not the actual damaged elements (e.g. the 18*th*, 19*th* and 20*th* elements) are also identified.

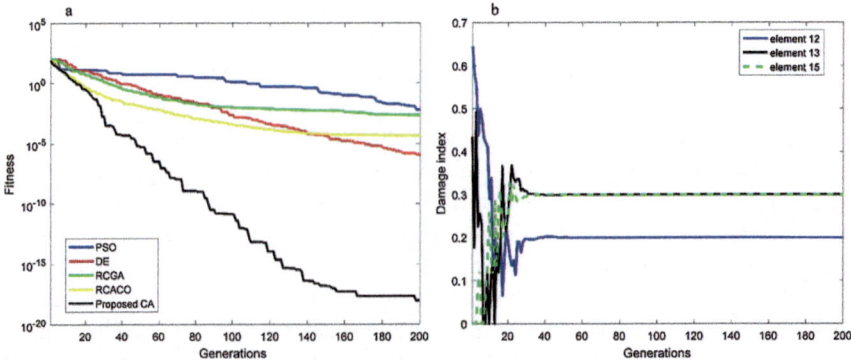

Fig. 4. Evolutionary process in case 2 of the beam: (a) the best fitness; (b) damage index.

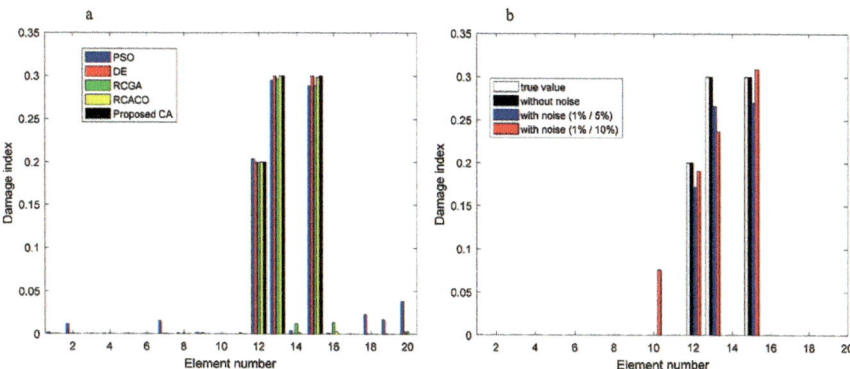

Fig. 5. Damage detection results in case 2 of the beam: (a) comparison of damage detection results without noise; (b) damage detection results of proposed CA.

Case Study 3: Effect of Measurement Noise

In practice, the measured frequencies and mode shapes are usually polluted by noise. The existence of the noise may have a negative influence on the accuracy of damage identification. In the numerical simulation, the noise on the frequency and mode shapes is simulated as follows:

$$f_i^{noise} = \left(1 + \eta_f(2rand[0,1] - 1)\right)f_i \qquad (5)$$

$$\Psi_{ij}^{noise} = (1 + \eta_m(2rand[0,1] - 1))|\Psi_{ij}| \qquad (6)$$

where f_i^{noise} is the ith frequency and Ψ_{ij}^{noise} is the jth component of the ith mode shape vector, both are contaminated by noise. η_f and η_m are the noise levels.

The last damage case is re-examined with two cases of noise levels, $\left(\eta_f = 1\%, \eta_m = 5\%\right)$ and $\left(\eta_f = 1\%, \eta_m = 10\%\right)$, on frequencies and mode shapes. In this case, the first six frequencies and the six first mode shapes are used. The identified results are illustrated in Fig. 5(b). For the first case of noise level, the presented method has detected exactly the location of damaged elements with the damage index $\alpha_{12} = 0.1721$, $\alpha_{13} = 0.2665$ and $\alpha_{15} = 0.2702$. For the second case of noise level, the proposed CA has detected the exact location of damaged elements with damage index relatively close to the true values ($\alpha_{12} = 0.1909$, $\alpha_{13} = 0.2371$ and $\alpha_{15} = 0.3090$), as depicted in Fig. 5(b). However, a false alarm element (the 10th element) with a low damage index $\alpha_{10} = 0.00759$, is also identified. From results, it can be concluded that even using frequencies and mode shapes contaminated by noise, the proposed CA still can locate exactly damaged elements and identify acceptably the damage extent. This indicates that the present method is relatively insensitive to noise in the measurement.

6 Conclusion

In this paper, a cultural algorithm is proposed, for the first time, to solve the damage detection problem using modal data. The objective function is expressed as the modal flexibility error residual. Modal flexibility is chosen due to its high sensitivity to damage. In the proposed cultural algorithm, the population space is modeled by the evolutionary programming approach. Three knowledge sources, which are situational, normative and domain knowledge sources, are included in the belief space. Domain knowledge source is applied as a repair operator whose purpose is to convert damaged elements, with very low stiffness reduction factor, into healthy ones. The core idea is to speed up the convergence of the algorithm and thus, reducing its computational cost. In order to verify the effectiveness of the proposed algorithm for structural damage detection, two illustrative test examples selected from the literature are considered. Numerical results demonstrate that the presented algorithm can identify exactly the locations and the extents of single and multiple damages successfully even under measurement noise. Furthermore, the proposed cultural algorithm performs the better convergence speed and higher accuracy than PSO, RCGA, RCACO and DE in all of the numerical examples.

References

Khoo, L.M., Mantena, P.R., Jadhav, P.: Structural damage assessment using vibration modal analysis. Struct. Health Monit. **3**, 177–194 (2004)

Alvandi, A., Cremona, C.: Assessment of vibration-based damage identification techniques. J. Sound Vib. **292**, 179–202 (2006)

Pandey, A.K., Biswas, M., Samman, M.M.: Damage detection from changes in curvature mode shapes. J. Sound Vib. **145**, 321–332 (1991)

Ali, M.Z., Awad, N.H., Suganthan, P.N., Duwairi, R.M., Reynolds, R.G.: A novel hybrid Cultural Algorithms framework with trajectory-based search for global numerical optimization. Inf. Sci. (Ny) **334–335**, 219–249 (2016). https://doi.org/10.1016/j.ins.2015.11.032

Ali, M.Z., Awad, N.H.: A novel class of niche hybrid cultural algorithms for continuous engineering optimization. Inf. Sci. (N.Y.) **267**, 158–190 (2014). https://doi.org/10.1016/j.ins.2014.01.002

Tomaszewska, A.: Influence of statistical errors on damage detection based on structural flexibility and mode shape curvature. Comput. Struct. **88**, 154–164 (2010)

Becerra, R.L., Coello, C.A.C.: A Cultural Algorithm for Solving the Job Shop Scheduling Problem, pp. 37–55. Springer, Berlin (2005). https://doi.org/10.1007/978-3-540-44511-1_3

Becerra, R.L., Coello, C.A.C.: A cultural algorithm for solving the job shop scheduling problem, pp. 37–55. Springer, Berlin (2005). https://doi.org/10.1007/978-3-540-44511-1_3

Begambre, O., Laier, J.E.: A hybrid particle swarm optimization-simplex algorithm (PSOS) for structural damage identification. Adv. Eng. Softw. **40**, 883–891 (2009). https://doi.org/10.1016/j.advengsoft.2009.01.004

Mares, C., Surace, C.: An application of genetic algorithms to identify damage in elastic structures. J. Sound Vib. **195**(2), 195–215 (1996)

Na, C., Kim, S.-P., Kwak, H.-G.: Structural damage evaluation using genetic algorithm. J. Sound Vib. **330**, 2772–2783 (2011)

Ding, Z.H., Huang, M., Lu, Z.R.: Structural damage detection using artificial bee colony algorithm with hybrid search strategy. Swarm Evol. Comput. **28**, 1–13 (2016). https://doi.org/10.1016/j.swevo.2015.10.010

Parloo, E., Guillaume, P., Overmeire, M.V.: Damage assessment using mode shape sensitivities. Mech. Syst. Signal Process. **17**, 499–518 (2003)

Guo, H.Y., Li, Z.L.: A two-stage method to identify structural damage sites and extents by using evidence theory and micro-search genetic algorithm. Mech. Syst. Signal Process. **23**, 769–782 (2009)

Gonçalves, M.B., de Cursi, J.E.S.: Parameter estimation in a trip model by random perturbation of a descent method. Transp. Res. Rec. **35**, 137–161 (2001). https://doi.org/10.1016/S0191-2615(99)00043-0

Ho, N.B., Tay, J.C.: GENACE: an efficient cultural algorithm for solving the flexible job-shop problem. In: Proceedings of the 2004 Congress Evolutionary Computation (IEEE Cat. No. 04TH8753), pp. 1759–1766. IEEE. https://doi.org/10.1109/cec.2004.1331108

Santos, J.V.A., Soares, C.M.M., Soares, C.A.M., Maia, N.M.M.: Structural damage identification in laminated structures using FRF data. Compos. Struct. **67**, 239–249 (2005)

Maeck, J., de Roeck, G.: Damage assessment using vibration analysis on the Z24-bridge. Mech. Syst. Signal Process. **17**, 133–142 (2003)

Kim, J.T., Ryu, Y.S., Cho, H.M., Stubbs, N.: Damage identification in beam-type structures: frequency-based method vs. mode-shape-based method. Eng. Struct. **25**(1), 57–67 (2003)

Chou, J.H., Ghaboussi, J.: Genetic algorithm in structural damage detection. Comput. Struct. **79**, 1335–1353 (2001)

Stutz, K.L.T., Castello, D.A., Rochinha, F.A.: A flexibility-based continuum damage identification approach. J. Sound Vib. **279**, 641–667 (2005)

Kang, F., Li, J., Xu, Q.: Damage detection based on improved particle swarm optimization using vibration data. Appl. Soft Comput. **12**, 2329–2335 (2012). https://doi.org/10.1016/j.asoc.2012.03.050

Khodabakhshian, A., Hemmati, R.: Multi-machine power system stabilizer design by using cultural algorithms. Int. J. Electr. Power Energy Syst. **44**, 571–580 (2013). https://doi.org/10.1016/j.ijepes.2012.07.049

Lopez, R.H., de Cursi, J.E.S., Lemosse, D.: Approximating the probability density function of the optimal point of an optimization problem. Eng. Optim. **43**, 281–303 (2011). https://doi.org/10.1080/0305215X.2010.489607

Friswell, M.I., Penny, J.E.T., Garvey, S.D.: A combined genetic and eigensensitivity algorithm for the location of damage in structures. Comput. Struct. **69**, 548–556 (1998)

Miguel, L.F.F., Miguel, L.F.F., Kaminski, J., Riera, J.D.: Damage detection under ambient vibration by harmony search algorithm. Expert Syst. Appl. **39**, 9704–9714 (2012). https://doi.org/10.1016/j.eswa.2012.02.147

Mohan, S.C., Maiti, D.K., Maity, D.: Structural damage assessment using FRF employing particle swarm optimization. Appl. Math. Comput. **219**, 10387–10400 (2013). https://doi.org/10.1016/j.amc.2013.04.016

Cacciola, P., Maugeri, N., Muscolino, G.: Structural identification through the measure of deterministic and stochastic time-domain dynamic response. Comput. Struct. **89**, 1812–1819 (2011)

Cornwell, P.J., Doebling, S.W., Farrar, C.R.: Application of the strain energy damage detection method to plate-like structures. J. Sound Vib. **224**, 359–374 (1999)

Reyes, L.C., Zezzatti, C.A.O.O., Santillán, C.G., Hernández, P.H., Fuerte, M.V.: A cultural algorithm for the urban public transportation. LNCS (including subseries LNAI and LNBI), vol. 6077, pp. 135–142 (2010). https://doi.org/10.1007/978-3-642-13803-4_17

Reynolds, R.G.: An introduction to cultural algorithms. In: Proceedings of the Third Annual Conference on Evolutionary Programming, pp. 131–139. World Scientific, River Edge, New Jersey (1994)

Fallahian, S., Seyedpoor, S.M.: A two stage method for structural damage identification using an adaptive neuro-fuzzy inference system and particle swarm optimization. Asian J. Civil Eng. **11**, 797–810 (2010)

Seyedpoor, S.M.: Structural damage detection using a multi-stage particle swarm optimization. Adv. Struct. Eng. **14**, 533–549 (2011)

Stutz, L.T., Tenenbaum, R.A., Corrêa, R.A.P.: The differential evolution method applied to continuum damage identification via flexibility matrix. J. Sound Vib. **345**, 86–102 (2015). https://doi.org/10.1016/j.jsv.2015.01.049

Doebling, S.W., Farrar, C.R., Prime, M.B.: A summary review of vibration-based damage identification. Shock Vib. Digest **30**, 91–105 (1998)

Ren, W.X., Chen, H.B.: Finite element model updating in structural dynamics by using the response surface method. Eng. Struct. **32**, 2455–2465 (2010)

Hasan, W.: Crack detection from the variation of the eigenfrequencies of a beam on elastic foundation. Eng. Fract. Mech. **52**, 409–421 (1995)

Yi, W.J., Liu, X.: Damage diagnosis of structures by genetic algorithm. Chin. J. Eng. Mech. **18** (2), 64–71 (2001)

Wang, W., Song, Y., Xue, Y., Jin, H., Hou, J., Zhao, M.: An optimal vibration control strategy for a vehicle's active suspension based on improved cultural algorithm. Appl. Soft Comput. J. **28**, 167–174 (2015). https://doi.org/10.1016/j.asoc.2014.11.047

Kanovic, Z., Rapaic, M.R., Jelicic, Z.D.: Generalized particle swarm optimization algorithm – theoretical and empirical analysis with application in fault detection. Appl. Math. Comput. **217**, 10175–10186 (2011)

Park, K.C., Reich, G.W., Alvin, K.F.: Structural damage detection using localized flexibilities. J. Intell. Mater. Syst. Struct. 9, 911–919 (1998). https://doi.org/10.1177/1045389X9800901107

Michalewicz, Z.: Genetic Algorithms + Data Structures = Evolution Program, 2nd edn. Springer, Heidelberg (1994)

Back, T., Rudolph, G., Schwefel, H.-P.: Evolutionary programming and evolution strategies: similarities and differences. In: Proceedings of Evolutionary Programming, pp. 11–22 (1993)

Kim, J.-H., Jeon, J.-Y., Chae, H.-K., Koh, K.-I.: A novel evolutionary algorithm with fast convergence. In: Proceedings of 1995 IEEE International Conference on Evolutionary Computation, Australia, vol. 1, pp. 228–238. IEEE (1995). https://doi.org/10.1109/ICEC.1995.489150

Yao, X., Liu, Y.: Fast evolutionary programming. In: Proceedings of the Fifth Annual Conference on Evolutionary Programming (1996)

Peng, B., Reynolds, R.G.: Cultural algorithms: knowledge learning in dynamic environments. In: Proceedings of the 2004 Congress on Evolutionary Computation (IEEE Cat. No. 04TH8753), vol. 2, pp. 1751–1758 (2004). https://doi.org/10.1109/CEC.2004.1331107

Reynolds, R.G.: Knowledge-based self-adaptation in evolutionary programming using cultural algorithms. In: Proceedings of 1997 IEEE International Conference on Evolutionary Computation (ICEC 1997), pp. 71–76 (1997). https://doi.org/10.1109/ICEC.1997.592271

Shi, Y., Eberhart, R.: A modified particle swarm optimizer. In: 1998 IEEE International Conference on Evolutionary Computation Proceedings. IEEE World Congress on Computational Intelligence (Cat. No. 98TH8360), pp. 69–73. IEEE (1998). https://doi.org/10.1109/ICEC.1998.699146

Storn, R., Price, K.: Differential evolution - a simple and efficient heuristic for global optimization over continuous spaces. J. Glob. Optim. **11**, 341–359 (1997). https://doi.org/10.1023/A:1008202821328

Fu, Y.Z., Lu, Z.R., Liu, J.K.: Damage identification in plates using finite element model updating in time domain. J. Sound Vib. **332**, 7018–7032 (2013). https://doi.org/10.1016/j.jsv.2013.08.028

Reynolds, R.G., Peng, B., Brewster, J.J.: Cultural swarms: knowledge-driven problem solving in social systems. In: SMC 2003 Conference Proceedings. 2003 IEEE International Conference on Systems, Man and Cybernetics. Conference Theme - System Security and Assurance (Cat. No. 03CH37483), vol. 4, pp. 3589–3594 (2003). https://doi.org/10.1109/ICSMC.2003.1244446

Reverse Engineering Techniques for Investigating the Vibro-Acoustics of Historical Bells

Vincent Debut[1,2(✉)], Miguel Carvalho[1,2], Filipe Soares[2], and José Antunes[1,2]

[1] Instituto de Etnomusicologia, Centro de Estudos em Música e Dança, Faculdade de Ciências Sociais e Humanas, Universidade Nova de Lisboa, Av. de Berna 26C, 1069-061 Lisbon, Portugal
vincentdebut@fcsh.unl.pt,
[2] Centro de Ciências e Tecnologias Nucleares, Instituto Superior Técnico, Universidade de Lisboa, Estrada Nacional 10 (Km 139.7), 2695-066 Bobadela LRS, Portugal
{vincentdebut,miguel.carvalho,
jantunes}@ctn.tecnico.ulisboa.pt,
filipedcsoares@gmail.com

Abstract. In this paper, we present an effective methodology for assessing the vibrational properties of real-life bells, using reverse engineering techniques. When struck by a clapper, bells vibrate in rather complicated ways, which result in complex sounds. Typically, to obtain pleasant sounds, bell founders tune the first five partials (vibration modes) according to specific frequency ratios, while also trying to control the amount of beats, which also affect the musical quality. In practice, many musically important aspects are strongly related to fine details of the bell geometry. In this work, we use scanning imaging technology to obtain precise 3D geometry bell data, and then assess the bell tuning features by combining the acquired 3D geometrical data with Finite Element modal computations. Our numerical results are compared with experimentally identified bell modes, attesting the feasibility and effectiveness of the proposed approach. This analysis strategy is particularly suited in the context of cultural heritage preservation, by providing new and comprehensive ways to characterize and describe historical bells. Moreover, it can also shed light when addressing bell casting and tuning techniques throughout times.

Keywords: Vibro-acoustics · Reverse engineering · 3D scanning Modal analysis · Bells

1 Introduction

This work was motivated by the need to enrich the musicological knowledge on a unique collection of 102 historical cast bronze bells, stemming from the largest surviving 18th century carillons in Europe. Although originally concerned with the assessment of the bell geometry using 3D scan imaging technology, the objective of

© Springer Nature Switzerland AG 2019
T. Fakhfakh et al. (Eds.): ICAV 2018, ACM 13, pp. 218–226, 2019.
https://doi.org/10.1007/978-3-319-94616-0_22

this paper centers on the development of an effective framework for assessing the bell vibrational features, using reverse engineering techniques. Enabling easy, accurate and fast measurements of objects with complex shapes, surface scanning technology is becoming widely used for many applications, ranging from quality control in the automotive industry (Kuş 2009) to healthcare applications (Treleaven and Wells 2007), or to support the safeguarding of cultural heritage (Levoy et al. 2000). It has benefited from the increasing computational capabilities to manipulate large dataset alongside with the advances in computer graphics and, now 3D scanning can provide detailed models of intricate objects ready to use for analysis purposes. In this work, the reverse engineering process involves a 3D geometrical capture of the bell geometry by manual scanning, the construction of a structural model by means of Finite Elements, and finally modal computations in order to estimate the modal frequencies and mode shapes of the bell. In spite of the specificity of the mechanical system of interest here, the developed approach is definitely very attractive for dynamical assessment of existing structural components in industry.

It is well known that, when struck by a clapper, bells vibrate in rather complicated ways, which result in complex sounds (Fletcher and Rossing 1998). Moreover, due to the bells axi-symmetry, their so-called "partials" appear in modal pairs, with identical natural frequencies but different orientations (thus their orthogonality). However, due to some unavoidable slight deviations from the perfect axi-symmetry, this modal degeneracy disappears and leads to some audible beats in the radiated sound, which affect our perception of bell quality.

Typically, to obtain a pleasant sound, bell founders tune the first five partials according to the specific frequency ratios 0.5, 1.0, 1.2, 1.5 and 2.0 (referred to the second mode pair, conventionally designated as "fundamental"), while also trying to control the amount of beating. In practice, this is achieved in two main steps: (1) through the design of the bell moulds, which allows a rough pre-tuning, and (2) by removing some of the bell material after casting, using a tool-shop lathe in order to correct the slight tuning errors. According to the amount and location of the removed metal, the bell modal frequencies may increase or decrease. One then understands that all the musically important aspects are strongly related to fine details of the bell geometry.

In this work, we demonstrate the efficiency of combining precise 3D geometrical data obtained using scanning imaging technology with modal computations, in order to assess the vibrational behaviour of three-dimensional shell-like structures. The feasibility of the methodology is illustrated on an historical bell of the 18th century, from the Witlockx carillon of the Mafra National Palace, near Lisbon, Portugal (Fig. 1), for which a full vibratory assessment was achieved from a sophisticated experimental modal identification technique previously developed (Debut et al. 2016a). The comparison of the results attest the overall feasibility and effectiveness of the reverse engineering approach developed in this paper, in particular for the analysis and subsequent re-tuning of ill-tuned bells.

Fig. 1. Photograph of the historical bell from the Witlockx carillon of the Mafra National Palace – approximately 20 cm high.

2   3D Bell Scanning

The precise metrology of structures with complex shapes typically poses considerable problems. Modern 3D scan imaging technology, combining sophisticated techniques of computationally intensive image acquisition with processing algorithms, enables a comparatively fast and very precise geometrical assessment of intricate objects. Figure 2 shows one such commercial device, used in the present work, the ARTEC EVA 3D handheld scanner. The assembly consists of one projector and two cameras, and uses structured light as sensing technology to capture the fine details of the objects. It captures shapes with an overall dimensional tolerance of 0.5 mm, as well as image colours, which are used to generate a texture map on the 3D rendering model. Basically, the scanner takes a sequence of images with different patterns of LED light projected onto the surface of the object. Once acquired the raw scan data for the entire object, a series of different post-processes, namely scans alignments, registration, surface reconstruction and rendering, is run either automatically or interactively, in order to create a complete 3D final model from the raw point data.

Figure 3 illustrates three aspects of the scanned bell geometry: the 3D mesh, which results from the transformation of the point data into polygons, the rendering model, and after mapping colour data onto it. These figures show the level of detail achievable.

Using such a device, when compared with conventional laser point-metrology, an order of magnitude improvement can typically be achieved in dimensional precision, while spatial coverage resolution is improved by several orders of magnitude. Moreover, the geometry data can be obtained in a fraction of the time, but post-processing for high-quality full rendering can be lengthy depending on the object size and the model complexity.

Fig. 2. 3D image scanning of the historical bell with the ARTEC EVA 3D handled scanner

Fig. 3. 3D digital models of the historical bell. Polygons mesh (left), full-rendering model with texture (middle) and with colour (right)

3 Finite Element Modal Computations

The scanned bell geometry can be readily organized in a specific file form to input most finite element computer programs for subsequent structural analysis. Here, the Finite-Element package Cast3M (2017) developed by CEA (France) has been used to generate the structural model and carry out the modal computation as in our previous work (Debut et al. 2016b).

The bell bronze alloy was modelled with parameters $E = 100.3$ GPa, $\rho = 8542$ kg m^{-3} and $v = 0.34$. The generated mesh, displayed in Fig. 4, consists of 30,000 solid elements (4-node tetrahedral), which appears fine enough for accurate modal computations based on preliminary convergence tests. Modal computations were performed assuming the bell in free conditions, and ignoring damping, which is a legitimate assumption for this system. Energy dissipation, which occurs as a consequence of both internal losses and acoustic radiation, is actually very low for bells, typically in the range 0.01–0.1% (Debut et al. 2016a), so that the eigenfrequencies remain almost unchanged compared to the conservative case, and the mode shapes approximate closely real modes. Since losses by sound radiation remain the primary cause of dissipation for bells (Rossing 2000), damping strongly affect the sound radiated. The weakly damped bell vibration results in a slow sound decay, where the amplitude of each radiated partial decreases at a specific rate due to the frequency dependence of the sound radiation phenomena.

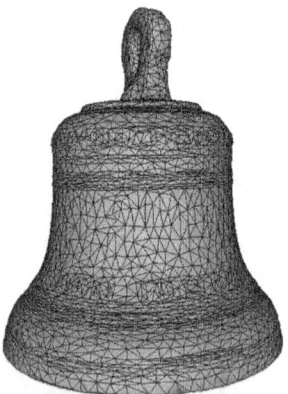

Fig. 4. Finite element mesh used for modal computations

The mode shapes ϕ_n and natural frequencies $f_n = \omega_n/(2\pi)$ were computed from the assembled mass and stiffness matrices $\mathbf{M}$ and $\mathbf{K}$, which stem from the standard eigen-computation $(\mathbf{K} - \omega_n^2\mathbf{M})\phi_n = \mathbf{0}$. The computed vibrational patterns and frequencies of the first five musically relevant modes are presented in Fig. 5. Note that although computed, rigid body modes, which correspond to translation and rotation motions of the undeformed structure, are not relevant here. As expected for typical axisymmetric structure, normal modes appear in degenerate pairs, so that one bell partial is constituted by a pair of modes. However, because slight asymmetries are present in the bell shape, one notices small differences in frequency between the two modal components of each partial, which lead to more or less intense beating when the bell is struck.

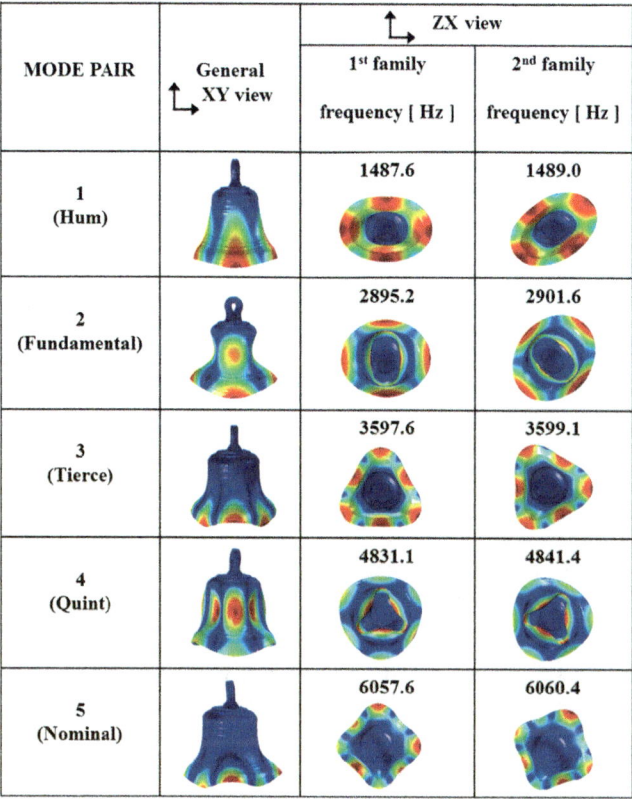

MODE PAIR	General XY view	ZX view	
		1st family frequency [Hz]	2nd family frequency [Hz]
1 (Hum)		1487.6	1489.0
2 (Fundamental)		2895.2	2901.6
3 (Tierce)		3597.6	3599.1
4 (Quint)		4831.1	4841.4
5 (Nominal)		6057.6	6060.4

Fig. 5. Mode shapes and modal frequencies of the relevant five mode pairs from FE modal computation of the scanned bell geometry

4 Experimental Modal Identification

For validation purposes, a full experimental modal identification of the bell, based on impact testing, was performed. The vibrational radial responses of the bell were recorded with three piezoelectric accelerometers glued on the outer rim of the bell, in the same horizontal plane, while impact excitations were performed on 32 locations regularly spaced near the rim – see Fig. 6. From the set of acquired impulse responses, modal identification was then performed using a custom developed program (Debut et al. 2016b) based on the Eigensystem Realization Algorithm (ERA) (Juang 1994), leading to estimate of the modal frequencies and modal damping values, as well as of the mode shapes.

The comparison of the FEM-computed and identified modal frequencies, reported in Table 1, confirms the effectiveness of the proposed approach. Once matching the modal frequency for the first mode, the errors of the modal frequencies predicted on the basis of the 3D scanned geometry and the finite element model proved consistently lower than 0.5%.

Fig. 6. Experimental test-rig for modal identification. Picture showing the bell suspended, with three accelerometers at the rim and the impact hammer

For the application of interest here, the internal tuning of the studied bell can be easily assessed, by calculating the frequency ratios of every partial relative to an arbitrary partial frequency (here the Fundamental). From the reported frequency ratios in Table 1, it appears that the founder achieved a satisfactorily tuning of the first two modes, while the tuning of the high-order modes remains less precise. One plausible speculation to explain this result is that Witlockx only attempted rigorous tuning of the Hum and Fundamental, being somehow aware of the decreasing sound intensity sensitivity of the ear toward very high frequencies (Campbell and Greated 1988).

Table 1. Computed and experimental modal frequencies of the bell

MODE PAIR	Modal frequencies (Hz) - Ratio f_n/f_2				Relative difference (%)
	FEM computations		Experimentally identified		
1 (Hum)	1487.6	0.51	1487.6	0.51	0
	1489.0	0.51	1490.8	0.52	0.12
2 (Fundamental)	2895.2	1.0	2891.8	1.0	0.12
	2901.6	1.0	2898.1	1.0	0.12
3 (Tierce)	3597.6	1.24	3593.8	1.24	0.11
	3599.1	1.24	3594.0	1.24	0.14
4 (Quint)	4831.1	1.67	4854.4	1.67	0.48
	4841.4	1.67	4855.9	1.68	0.30
5 (Nominal)	6057.6	2.09	6048.2	2.09	0.16
	6060.4	2.09	6060.8	2.09	0.01

Interestingly, this also reflects the difficulty of bell founders of casting and tuning small bells, which is an aspect of bell founding which is still challenging today (Lehr 2000). Finally, by comparing the modal frequencies between the doublet pair, one can notice that the beating frequency is rather large for the Fundamental, of approximately 6 Hz, which certainly has an important negative effect in terms of pitch perception and sound quality.

5 Conclusions

This paper presented a consistent and effective methodology for inferring the dynamical features of complex structures, based on scan data coupled to finite element computations. Application of the technique was presented for an historical bell from the 18th century, and convincing illustrative results were produced. The reverse engineering process involves the 3D geometrical capture of the bell geometry by manual scanning, the construction of a rendering model from the raw scan data, and finally, Finite Element modal computations in order to estimate its modal frequencies and mode shapes. As a result, an objective assessment of the tuning of this rare 18th century bell has been successfully achieved.

This analysis strategy, which can be applied in many different fields, seems particularly suited to support the preservation of cultural heritage. Besides the tuning assessment presented here, the high-resolution of the scans enables a precise, detailed and textured 3D image of the bell structure, for which no substantial information is today available. Besides documenting such a rare artefact, a careful analysis of the rendering model can also explain subtle details for the sound radiated, and might point toward other information to appreciate the practices and skills of the bell founder.

Acknowledgements. This work was supported by Santander-Totta/Universidade NOVA, through the project "Singing bronze: material sciences and acoustic engineering advanced techniques toward the preservation of the Mafra carillon bells". The authors thank the National Palace of Mafra for lending the studied bell, as well as Diogo Alves and Vasco Completo from NOVA University for scanning.

References

Campbell, M., Greated, C.: The Musician's Guide to Acoustics. Oxford University Press, Oxford (1988)

CAST3M: Finite Element Software. Commissariat à l'Energie Atomique et aux Energies Alternatives (CEA-Saclay, France) (2017). http://www-cast3m.cea.fr/

Debut, V., Carvalho, M., Antunes, J.: An objective approach for assessing the tuning properties of historical carillons. Appl. Acoust. **101**, 78–90 (2016a)

Debut, V., Carvalho, M., Figueiredo, E., Antunes, J., Silva, R.: The sound of bronze: virtual resurrection of a broken 13th century bell. J. Cult. Herit. **19**, 544–554 (2016b)

Fletcher, N.H., Rossing, T.D.: The Physics of Musical Instruments. Springer, New York (1998)

Juang, J.: Applied System Identification. PTR Prentice-Hall, New Jersey (1994)

Kuş, A.: Implementation of 3D optical scanning technology for automotive applications. Sensors **9**(3), 1967–1979 (2009)

Lehr, A.: The geometrical limits of the carillon bell. Acta Acust. **86**, 543–549 (2000)

Levoy, M., Pulli, K., Curless, B., Rusinkiewicz, S., Koller, D., Pereira, L., Ginzton, M., Anderson, S., Davis, J., Ginsberg, J., Shade, J., Fulk, D.: The digital michelangelo project: 3D scanning of large statues. In: Proceedings of the ACM SIGGRAPH Conference 2000 (2000)

Rossing, T.D.: Science of Percussion Instruments. World Scientific, Singapore (2000)

Treleaven, P., Wells, J.C.K.: 3D body scanning and healthcare applications. Computer **40**(7), 28–33 (2007)

Experimental Investigation of Normal/Lateral Excitation Direction Influence on the Dynamic Characteristics of Metal Mesh Isolator

Fares Mezghani[1,2(✉)], Alfonso Fernandez Del Rincon[2],
Mohamed Amine Ben Souf[1], Pablo Garcia Fernandez[2],
Fakher Chaari[1], Fernando Viadero Rueda[2], and Mohamed Haddar[1]

[1] Laboratory of Mechanics, Modeling and Production,
National School of Engineering of Sfax, University of Sfax, Sfax, Tunisia
fares.mezghani@gmail.com
[2] Laboratory of Structural and Mechanical Engineering, Superior Technical
School of Industrial Engineering and Telecommunications,
University of Cantabria, Santander, Spain

Abstract. Vibrations, considered one of the major problems in the engineering applications, are analyzed to predict their detrimental effects on the equipment and structures. The metal mesh isolator has become widely applied to mitigate the disturbing vibration due to its special production techniques. The metal mesh isolator is a kind of novel style porous damping material that is manufactured via a process of wire-drawing, weaving and compression molding. The influencing laws of the manufacturing parameters including the relative density and the working condition together with the excitation direction dependence should be taken into account in the characterization of the metallic wires material. In this paper, the mechanical properties of three models with different relative density will be investigated under different preload masses and for three acceleration levels. A number of experiments can be examined by changing the direction of excitation in order to describe the compression and non-compression molding direction effect on the dynamic behavior. A modal analysis is performed using the rational fraction polynomial method to determine the stiffness and the damping ratio from the measured transmissibility data. According to the reported experimental results, the major factor affecting the mechanical characteristics (stiffness and damping) is the sliding friction that exists at the contact-points between wires.

Keywords: Metal mesh isolator · Direction effect
Rational fraction polynomial method · Stiffness and damping ratio

1 Introduction

Metal mesh vibration isolator is a form of porous material manufactured from metallic wires. Owing to their outstanding properties including the significant damping capacity and excellent mechanical resistance, metal mesh isolator are suitable for many engineering applications. Metal mesh damper is originally used as metal mesh damper of inertial platform (Wang et al. 2004), damper for fuel nozzles in gas engines

© Springer Nature Switzerland AG 2019
T. Fakhfakh et al. (Eds.): ICAV 2018, ACM 13, pp. 227–234, 2019.
https://doi.org/10.1007/978-3-319-94616-0_23

(Ma et al. 2008) and sound absorption (Wu et al. 2009). Furthermore, the different shapes and structural parameters allow this material to be widely used as vibration isolators especially in the industrial and defense equipment, naval vessels, aircraft engines and other sensitive equipment (Di Massa et al. 2013; Chaudhuri and Kushwaha 2008). Therefore, a study on the properties of metal mesh dampers is required.

Ma et al. (2008) have studied the effects of structural parameters including the wire diameter, helix dimension, and relative density on the compressive properties of EMWM. Wang et al. (2010) have focused on the nonlinear static and dynamics properties of the MR damper and have experimentally studied the structure effects on the static properties. Zhang et al. (2013) have tested EMWM samples made from nickel based super alloy, and shown the relative density as one of the most important factors affecting the tangent modulus, Poisson's ratio, and damping of EMWM. Mezghani et al. (2017) have analyzed the relative density influence on the compression and dissipative behavior of the metal mesh isolator. Metal damper samples with different relative densities have been produced and subjected to quasi-static compression loading. Experimental analysis indicated that the porosity affects the stiffness but has an opposite effect regarding the loss factor. However, the previous research work on the metallic wires material has not focused on the excitation direction dependence, which exhibits a significant influence in the metal damper's performance.

In this paper, the mechanical properties of three models with different relative density will be investigated under different preload masses and for three acceleration levels. Furthermore, a significant number of experiments can be examined by changing the direction of excitation in order to describe the compression and non-compression molding direction effect on the dynamic behavior of the metallic wires mesh.

2   Experimental Apparatus

2.1   Metal Mesh Isolator Description

The metal mesh isolator essentially consists of a cushion of stainless steel, woven using a knitting machine, rolled and/or pressed into the required geometric shape via a press mold in order to achieve the desired geometric shape, so that different geometries can be manufactured depending on the application by changing the mold process.

A hollow cylinder wire mesh specimen, shown in Fig. 1, was used for this research. It was produced with an inner diameter of 10 mm, an outer diameter of 48 mm, and an axial thickness of 27 mm.

Three models of metal mesh isolators (A, B, C) with different relative density as shown in Fig. 2, have been produced and selected as a test elements for an experimental investigation of the nonlinear dynamic proprieties. Each model is composed by nine samples following the same production process. The relative density of the three samples for the studied metal mesh isolator is listed in the Table 1.

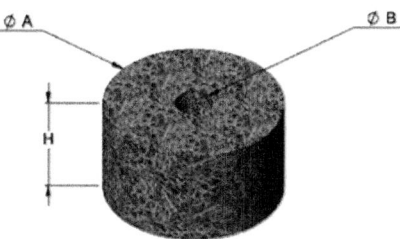

Fig. 1. Metal mesh specimen dimension

Fig. 2. Three models of isolators

Table 1. The relative density of each model.

Model	Relative density (%)
A	21.41
B	24.93
C	32.47

2.2 Experiment Set-Up

The experiments are performed in the electro-dynamic shaker Gearing and Watson V400, connected to an amplifier DSA4-8k. The experimental set up is displayed in Fig. 3. The exciter was positioned vertically (Fig. 3a)/horizontally (Fig. 3b) and was controlled by a USB laser system through the Dactron associated software for data acquisition and analysis. Two tri-axial accelerometers (ENDEVCO, model 65M100) are placed: one on the shaker table and another on the mass plate. The test rig is designed to measure the normal/lateral direction responses using four vibration isolators of the same model to avoid error due to the unbalance.

In order to study the mounts behavior under different structural conditions, three upper masses were tested: M_1, M_2 and M_3, which dependent on the number of plates. The mass values used during the test are given in Table 2.

(a)   **(b)**

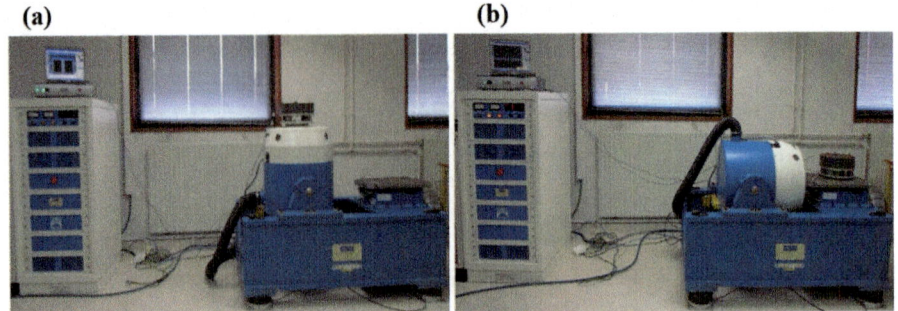

Fig. 3. Test machine: (a) normal (b) lateral excitation direction test

Table 2. The mass values.

	Number of plates	Mass (g)
M_1	1	5686.2
M_2	4	17552.6
M_3	7	29270.3

3 Parameter Estimation from Transmissibility Measurements of the Metal Mesh Isolator Using Rational Fraction Polynomials

In this section, the influence of parametric factors on the dynamic behavior of the nonlinear metallic wire isolator are experimentally investigated. The stiffness and damping ratio are extracted from the transmissibility measurements using the ration fraction polynomials method to characterize the mechanical properties and damping capacity of this porous material.

3.1 Experimental Methodology

Commonly, the characterization of material's damping capacity consist in estimating either its inverse quality factor or loss factor via the half-power bandwidth (HPB) (Ma et al. 2017). The convenient HPB formula provides a method used to determine the inverse quantity factor for linear system. For nonlinear system that exhibits a jump discontinuities in their response, the use of this method is invalid since one of the half power points ceases to be observed. To overcome this issue, the quantity factor is obtained from the measurements by fitting the response curve with the Lorentzian function (Antonio and Pastoriza 2009). However, this estimation lead to be poor for strong nonlinearity and when the unknown parameters depend on the motion amplitudes. Davis (2011) provided a new formula derived from the HPB method and used in the same manner but with the physical validity significantly improved accuracy.

The common feature of these aforementioned methods is that the physical constitutive model is not taken into consideration and the dynamic mechanical properties are not reflected in these methods. The Rational Fraction Polynomial (RFP) method (Richardson and Formenti 1982) is one of one of the traditional techniques in the modal analysis applications for identifying the modal parameters of the predominant modes. Curve fitting has been performed on the measurement to estimate the natural frequency and the damping ratio. The transfer function can be given in partial fraction form as in Eq. (1).

$$H(\omega) = \sum_{r=1}^{N} \frac{A_r + i\omega B_r}{\omega_r^2 - \omega^2 + i2\xi_r\omega_r\omega} \tag{1}$$

Where ω_r represents the natural frequency of the mode r, ξ_r is the damping ratio and A_r and B_r are constants and are residues corresponding to mode r. All these parameters can be determined through a series of matrix manipulations and operations. This process is further explained in details in Ewins (1984), Maia and Ewins (1989). The Rational Fraction Polynomial RFP method is employed to generate the most accurate damping ratio estimate (Schwarz and Richardson 1999) and provide very good estimates of the dynamic parameters.

A typical RFP model for a single-degree of freedom system is described in Eq. (2).

$$H(\omega) = \frac{A}{i\omega - \lambda} + \frac{A^*}{i\omega - \bar{\lambda}} \tag{2}$$

where $\lambda = \sigma + i\omega_d$; ω_d is the damped natural frequency.

3.2   Results and Discussion

3.2.1   Excitation Amplitude Effects

Figure 4 shows the stiffness variation as well as damping altering versus excitation level. These results were obtained with different modeling directions for the metallic wire-samples that are belonging to model-B. For both of excitation-directions, the stiffness decreased as the acceleration-amplitude increased. It is also to notice that the stiffness decreased when passing from normal load-direction to lateral load-direction, as shown in Fig. 4a.

Consequently, one can affirm that the excitation level in the compression direction is more efficient. In the other hand, the influence of the excitation direction is differently obvious in the variation of the damping ratio as shown in Fig. 4b. Indeed, the damping ratio decreased with the increase of the acceleration amplitude for the normal load direction while it increased for a lateral load-direction. Furthermore, the damping-ratio's variation range appears to be between 0.25 and 0.15 for compression direction. The values are smaller than those of non-compression excitation and are ranging between 0.25 and 0.6.

3.2.2   Preloading Effects

Similarly, the variation of the stiffness behavior of the model-B as function of the preload using both excitation direction under higher level of acceleration (3 m/s^2) are

(a)   **(b)**

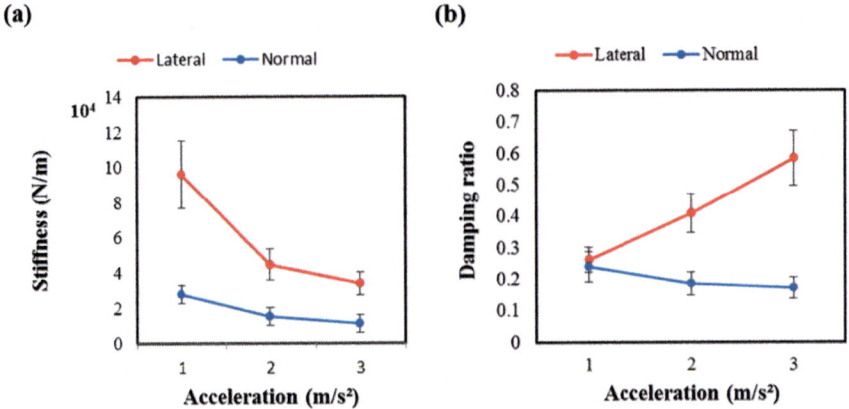

Fig. 4. The dependence of the excitation amplitude on the dynamic behavior in different excitation direction

summarized in Fig. 5a. The behavior of the preload-dependent damping ratio for normal and lateral direction are shown in Fig. 5b. The results show the significant influence exerted by the excitation direction over the damping ratio behavior of the third metal mesh model. One can observe an evident increase of the damping ratio in the lateral direction for increasing values of the preload, passing from a preload mass of 5.86 kg to the highest one of 29.27 kg. For the normal direction, the variation of the damping ratio is, however, slightly changed.

(a)   **(b)**

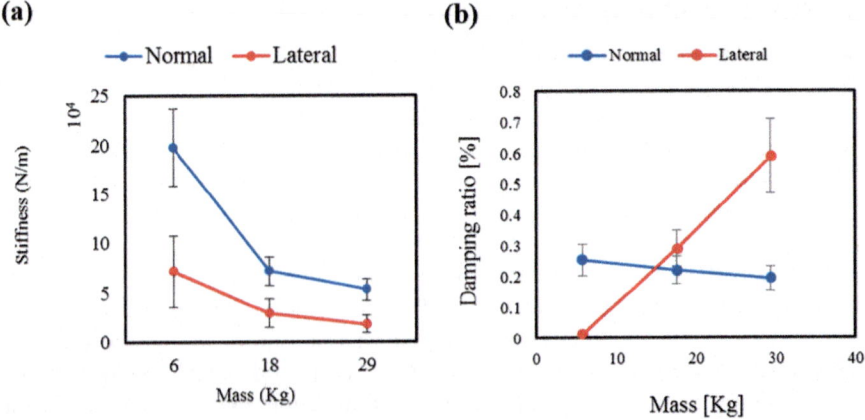

Fig. 5. The dependence of the preload on the dynamic behavior in different excitation direction

3.2.3   Relative Density Effects

Three metal mesh models with different relative density, as shown in the Fig. 6, have been tested with the same preload and excitation level conditions in the normal and lateral excitation directions. The results, related to the stiffness variation (Fig. 6a), reveal that the lateral excitation direction exhibits similar effect comparing to the normal direction with the increase of the relative density, otherwise, the increase in the contact between wires. Contrary to the case of the stiffness, the damping ratio decreases for the increase of the relative density in the lateral direction (Fig. 6b), however, it is slightly increased from 0.18 to 0.2 when the relative density vary from 21.41 to 32.47%. Meanwhile, it is worth notice that the damping ratio for the lateral direction present higher values compared to the normal direction ones.

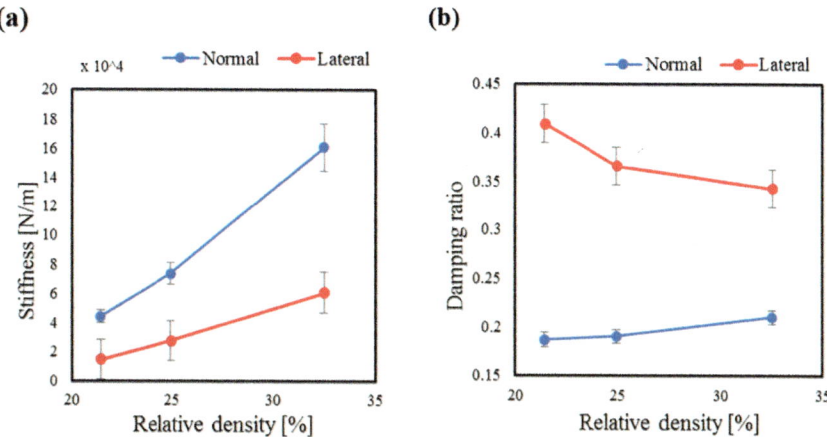

Fig. 6. The dependence of the relative density on the dynamic behavior in different excitation direction

4   Conclusion

A modal-analysis was performed in order to determine the stiffness and damping variation during the normal/lateral excitation-direction's tests. According to the reported experimental results, it is obvious that the mechanical properties of the metallic wires' mesh depend on the excitation direction. Furthermore, it can be concluded that the relative density and the working condition, including the pre-compression and the excitation level, have a significant influence on the global mechanical properties of the metallic mesh-wires. With regard to the previously cited influencing parameters, the major factor affecting the mechanical characteristics (stiffness and damping) is the sliding friction that exists at the contact-points between wires.

The adopted method, in this paper, is far from being suitable for stiffness and damping determination, since the model is supposed to be locally linear. Hence, a nonlinear identification method should be adopted.

References

Chaudhuri, S., Kushwaha, B.: Wire rope based vibration isolation fixture for road transportation of heavy defence cargo. In: Vibration Problems ICOVP-2007, pp. 61–67. Springer, Dordrecht (2008)

Di Massa, G., Pagano, S., Rocca, E., Strano, S.: Sensitive equipments on WRS-BTU isolators. Meccanica **48**(7), 1777–1790 (2013)

Davis, W.O.: Measuring quality factor from a nonlinear frequency response with jump discontinuities. J. Microelectromech. Syst. **20**(4), 968–975 (2011)

Antonio, D., Pastoriza, H.: Nonlinear dynamics of a micromechanical torsional resonator: analytical model and experiments. J. Microelectromech. Syst. **18**(6), 1396–1400 (2009)

Ewins, D.J.: Modal Testing: Theory And Practice, vol. 15. Research Studies Press, Letchworth (1984)

Ma, Y., Zhang, Q., Zhang, D., et al.: Experimental investigation on the dynamic mechanical properties of soft magnetic entangled metallic wire material. Smart Mater. Struct. **26**(5), 055019 (2017)

Ma, Y., Wang, H., Li, H., et al.: Study on metal rubber materials characteristics of damping and sound absorption. In: ASME Turbo Expo 2008: Power for Land, Sea, and Air, pp. 477–486. American Society of Mechanical Engineers (2008)

Maia, N.M.M., Ewins, D.J.: A new approach for the modal identification of lightly damped structures. Mech. Syst. Signal Process. **3**(2), 173–193 (1989)

Mezghani, F., Del Rincon, A.F., Souf, M.A.B., Fernandez, P.G., Chaari, F., Rueda, F.V., Haddar, M.: Experimental investigation on the influence of relative density on the compressive behaviour of metal mesh isolator. In: International Conference Design and Modeling of Mechanical Systems, pp. 941–947. Springer, Cham (2017)

Richardson, M.H., Formenti, D.L.: Parameter estimation from frequency response measurements using rational fraction polynomials. In: Proceedings of the 1st International Modal Analysis Conference, vol. 1, pp. 167–186. Union College Schenectady, New York (1982)

Schwarz, B.J., Richardson, M.H.: Experimental modal analysis. CSI Reliab. Week **35**(1), 1–12 (1999)

Wang, H., Rongong, J.A., Tomlinson, G.R., Hong, J.: Nonlinear static and dynamic properties of metal rubber dampers. Energy **10**, 1 (2010)

Wang, J.M., Pei, T.G.: Research on nonlinear characteristic of a new type of metal rubber damper for inertial platform. J. Astronaut. **25**, 256–311 (2004)

Wu, G., Ao, H., Jiang, H., Izzheurov, E.A.: Calculation of resonant sound absorption parameters for performance evaluation of metal rubber material. Sci. China Ser. E: Technol. Sci. **52**(12), 3587–3591 (2009)

Zhang, D., Scarpa, F., Ma, Y., Boba, K., Hong, J., Lu, H.: Compression mechanics of nickel-based superalloy metal rubber. Mater. Sci. Eng. A **580**, 305–312 (2013)

Water-Hammer Control in an Actual Branched Cast Iron Network by Means of Polymeric Pipes

Lamjed Hadj Taieb[1](✉), Med Amine Guidara[2], Noura Bettaieb[2],
Sami El Aoud[2], and Ezzeddine Hadj Taieb[2]

[1] College of Engineering, Prince Sattam Bin Abdul-Aziz University, Al-Kharj,
Kingdom of Saudi Arabia
l.hadjtaieb@psau.edu.sa
[2] Research Laboratory "Applied Fluid Mechanics, Process Engineering and
Environment", National Engineering School of Sfax, 3038 Sfax, Tunisia
elaoudsa@yahoo.fr

Abstract. The purpose of this paper is to numerically investigate the impact of replacing existing cast iron pipes of an actual branched network with High-Density Polyethylene HDPE pipes on damping and dispersing pressure waves created by water hammer phenomenon. A transient solver based on the Kelvin-Voigt formulations was developed. The numerical model takes into consideration the pipe wall viscoelastic behavior of polymeric pipes. The numerical resolution method of characteristics MOC with specified time intervals was adopted to solve the nonlinear, hyperbolic, partial differential equations that govern the unsteady flow generated in the network. The reliability of the numerical model was validated with experimental results from the literature. Flow disturbances in the branched network were generated due to the simultaneous fast closure of both downstream valves. Three different configurations of control strategies were suggested. The proposed strategies were based on implementing one or two high-density polyethylene (HDPE) pipes in the sensitive regions where the highest pressure perturbations took place. A comparison between the performances of the different control strategies was performed. Obtained results showed that even with one HDPE pipe implemented in the network, the positive and negative pressure peaks were reduced remarkable. Furthermore, the results have also shown that the risk of cavitation has been completely avoided using the control strategies.

Keywords: Hydraulic transient · Water hammer control · Viscoelastic model
MOC

1 Introduction

Hydraulic transient events are the response of the fluid under pressure-flow disturbances. These events take place in hydraulic networks such as water supply, gas-oil distribution networks.... In most cases, these disturbances may destroy or create great

© Springer Nature Switzerland AG 2019
T. Fakhfakh et al. (Eds.): ICAV 2018, ACM 13, pp. 235–244, 2019.
https://doi.org/10.1007/978-3-319-94616-0_24

damages in the hydraulic equipment of the network (Almeida and Koelle 1992; Chaudhry 1979; Wylie et al. 1993).

The unexpected changes in flow regimes through the pipe, frequently caused by fast opening and closing of valves, sudden stopping and starting of pumps, etc., result in instantaneous pressure changes known as the Water hammer phenomenon. The intensity of this phenomenon is more important in case of a liquid subjected to brutal variations in the flow. Thus, with the continuous growth and complexity of hydraulic networks, the urge to protect these facilities has never been more important. Several water hammer control devices such as flywheels, relief valves, air tanks, chimneys and balance balls protection are used to protect hydraulic systems (Pejovic 1987; Riasi and Nourbakhsh 2011). These devices contribute in decreasing the transient impact on the network. However, in many cases, due to economic and technical reasons, the use of such protective devices is not always possible. To overcome these shortcomings (Massouh and Comolet 1984) have first presented a solution consisting on adding a short section made of rubber to a steel piping system. It has been demonstrated that this addition to the rigid network alleviates the excessive pressure surge initiated by water hammer. Gargouri et al. (2008) investigated the effect of adding polymeric pipes to a real existing network. Results have shown remarkable decrease and damping of the transient pressure surges. However, these studies are limited to the elastic behavior of the pipes. In this context, Triki (2016) has investigated the possibility to lower pressure surges in a rigid pipe by an addition of a short section made of polymeric materials. The results were based on the model presented by Ramos et al. (2004). The obtained results have shown the efficiency of this solution on the pressure surge dispersion and damping. Triki (2017) has investigated the branched polymeric penstock control technique. It has been shown that this strategy contributes in a smaller increase of the period compared to the inline section technique.

Yet, in all previous mentioned studies, investigations were only carried on simple reservoir-pipe-valve systems. However, attention should rise to study the effectiveness of these control strategies in branched, more complicated networks.

The objective of this work is to numerically investigate the protection strategies of existing water facilities against water hammer phenomenon by adding a deformable segment in series along the pipeline. Firstly, a mathematical model was developed through the simplified formulations for the transient flow of fluid in quasi-rigid and viscoelastic cylindrical pipes. The characteristics method MOC is used for the numerical solution of the obtained partial differential equations. A generalized Kelvin-Voigt model is adopted to simulate the viscoelastic behavior of the rheological pipe wall behavior. Various cases are discussed to demonstrate the effectiveness of the protection strategies of water facilities by means of polymeric segments, nevertheless, the study of the effect of some parameters on the reliability of this water hammer control strategy.

2 Mathematical Model

2.1 Governing Equations

Mathematical modeling of unsteady flows in closed pipes is based essentially on the continuity and momentum equations. This set of two partial differential equations is adopted from the classic water hammer model developed by Almeida and Koelle (1992), Wylie and Streeter (1978). However, this model cannot accurately predict the transient response of polymeric pipes. To take into account the rheological behavior of such pipes, an additional term is added to the continuity equation (Covas et al. 2005), as follows:

$$\frac{\partial H}{\partial t} + \frac{a_0^2}{gA}\frac{\partial Q}{\partial x} + \frac{2a_0^2}{g}\frac{\partial \varepsilon_r}{\partial t} = 0 \tag{1}$$

$$\frac{\partial Q}{\partial t} + gA\frac{\partial H}{\partial x} + h_f = 0 \tag{2}$$

where H is the piezo-metric head; x is space co-ordinate; A is the cross sectional area of the pipe; h_f is the steady state friction term given by:

$$h_f = \frac{f_s}{2gD}\frac{Q|Q|^{n-1}}{A^n} \tag{3}$$

where f_s represents the Darcy-Weisbach friction factor, D is the inner pipe diameter, n is the exponent of the flow.

The last element of Eq. (1) represents the time-derivative retarded strain with ε_r is the retarded component of the strain. This derivative cannot be computed directly, further discretization are needed in this case. The generalized Kelvin-Voigt (KV) model is used to describe the pipe wall rheological behavior (Aklonis et al. 1972).

The KV model presented in Fig. 1 is a series combination of a compliance J_0 and Nkv Kelvin-Voigt elements, solids of viscosity μ_i and creep compliances $J_i = 1/E_i$. Thus, the creep function used in this model is given by Eq. (4) (Liu 2007).

$$f(t) = J(t) = J_0 + \sum_{i=1}^{N_{kv}} J_i\left(1 - e^{\frac{-t}{\tau_i}}\right) \tag{4}$$

where $\tau_i = J_i\mu_i$ is the delay time of the ith solid of Kelvin-Voigt. The parameters of the Kelvin-Voigt model could be defined either based on the experimental data from the study of Covas et al. (2004) or by calculating the creep function using the creep–compliance experimental data (Urbanowicz et al. 2016).

The expression relating the linear viscoelastic strain and the stress is given by:

$$\varepsilon(t) = \sigma(t)J(0) + \int_0^t \sigma(t-s)\frac{dJ}{ds}(s)ds = (\sigma * dJ)(t) = (J * d\sigma)(t) \tag{5}$$

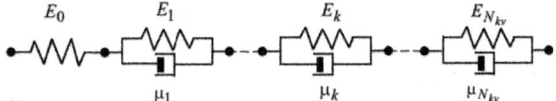

Fig. 1. Generalized Kelvin–Voigt model

2.2 Numerical Scheme

The Method of characteristics is used to solve the previous set of partial differential equations. To verify the stability of this method, the Courant-Friedrich-Lewy stability condition must be satisfied. This condition describes the propagation of the flow features along the characteristic lines, $dx/dt = V \pm ao$.

$$\frac{dH}{dt} \pm \frac{a_0}{gA}\frac{dQ}{dt} + \frac{2a_0^2}{g}\left(\frac{\partial \varepsilon_r}{\partial t}\right) \pm a_0 h_f = 0 \tag{6}$$

Equation (6) are valid along the characteristic lines $dx/dt = V \pm ao$. To model unsteady flows in rigid and quasi-rigid pipes, linear-elastic case, the element representing the retarded strain-time derivative is eliminated.

In case of water engineering problem, the fluid velocity is neglected compared to the pressure wave speed. Thus, the set of Eq. (6) is simplified by neglecting convective terms. These equations can be solved by the following scheme:

$$[H(x,t) - H(x \mp \Delta x, t - \Delta t)] \pm \frac{a_0}{gA}[Q(x,t) - Q(x \mp \Delta x, t - \Delta t)]$$
$$+ \frac{2a_0^2 \Delta t}{g}\left(\frac{\partial \varepsilon_r}{\partial t}\right) \pm a_0 \Delta t . h_f = 0 \tag{7}$$

with $dx/dt = \pm a_0$, the characteristic lines are approximately straight. The second term of Eq. (4) is derived in order to compute the time-derivative of the retarded strain (Covas et al. 2005). Further equations are required to fully compute the pressure and discharge through the pipes. These equations are called the boundary conditions. A detailed representation of the latter are presented in Chaudhry (2014).

To validate the previous developed model, the Reservoir-pipeline-valve system adopted for experimental investigations by Covas et al. (2004) was considered for numerical transient investigations. The test facility contains a single Polyethylene pipeline system with a total length of 271.5 m. The chosen material was high-density polyethylene SDR11 PE100 NP16 with a 6.2 mm wall thickness and a 63 mm nominal diameter (ND). The steady stat discharge flow was $Q0 = 1.0$ l/s. The flow disturbances in the Reservoir-pipe system were initiated by the closure of the ball valve located at the downstream end. The creep function $f(t)$ is expressed by a five-element KV model. The creep coefficient and the retardation times are presented in Table 1.

Table 1. Creep function parameters

Creep coefficients J_k $\left(10^{-10}\ \mathrm{Pa}^{-1}\right)$	Value	Retardation time τ_k (s)	Value
J_0	6.99		
J_1	1.057	τ_1	0.05
J_2	1.054	τ_2	0.5
J_3	0.9051	τ_3	1.5
J_4	0.2617	τ_4	5
J_5	0.7456	τ_5	10

In Fig. 2, the piezo-metric head variations in two positions T1 and T5 (distant 271 m and 116.42 m respectively from the upstream reservoir), are compared with those measured experimentally by Covas et al. (2004).

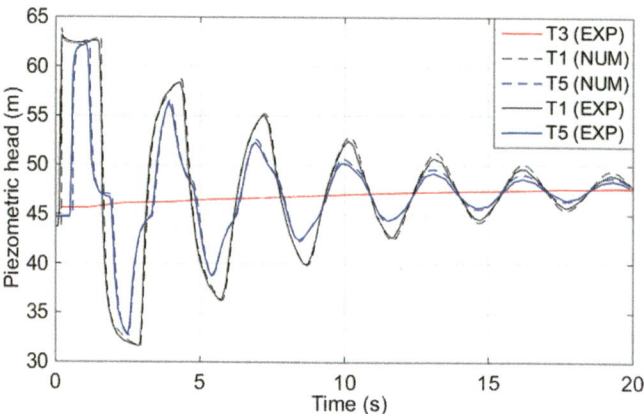

Fig. 2. Pressure variations at location T1, T5

It can be observed that the results obtained from the non-conventional water hammer model are in good agreement with the experimental ones. Thus, the previously developed model could be adopted to simulate pressure variations in hydraulic pipeline systems with both elastic and viscoelastic rheological behavior.

3   Results and Discussions

The southern Tunisian network, presented in Fig. 3(a), was adopted for the numerical investigations. The Water supply network consists of an upstream reservoir located at node 1 the Lassifer feeding two downstream reservoirs located at Guellala and Zarzis, node 12 and 17 respectively.

The initial conditions of the network are the steady state data through the system provided by the Tunisian company of water supply and exploitation SONEDE.

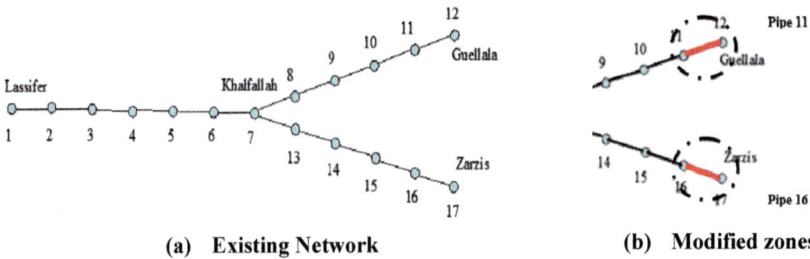

(a) **Existing Network**   (b) **Modified zones**

Fig. 3. South Tunisia water supply network

Transient events are created by sudden disturbances in the flow rate at nodes 12 and 17. Both valves are closed, simultaneously, instantaneously. The upstream boundary is represented by the constant level reservoir $H_1(t) = 102.3$ m.

The effectiveness of the water hammer control strategy is investigated for instant total closure of both valves at node 12 and 17 (Guellala) and (Zarzis) respectively. Fast or instantaneous closure of the valves is likely to cause enormous pressure rise at the valve level followed by a pressure wave traveling throughout the network. Since pipe 11 and 16 are directly related to the downstream valves, they are considered the regions where high pressure perturbations may occur. In this light, control strategies based on replacing pipe 11 and 16 with HDPE pipes, as presented in Fig. 3(b), are suggested. Three configurations were proposed to choose the most economical yet reliable solution.

(a) Only pipe 11 is replaced with a HDPE pipe (pipe length 7690 m)
(b) Only pipe 16 is replace with a HDPE pipe (pipe length 4650 m)
(c) Both pipe 11 and pipe 16 are replaced with HDPE pipes

The pressure variations were plotted at three different locations: at node 7 which corresponds to the branching junction and node 12 and 17 at the level of the valves. The pressure response of the unprotected network is compared to the response of the protected network under the three different control configurations. Figure 4 presents the pressure variation at the junction level. Node 7 is located approximately 35.5 km away from node 12 and 23.75 km away from node 17. This explains the delay in the pressure rise as the pressure waves generated at the downstream ends need some time to travel along the pipelines. However, in case of the unprotected network, the first pressure wave that reached node 7 caused an immediate rise in the pressure by approximately 7 bars. As the pressure waves continue to reach the junction level, a maximum of pressure rise of approximately 9 bars is observed.

For the protected network under configuration (a), the immediate overpressure was less than 2 bars while the maximum reached overpressure was approximately about 3.7 bars. In case of control configuration (b), the junction level is subjected to an instantaneous rise of 5 bars. Following the continuous arrival of pressure waves, the pressure rise reached a maximum of 5 bars. However, with HDPE pipes replacing both pipes 11 and 16, no instantaneous pressure rise was observed at node 7. The pressure smoothly rises reaching a maximum rise of 3.2 bars.

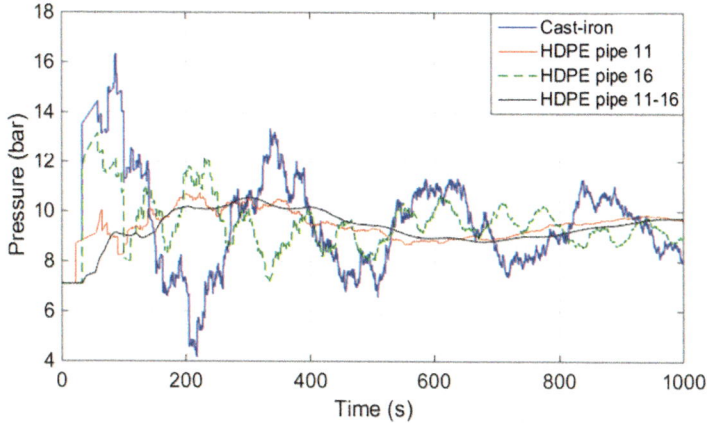

Fig. 4. Pressure variations at node 7 with different control strategies

A further practical consideration should be given to the dependency of the value of the first pressure peak upon the length of the polymeric replaced pipe. This explains why control strategy (a) presented better first pressure peak damping than configuration (b) as the difference in length between pipe 11 and pipe 16 is almost 3 km. Nevertheless, the decrease in the magnitude of the first pressure wave is almost the same for both control strategies (a) and (c). In addition to the decrease of the first pressure peak, an increase in the period of wave oscillations was also observed in the response of protected network under configuration (a) and (c). These results were physically well predicted regarding to the fact that High-density polyethylene pipes are capable of producing more damping and reduction of pressure waves.

Figures 5 and 6 compare the pressure variations at node 12 and 17 calculated for the existing hydraulic network and with those predicted in the protected system. From the first glance, one can clearly notice that the unprotected network is subjected to a pressure rise followed by a pressure drop. Noting that the vapor pressure of water at 25 °C is 0.03169 bar, the pressure at node 12 in this case fell under the liquid vapor pressure. When subjected to higher-pressure waves, the bubbles filled with vapor explode and can generate an intense shock wave. These events are defined as the cavitation phenomenon. However, with one or more HDPE pipes implemented in the network, the cavitating flow may be avoided.

According to Figs. 5 and 6, the most severe transient events were observed in the unprotected existing system. The change in the downstream boundary conditions yield both dispersion and damping of pressure waves. In other terms, the pressure variations exhibited a large pressure increase followed by a pressure decrease to the saturated liquid pressure followed by a pressure rise forcing the cavity to collapse. By comparing the response of the network under the different proposed control strategies, one can easily notice that configuration (b), where only pipe 16 was replaced with a HDPE, presented the least practical results. Pressure variations at node level 12 presented almost the same patterns as those of the unprotected network with only a slight

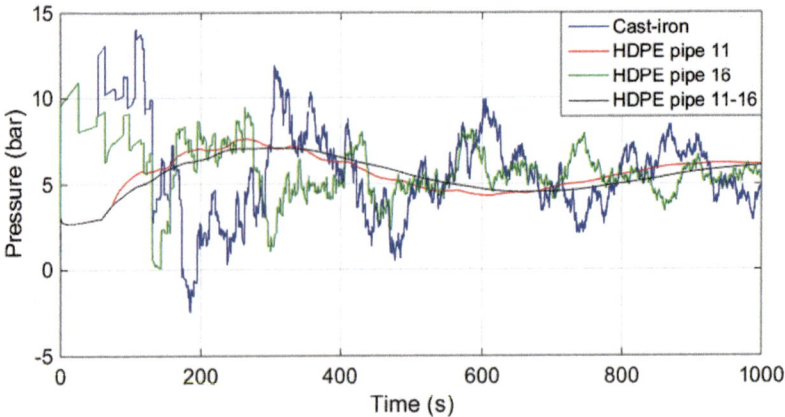

Fig. 5. Pressure variations at node 12 with different control strategies

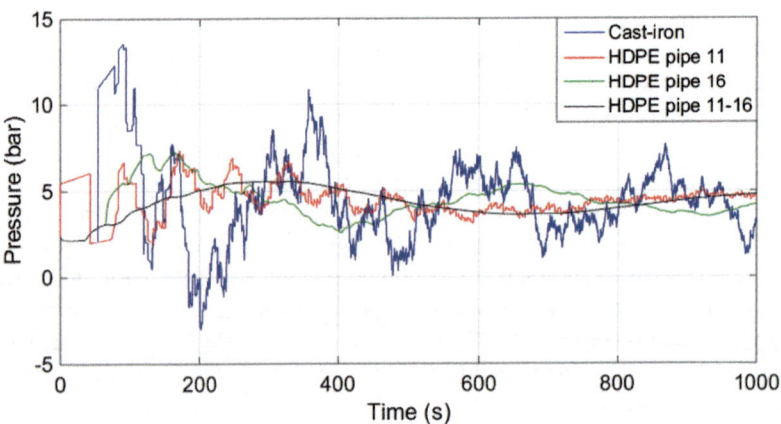

Fig. 6. Pressure variations at node 17 with different control strategies

difference in oscillation period. Meanwhile, control strategy (a) and (c) presented the same performances as observed at the junction level.

However, pressure variation at node 17 experienced, under control strategy (a), relatively fast pressure wave oscillations with low magnitude compared to the unprotected network performance. Overall, the numerically investigated performance of the existing network under three different configurations presented better results than the case of the unprotected network.

4 Conclusion

Different water hammer control configurations implemented in a real branched water supply network were numerically investigated. The control strategies consisted on replacing one or more existing cast iron pipes with high-density polyethylene pipes. Pipes located at the most sensitive regions of the branched water supply network where transient events are likely to occur following valve closures where considered for the replacement. Subjecting the network to transient events, the numerical obtained performances for different cases were compared.

The protected network presented better performances than the real existing one with cast iron pipes under the same transient events. Obtained results were physically interpreted by the fact that high density polyethylene pipes result in an important pressure wave attenuation.

Moreover, damping and dispersion of pressure waves were found to be different from one control strategy to another. The magnitude of the first pressure peak was found to be highly sensitive to the length of the replaced polymeric pipe.

References

Aklonis, J.J., MacKnight, W.J., Shen, M.: Introduction to Polymer Viscoelasticity. Wiley, New York (1972)

Almeida, A.B., Koelle, E.: Fluid Transients in Pipe Networks. Computational Mechanics Publications, Southampton (1992)

Boulos, P.F., Karney, B.W., Wood, D.J., Lingireddy, S.: Hydraulic transient guidelines. Am. Water Works Assoc. **97**, 111–124 (2005)

Chaudhry, M.H.: Applied Hydraulic Transients. Van Nostrand Reinhold Company, New York (1979)

Chaudhry, M.H.: Applied Hydraulic Transients, 3rd edn. Springer, New York (2014). https://doi.org/10.1007/978-1-4614-8538-4

Covas, D., Stoianov, I., Mano, J.F., Ramos, H., Graham, N., Maksimovic, C.: The dynamic effect of pipe-wall viscoelasticity in hydraulic transients. Part II. Model development, calibration and verification. J. Hydraul. Res. **43**(1), 56–70 (2005). https://doi.org/10.1080/00221680509500111

Covas, D., Stoianov, I., Ramos, H., Graham, N., Maksimovic, C.: The dynamic effect of pipe-wall viscoelasticity in hydraulic transients. Part I. Experimental analysis and creep characterization. J. Hydraul. Res. **42**(5), 517–532 (2004). https://doi.org/10.1080/00221686.2004.9641221

Gargouri, J., Hadj-Taieb, E., Thirriot, C.: Influence de l'élasticité de la paroi sur l'évolution des ondes de pression dans les réseaux de conduites. Mecanique et Industrie **9**, 33–42 (2008). https://doi.org/10.1051/meca

Liu, H.: Material Modelling for Structural Analysis of Polyethylene (2007)

Massouh, F., Comolet, R.: Study of a water-hammer protection system in fine. La Houille Blanche (1984). https://doi.org/10.1051/lhb/1984023

Pejovic, S.: Guidelines to hydraulic transient analysis (1987)

Ramos, H., Covas, D., Borga, A., Loureiro, D.: Surge damping analysis in pipe systems: modelling and experiments. J. Hydraul. Res. **42**(4), 413–425 (2004). https://doi.org/10.1080/00221686.2004.9641209

Riasi, A., Nourbakhsh, A.: Influence of surge tank and relief valve on transient flow behaviour in hydropower stations Alireza. In: Proceedings of 11th International Conference on Fluid Power. ASME-JSME-KSME Fluids Engineering Division, New York (2011)

Triki, A.: Waterhammer control in pressurized-pipe flow using an in-line polymeric short-section. Acta Mech. **227**(3), 777–793 (2016). https://doi.org/10.1007/s00707-015-1493-13

Triki, A.: Water-hammer control in pressurized-pipe flow using a branched polymeric penstock. J. Pipeline Syst. Eng. Pract. **8**(4), 1–9 (2017). https://doi.org/10.1061/(ASCE)PS.1949-1204.0000277

Urbanowicz, K., Firkowski, M., Zarzycki, Z.: Modelling water hammer in viscoelastic pipelines: short brief. J. Phys. Conf. Ser. **760**(1), 12037 (2016). https://doi.org/10.1088/1742-6596/760/1/012037

Wylie, E.B., Streeter, V.L.: Fluid Transients_Streeter.pdf (1978)

Wylie, E.B., Streeter, V.L., Lisheng, S.: Fluid Transients in Systems (1993)

Materials Behavior in Dynamic Systems

Reliability Based Design Optimization of Shape Memory Alloy

Fatma Abid[1,2]($\boxtimes$), Abdelkhalak El Hami[1]($\boxtimes$), Tarek Merzouki[3]($\boxtimes$),
Hassen Trabelsi[2]($\boxtimes$), Lassaad Walha[2]($\boxtimes$), and Mohamed Haddar[2]($\boxtimes$)

[1] Laboratory of Mechanics of Normandy LMN,
National Institute of Applied Sciences of Rouen, Cedex, France
{fatma.abid,abdelkhalak.elhami}@insa-rouen.fr
[2] Laboratory of Mechanics, Modeling and Manufacturing LA2MP,
Mechanical Engineering Department, National School of Engineers of Sfax,
Sfax, Tunisia
hassen.trabelsi@outlook.fr, walhalassaad@yahoo.fr,
mohamed.haddar@enis.rnu.tn
[3] Laboratory of System Engineering of Versailles,
University of Versailles Saint Quentin in Yvelines, Versailles, France
tarek.merzouki@uvsq.fr

Abstract. "Smart Materials" become more and more used due to their physical properties compared to other materials. Shape memory alloys (SMA) could be classified as one of them. Such material is characterized by the ability to remember its original shape after deformation. SMA provides high structural performance. However, this kind of material increases the cost of structures. Thus, optimization techniques have been applied in order to obtain minimum the volume structure. As SMA requires a global optimization tool, the evolutionary algorithms such as particle swarm optimization (PSO) is well suited. In deterministic optimization, the uncertainties of the system parameters are not taken into consideration. As a result, the optimal design obtained does not ensure the target reliability level. Thus, reliability based design optimization (RBDO) method was applied to provide an enhanced design. However, classical RBDO leads to high computational time. So, the purpose of this paper is to optimize SMA structure taking into consideration uncertainties in the structural dimensions. Therefore, a proposed RBDO methodology based on safety factors derived from Karush Kuhn Tucker (KKT) methodology coupled with PSO is developed. Compared to deterministic optimization, the proposed method guarantees the target reliability level of the structure but requires little extra computational effort.

Keywords: Shape memory alloy (SMA)
Reliability based design optimization (RBDO)
Particle swarm optimization (PSO) · Karush Kuhn Tucker (KKT)

1 Introduction

In recent decades, there are increasing interest in the use of "Smart Materials". Shape Memory Alloy (SMA) could be classified as one of them (Lagoudas 2008). Such material is characterized by the ability to remember its original shape after deformation.

© Springer Nature Switzerland AG 2019
T. Fakhfakh et al. (Eds.): ICAV 2018, ACM 13, pp. 247–256, 2019.
https://doi.org/10.1007/978-3-319-94616-0_25

SMA provides high structural performance. However, this kind of material increases the cost of such structures. As a result, Optimization techniques have been applied in order to obtain minimum volume structure. As Shape Memory Alloy requires a global optimization tool, the Genetic Algorithms (GA) (Strelec et al. 2003) and Particle Swarm Optimization (PSO) are well suited.

In deterministic design optimization, the uncertainties of the system parameters are not taken into consideration. As a result, the optimal design obtained does not ensure the target reliability level. In fact, the resulting optimum solution may lead to a high risk of failure. Thus, Reliability Based Design Optimization (RBDO) method is applied to provide an enhanced design with high level of confidence (Kharmanda et al. 2014). The main RBDO formulations can be classified into three categories namely two level approach, decoupled approach and single loop approach (Aoues and Chateauneuf 2010). At first, the two level RBDO approach considers the reliability assessment inside the optimization loop. In fact, the outer loop of this approach deals with optimization while the inner loop deals with reliability evaluation. The main drawback of this approach is the numerical effort required to solve the system. The main objective of this work is the RBDO of the SMA which leads to the problem of coupling RBDO and global optimization techniques. The main advantages of such a procedure is that it requires little additional computational time compared to standard deterministic optimization and at the same time it guarantees the target reliability level of the structure. Thus, this work presents the RBDO methodology based on safety factors that are derived from Karush Kuhn Tucker (KKT).

2   Optimization of Shape Memory Alloy Plate

The case of a shape memory alloy square plate with a quarter circle excised from the corner is simulated by finite element process. The characteristics of the mentioned material are described in the Table 1. This plate is clamped at its right boundary and its lower boundary and loaded at its left boundary by a displacement. The design variables of this example are both the radius of the hole r and the thickness of the plate h (Fig. 1).

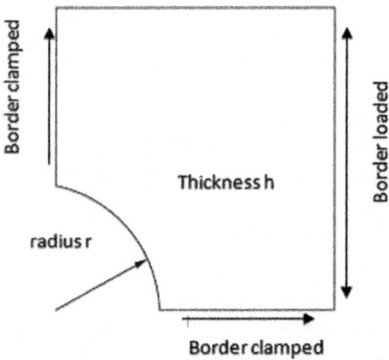

Fig. 1. Reliable design of the SMA plate

Table 1. Material parameters

Constant	Meaning	Value
h (MPa)	Hardening parameter	5000
T_0 (K)	Reference temperature	253.15
R (MPa)	Elastic limit	45
β (MPa K^{-1})	Temperature scaling parameter	75
ε_L	Maximum transformation strain	0.33
E_m (MPa)	Martensite modulus	70000

2.1 Deterministic Optimization of Shape Memory Alloy Plate

The plate is optimized to minimize its volume under the Von mises constraint. As mentioned before, h and r are considered as the design variables. For convenience of notation, these design variables are grouped into design vector $d = [r, h]$. The RBDO of the problem is written as follows:

$$Minimize : f(d) = (l^2 - \frac{\pi . r^2}{4}) . h$$

$$Subjet\ to :$$

$$\begin{cases} P_f(G(d) \leq 0) \leq P_f^{allowed} \\ 15\ \text{mm} \leq r \leq 25\ \text{mm} \\ 5\ \text{mm} \leq h \leq 15\ \text{mm} \end{cases} \quad (1)$$

As a result, the optimization algorithm aims at finding the radius r and the thickness h that minimize the plate and not let the structure fail according to the chosen failure mode. The solution of the deterministic shape memory alloy optimization problem will be noted $d^{optimal}$ in the following. However, the uncertainties of the dimensions are taken into account. Consequently, the deterministic variables r and h are replaced by random variables $X = [X_r, X_h]$. In accordance with this modification, the deterministic design vector d is replaced by the reliability design vector m, which is formed by the mean value m_r of the random radius X_r and the mean value m_h of the random thickness X_h: $m = [m_r, m_h]$.

In the next step, the classical RBDO methodology is presented in order to treat the constraint in a probabilistic sense.

2.2 Reliability Based Design Optimization of Shape Memory Alloy Plate

The converting of the deterministic shape memory alloy optimization problem into an RBDO problem leads to looking for $m^{reliable}$. The solution of the equation is:

$$Minimize : f(m)$$

$$subject\ to :$$

$$\begin{cases} P_f(G(m, X) \leq 0) \leq P_f^{allowed} \\ 15\ \text{mm} \leq m_r \leq 25\ \text{mm} \\ 5\ \text{mm} \leq m_h \leq 15\ \text{mm} \end{cases} \quad (2)$$

With $G(m, X)$ is the limit state function of random variables X, P_f is the probability of failure and $P_f^{allowed}$ is the maximum allowed probability of the structure. P_f is written as follows:

$$P_f(m) = P_f(G(m, X) \leq 0) = \int_{G(m,X)} f(X)dX \tag{3}$$

Where $f(X)$ is the joint probability density function of X. $P_f(m)$ and $P_f^{allowed}$ can be expressed using the FORM of the RIA using the following forms:

$$\begin{cases} P_f(m) \approx f(-\beta(m)) \\ P_f^{allowed} \approx f(-\beta^{target}) \end{cases} \tag{4}$$

Where $\phi(.)$ is the standard normal cumulative distribution function, $\beta(m)$ is the reliability index of the problem and β^{target} represents the target reliability index of the problem. As a result, the use of the previous relations to describe the probabilistic constraint leads to the equation:

$$\begin{aligned} &Minimize: f(m) \\ &subject\ to: \\ &\begin{cases} \beta^{target} \leq \beta(m) \\ 15\ mm \leq m_r \leq 25\ mm \\ 5\ mm \leq m_h \leq 15\ mm \end{cases} \end{aligned} \tag{5}$$

In order to measure the reliability index $\beta(m)$, the original random vector X (physical space) is transformed into a standard Gaussian vector U (standard space). First, we introduce the vectors x and u which are the realizations of the random vectors X and U. Consequently, the transformation between the two spaces is expressed as follows:

$$u = T(m, x) \quad or \quad x = T^{-1}(m, u) \tag{6}$$

The limit state function is presented as follows:

$$G(m, X) = G(m, T^{-1}(m, U)) = g(m, U) \tag{7}$$

With $g(m, U)$ represents the limit state function in the standard space. Then, the reliability index $\beta(m)$ can be obtained through the resolution of the following optimization problem in the U space:

$$\begin{cases} For\ a\ given\ m: \\ minimze\ \|u\| \\ subject\ to: \ g(m, u) \leq 0 \end{cases} \tag{8}$$

With the solution of the optimization $u^*(m)$ represents the Most Probable Failure Point
(MPP) and the reliability index $\beta(m) = \|u^*(m)\|$ represents the distance of the origin
between the MPP and the origin of the standard space.

In its classical form, the RBDO consists in two optimization problems: structural
optimization and reliability analysis. The solving of classical RBDO problem by any
sequential approximation leads to:

$$For\ k = 1, .., n_{iteration}$$

$$minimize\ f(m^k)$$

$$subject\ to :$$

$$\begin{cases} \beta^{target} \leq \beta(m^k) \\ 15\ mm \leq m_r^k \leq 20\ mm \\ 5\ mm \leq m_h^k \leq 15\ mm \end{cases} \tag{9}$$

With $n_{iteration}$ is the required number of iterations to solve the RBDO problem. Such a
method gives high computational time.

A methodology consists in the coupling of PSO method and the safety factors
derived from KKT optimally condition is used. This method eliminates the requirement
of reliability analysis. Thus, it is suitable for the optimization of shape memory alloy
materials.

3 Proposed RBDO Methodology and Its Coupling with Global Optimization Algorithms

The aim of this section is to couple reliability analysis with global optimization
algorithms in order to calculate the RBDO of shape memory alloy plate, a new RBDO
methodology that differs from the one described in Sect. 2 is given here.

The main purpose is to estimate the Most Probable Point (MPP) of an optimal
design (approximate u^* or x^* of $m^{optimal}$) obtained with any global optimization algorithm. Then, to calculate the safety factors S_f applied to such points in order to obtain
the final design $m^{reliable}$ that guarantees the target reliability level of the structure. Such
a methodology is described in the following.

We deduce the safety factors from the KKT optimality conditions of the RIA which
is written as follows (Lopez et al. 2011):

$$\begin{cases} \nabla_u(\|u^*(m)\| + \lambda\nabla_u g(m, u^*(m)) = 0 \\ \lambda g(m, u^*(m)) = 0 \\ \lambda \geq 0 \\ g(m, u^*(m)) \leq 0 \end{cases} \tag{10}$$

where λ is the Lagrange multiplier allowing to take into account of the restriction
$g(m, u^*(m)) \leq 0$. Then, Eq. (3) can be written as:

$$\begin{cases} \dfrac{u^*(m)}{\|u^*(m)\|} = \dfrac{\nabla_u g(m,u^*(m))}{\|\nabla_u g(m,u^*(m))\|} \\ g(m,u^*(m)) = 0 \end{cases} \tag{11}$$

The gradient of the constraint can be written in the physical space as mentioned in the equation:

$$\nabla_u g(m, u^*(m)) = (\nabla_m x)^T (\nabla_x G(m, x^*(m))) \tag{12}$$

With $\nabla_m x$ represents the jacobian of the transformation between the two space.

To calculate the RBDO of a given structure, The target reliability level must be attained. For that reason, the optimum has to verify:

$$\|u^*(m^{reliable})\| = \beta^{target} \tag{13}$$

Then, we substitute the Eq. (13) into the Eq. (12):

$$u^*(m^{reliable}) = -\beta^{target} \frac{(\nabla_m x)^T (\nabla_x G(m^{reliable}, x^*(m^{reliable})))}{\|(\nabla_m x)^T (\nabla_x G(m^{reliable}, x^*(m^{reliable})))\|} \tag{14}$$

In this paper, the random design variables of the problem are Gaussian. The transformation from the physical space to normal one is written as follows:

$$u_i = \frac{x_i - m_i}{s_i} \tag{15}$$

With s_i is the standard deviation of the i-th random variable. Considering that x_i^* is related to m_i using the following equation:

$$m_i = S_{f_i}.x_i^* \tag{16}$$

Substituting the Eqs. (16) and (15) into equation gives (14):

$$S_{f_i} = 1 + \beta^{target} \frac{s_i}{x_i^*(m^{reliable})} \frac{[(\nabla_m x)^T (\nabla_x G(m^{reliable}, x^*(m^{reliable})))]_i}{\|(\nabla_m x)^T (\nabla_x G(m^{reliable}, x^*(m^{reliable})))\|} \tag{17}$$

With S_{f_i} is the safety factor of the i-th design variable (i = 1, ..., Np) with Np is the number of design variables.

4   Results and Discussions

The numerical performance of classical RBDO method and the proposed one are studied through two examples: (1) mathematical problem of short column design and (2) shape memory alloy plate. For this benchmark, the RBDO methods was implemented in MATLAB environment. For the finite element analysis, ANSYS software was used.

4.1 Short Column Design

A short column, having a rectangular cross section with dimensions B and H, is optimized in order to minimize its cross section. The column is subjected to bending moments $M_1 = 250$ kNm, $M_2 = 125$ kNm and axial force $F = 2500$ kN. A limit state function is written in terms of the design vector $d = (B, H)$:

$$G(d) = 1 - \frac{4M_1}{BH^2Y} - \frac{4M_2}{HB^2Y} - \frac{F^2}{(BHY)^2} \qquad (20)$$

Where $Y = 40$ MPa is the yield stress of the column material. In this problem, uncertainties in the dimensions B and H of the column are considered. The variables have normal distributions with standard deviations of 0.03 m. The RBDO problem may be stated as follows (Lopez et al. 2011).

$$Minimize f(m) = B.H$$

subject to :

$$\begin{cases} \Pr(G(m,x) \leq 0) \leq P_f^{allowed} \\ 0 \leq B \\ 0.5 \leq \frac{B}{H} \leq 2 \end{cases} \qquad (21)$$

The RBDO of the column was studied by three different methods: the proposed method, RIA as well as PMA. The target reliability index β^{target} of structure is equal to 3. The proposed strategy based on safety factors requires only a few more evaluations after the deterministic optimization. Classical RBDO performs complete optimization many times. The computational time (TC) is used in order to compare the proposed method as well as RIA and PMA. We notice that the results obtained by RIA and PMA were the same. Besides, the final area was smaller than that achieved by the proposed method. We can conclude that although the deterministic optimization performed in the proposed method obtained a smaller area than the one calculated by PMM of RBDO RIA or PMA. When we apply the safety factors, the final result was higher than safe design calculated by RIA or PMA. Because of the use of finite element analysis for the calculation of finite state function, the computational time reduction is very significant. In the next section, shape memory alloy (SMA) RBDO problem is solved to demonstrate the effectiveness of the proposed methodology (Table 2).

Table 2. RBDO results of short column design

$\beta = 3$	Safety factor		RBDO RIA		RBDO PMA	
	B	H	B	H	B	H
x^*	0.2543	0.5065	0.3056	0.4249	0.3057	0.4247
$m^{reliable}$	0.3201	0.5723	0.3765	0.4803	0.3765	0.4803
Area (m^2)	0.1832		0.1808		0.1808	
TC (s)	1.0825		2.0123		2.9359	

4.2   Shape Memory Alloy Plate

The numerical results described in Sect. 2 are presented in this section. First, let us consider a strain driven uniaxial traction compression tensile test. The material parameters presented in Table 1 are considered in the following example. The SMA is characterized by two solid phases: the austenic phase (A) which is stable at high temperature (T > A_f) (austenite finish transformation temperature) and the martensitic phase (M) which is stable at low temperature (T < M_f) (martensite finish transformation temperature). Besides, the martensite can be divided into two configurations: (i) the stress free martensite which is formed by a twinned multivariant crystallographic structure and (ii) the stress induced martensite which is formed by a detwinned crystallographic structure with a single variant (S). Figure 2 shows a hysteresis loop. In fact, the curve can be divided to five parts: (1) Elastic deformation of the austenite, (2) the transformation from the austenite to single variant martensite (upper plateau), (3) Elastic deformation of the single variant martensite, (4) elastic strain recovery, (5) the transformation from single variant martensite to the austenite (lower plateau).

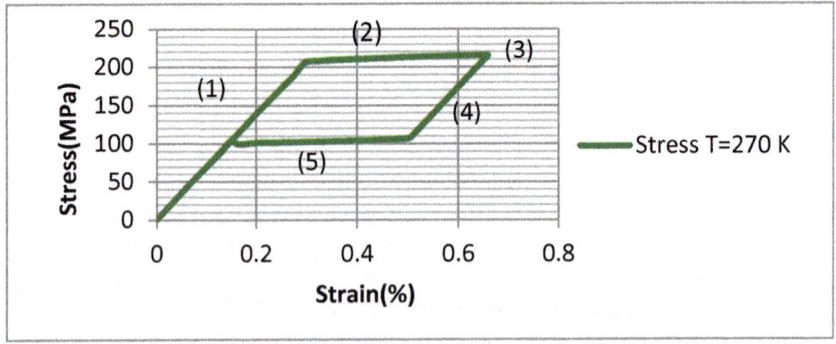

Fig. 2. Pseudo-elastic effect of the SMA plate

Figure 3 shows the corresponding stress strain curves at three distinct temperatures. Hysteresis loops for both traction and compression are presented. We can notice that the raise of the temperature increases the distance between the loop. A convergence study leads to a mesh with 27 elements and 108 nodes. Stresses are evaluated on Gaussian integration points. The optimization problem was solved using the PSO algorithm. The Von Mises stress is used for the evaluation of the limit state function G (h, r) starting from an initial design $(h, r) = (25, 5)$. A Sequential Quadratic Programming (SQP) optimization algorithm leads to an optimal conception $(h^*, r^*) = (24.014, 9.8)$. The reliability study has been performed with a target reliability index equal to 3. All the final designs are presented in Table 3.

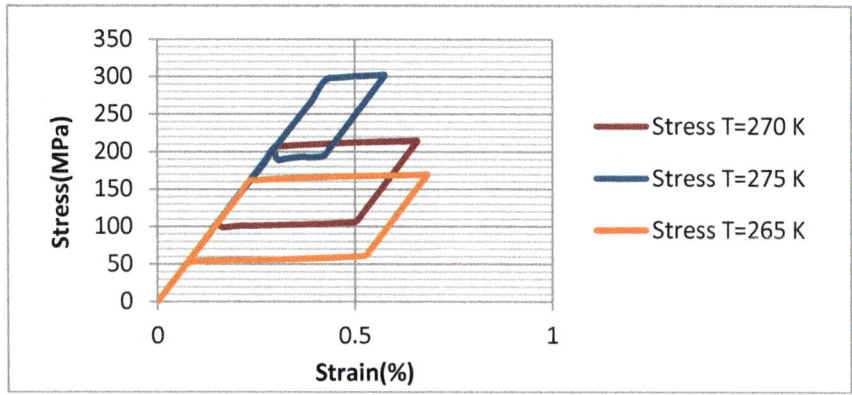

Fig. 3. Effect of temperature in the hysteresis loop

Table 3. RBDO results of SMA plate

		Distribution	Design point	Safety factor	Optimal point
Design variables	r (mm)	Normal	20	1.2007	24.014
	h (mm)	Normal	10	0.98	9.8
Objective function (mm³)	V		96,858		93,561
Constraint (MPa)	σ_{max}		219.29		218.86

5   Conclusion

In this work, the optimization of shape memory alloy was performed taking into account uncertainties. Safety factors based on KKT method were proposed as RBDO methodology. The optimization tools of the PSO method were applied owing to their ability to handle global optimization problems. The proposed safety factors based on RBDO methodology was employed and validated in the optimization of shape memory alloy. Even if the mechanical model used in this paper is simple, the RBDO methodology can be extended to complex one. To overcome such limitations, further research has to be done.

References

Aoues, Y., Chateauneuf, A.: Benchmark study of numerical methods for reliability-based design optimization. Struct. Multidiscip. Optim. **41**(2), 277–294 (2010). https://doi.org/10.1007/s00158-009-0412-2

Kharmanda, G., Ibrahim, M.H., Al-Kheer, A.A., Guerin, F., El-Hami, A.: Reliability-based design optimization of shank chisel plough using optimum safety factor strategy. Comput. Electron. Agric. **109**, 162–171 (2014). https://doi.org/10.1016/j.compag.2014.09.001

Lagoudas, D.C. (ed.): Shape Memory Alloys: Modeling and Engineering Applications. Springer, New York (2008). https://doi.org/10.1007/978-0-387-47685-8

Lopez, R.H., Lemosse, D., de Cursi, J.E.S., Rojas, J., El-Hami, A.: An approach for the reliability based design optimization of laminated composites. Eng. Optim. **43**(10), 1079–1094 (2011). https://doi.org/10.1080/0305215X.2010.535818

Strelec, J.K., Lagoudas, D.C., Khan, M.A., Yen, J.: Design and implementation of a shape memory alloy actuated reconfigurable airfoil. J. Intell. Mater. Syst. Struct. **14**(4–5), 257–273 (2003). https://doi.org/10.1177/1045389X03034687

Relaxation of Residual Stresses Induced by Ultrasonic Shot Peening Due to Cyclic Loading

Sondess Manchoul[1,2]([✉]), Raoutha Seddik[1,2], Rabii Ben Sghaier[1,3], and Raouf Fathallah[1,2]

[1] UGPMM, ENIS, Route de Soukra Km 2.5, BP 1173-3038 Sfax, Tunisia
manchoulsondess@yahoo.fr
[2] National Engineering School of Sousse, University of Sousse,
BP 264 Erriadh, 4023 Sousse, Tunisia
[3] Higher Institute of Applied Sciences and Technology of Sousse (ISSATS)
CitéTaffala (Ibn Khaldoun), University of Sousse, 4003 Sousse, Tunisia

Abstract. Conventional Shot-Peening is one of the popular surface enhancement processes. It consists on projecting small shots at the surfaces of the metallic components. Ultrasonic Shot-Peening is based on the same principle. The differences between both mechanisms were: the size of shot (from 0, 25 and 1 mm for Conventional Shot-Peening, and 1 to 8 mm for Ultrasonic Shot-Peening) and the velocity (from 20 to 150 m/s for Conventional Shot-Peening, and 3 to 20 m/s for Ultrasonic Shot-Peening). Another difference is the mechanism used for projecting the shots. In Ultrasonic Shot-Peening process the shots, confined in a closed chamber, are projected by sonotrode vibration on the treated specimen that is fixed on the top of this chamber. So, during the Ultrasonic Shot-Peening, the shots can be recovered after the treatment. In this paper, we propose three dimensional finite element model of Ultrasonic Shot-Peening which enable predicting the residual Stresses generated by this process on a semi-infinite target after a repetitive impacts. Moreover, this model is used to evaluate the residual stresses relaxation in AISI 316L target under cyclic tensile loading. The numerical results are validated by comparing the residual stress profile induced by the numerical model with the experimental findings.

Keywords: Ultrasonic shot peening · Compressive residual stresses
Number of impacts · Relaxation

1 Introduction

During the last decade, a significant progress has contributed to the development of cold surface treatments used to improve mechanical properties of treated materials. Recently, new mechanisms of peening processes have been performed to generate Compressive Residual Stress (CRS) at surface, for instance Ultrasonic Shot Peening (USP), Water Jet-Peening (WJP), Ultrasonic Impact Treatment (UIT) and Surface Mechanical Attrition Treatment (SMAT). In this study we have focused on the USP mechanism which allows enhancing the fatigue performance of treated parts. USP consists of impacting the surface of the specimen by spherical hard shots, by the use of sonotrode vibration. It has the advantage to introduce deeper CRS and lower

© Springer Nature Switzerland AG 2019
T. Fakhfakh et al. (Eds.): ICAV 2018, ACM 13, pp. 257–265, 2019.
https://doi.org/10.1007/978-3-319-94616-0_26

roughness, compared to conventional shot peening (CSP) treatment. As the CRS present the key parameter of improving the fatigue behaviour of peened parts, the majority of numerical studies related to USP have been focused upon prediction the CRS profiles. In this context, Chaise et al. (2012) performed a specific USP pattern to compute CRS after normal impacts. Rousseau et al. (2015) examined the effect of shots' number in the USP process, proving that an increase in the shots quantity leads to a significant concentration of the CRS into the peened target.

Unfortunately, the CRS fields may relax due to subsequent mechanical cyclic loading. Whereas, these relaxed CRS could stabilize after few numbers of cycles. Hence, so as to predict correctly the fatigue strength of peened components it is crucial to take into account these stabilize residual stresses (RS). Experimental analyses prove that RS relaxation is affected by different factors which are: the amplitude, the type, and the number of applied cycles of the cyclic loading (Dalaei et al. 2011; Zaroog et al. 2011). In this present work, a three dimensional (3D) finite element (FE) pattern is proposed to simulate the USP process. This model accounts the relevant parameters of peening process, the cyclic elastic-plastic law coupled with superficial damage well as the surface contact conditions. This model is utilized to evaluate the RS distribution induced by USP after repetitive impacts for different velocities of impact. Thereby, we propose to predict the change of the RS relaxation due to mechanical cyclic loading.

2   Ultrasonic Shot Peening Finite Element Model

In order to predict the RS after USP, a three-dimensional model is carried out using ABAQUS/Explicit code. The USP finite element model is composed of elastic-plastic shots which impact the surface of the semi-infinite target with impingement angle 90°. The target used in this study is modeled as a rectangle 5 mm × 5 mm × 15 mm. It is meshed using hexagonal elements with reduced integration elements (C3D8R). To improve the precision of the numerical results, a refiner mesh is used in the Central area of the peened-target. For the BC, the bottom surface of the part is fully fixed. A co-efficient of friction is introduced to define the contact between the uniform spherical shots and the surface of the target. A reference area (Fig. 1a) chosen in this work is defined by taking the equilateral triangle formed by the centers of three successive shots. This area is considered instead of the whole surface in order to reduce the number of used shots. The peening coverage rate T can be determined as the ratio between the affected surfaces (impacted by shots) over the full representative area (surface of the equilateral triangle). Therefore, the coverage rate T can be expressed as follows:

$$T = \frac{2\pi}{\sqrt{3}} \left(\frac{a^*}{d}\right)^2 \tag{1}$$

Where T is the coverage rate, a* is the radius of indentation resulted from a mono-impact and d defines the distance between two adjacent shots. This radius a* can be determined numerically after the first impact as shown Fig. 1b.

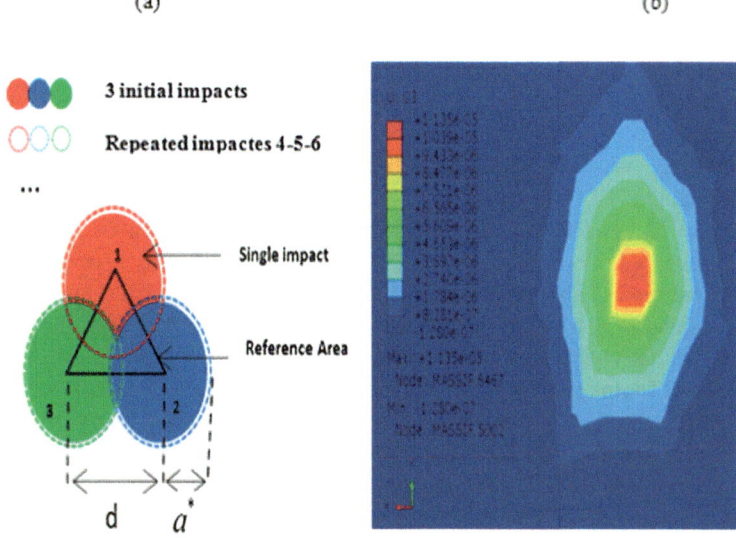

Fig. 1. (a) The reference area (b) the indentation created by a single shot impact

In this present study, the coverage is adjusted by defining the distance between the three shots in the triangular pattern. Indeed, with the first 3 impacts we can achieve the percentage of the desired coverage; and when we promote the number of impacts, the surface coverage rate will automatically increase. Hence, the full coverage (100%) is reached by the 3 first impacts while 6 and 9 impacts define respectively the 200% and 300% of the surface coverage rate.

An elasto-plastic material constitutive law with nonlinear kinematic and isotropic hardening of Lemaitre and Chaboche (2002) is used in this simulation.

3   Relaxation of Residual Stresses

The enhancement the fatigue behavior of ultrasonic-peened part is linked, principally, to the generation of CRS field into the treated part. However, these CRS may decrease significantly due to mechanical cyclic loading. This relaxation is generally linked to the accumulation of a plastic-strain with subsequent cyclic loading.

This relaxed CRS has a significant influence on the fatigue performance of the ultrasonic-peened parts. Indeed, investigating the fatigue behaviour without considering the residual stresses relaxation results in unreliable results. Therefore, we devote a great attention in the present study to the effect of cyclic loading on the redistribution of residual stress. The obtained results are devised into three different steps concerned with:

(i)   Change of residual stresses profiles during cyclic tensile loading using a load ratio $R_\sigma = -1$
(ii) Effect of number of applied cycles
(iii) Effect of the amplitude of cyclic loading.

4 Application

The present application is performed on the AISI 316L material which is considered as an elasto-plastic material with nonlinear kinematic and isotropic hardening. The mechanical properties of the studied AISI 316L material are: $E = 196$ GPa, $\sigma_y = 220$ MPa, $v = 0.29$, $C = 30000$ MPa, $\gamma = 60$, $Q = 150$ MPa and $b = 1$ (Laamouri et al. 2013). We propose to evaluate the influence of the number of impacts (i.e. 3, 6, and 12 impacts) for two velocities (4 m/s and 8 m/s) on the residual stresses profiles induce by USP process. In this present study, the diameter D of shots is fixed at 4 mm. As mentioned above; the indentation radius a^* is calculated numerically after the first indentation of a single shot. Thereafter, it is introduced as an input value to this Eq. 1, so as to determine the distance d between two shots ensuring a surface coverage equal to 100%. In this present simulations, the values of a^* are evaluated for two cases: (i) for the case of the impact velocity 8 m/s, the indentation radius a^* is equal to 0.325 mm and the distance d is fixed at 0.619 mm (ii) for the case of velocity 4 m/s, the radius a^* is equal to 0.21 mm and the distance d adopted to achieve a 100% of the coverage is equal to 0.41 mm.

The present approach is performed for purely cyclic tension loading with a load ratio of R = −1. The applied cyclic stress tensor below the target is given as follows:

$$\underline{\underline{\sigma}}_{app}(t,z) = \begin{pmatrix} \sigma_{xx}^a(z)\sin(\omega t) & 0 & 0 \\ 0 & 0 & 0 \\ 0 & 0 & 0 \end{pmatrix} \tag{2}$$

where σ_{xx}^a is the amplitude of the alternate stress within the x-direction. In this study $\sigma_{xx}^a = 350$ MPa.

As the RS values in the Z-direction are negligible compared to those obtained in X and Y directions, the biaxial RS tensor are given is:

$$\underline{\underline{\sigma}}_R(z) = \begin{pmatrix} \sigma_{Rxx}(z) & 0 & 0 \\ 0 & \sigma_{Ryy}(z) & 0 \\ 0 & 0 & 0 \end{pmatrix} \tag{3}$$

4.1 Validation of the USP Finite Element Model

The precision of the numerical USP model is validated by using the same ultrasonic-peening conditions adopted by Li (2011) in his experimental investigations: (i) 100 Cr6 steel shots with 4 mm of diameter and a velocity of impact equal to 4 m/s (iii) angle of impact 90°, and (vi) 100% of surface coverage rate. Figure 2 demonstrates a good harmony between the numerical and investigated X-ray CRS profiles induced by the USP especially for the outer layers. While for deeper layers, a gap between the RS predicted and the experimental analysis is noted. This little deviation is due to the uncertainties of the X-ray diffraction measurements in inner depths.

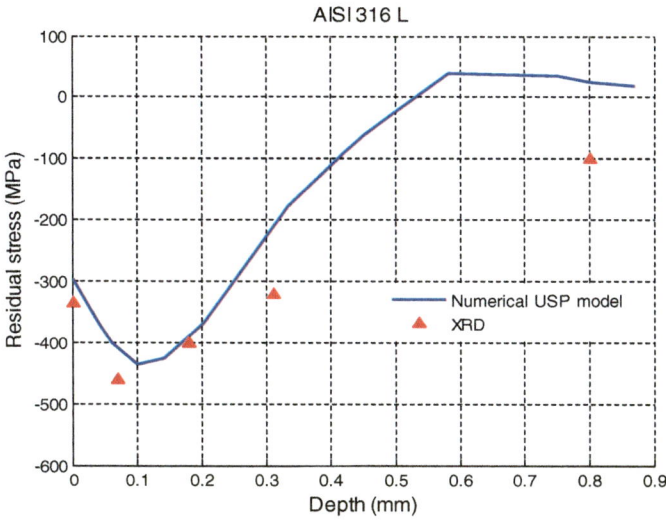

Fig. 2. Predicted and the analyzed X-ray CRS profiles of ultrasonic shot-peened AISI 316 L part

5   Results and Discussion

5.1   Prediction of RS Profiles Before and After Relaxation

Figures 3 and 4 depict the profiles of RS reached for three ultrasonic peening conditions (i.e. 3, 6, and 12 impacts) using two velocities of impact 4 ms^{-1} and 8 ms^{-1}. These figures, obtained before relaxation, reveal that:

For 4 ms^{-1} (Fig. 3), raising the number of impacts from 3 to 12 impacts changes the CRS at the surface from -180 MPa to -390 MPa. In addition, the maximum of CRS goes down significantly proving the beneficial effect of promoting the impact number. Moreover, Fig. 3 illustrates that varying the number of impacts results in a deeper thickness of the CRS layers.

However, for 8 ms^{-1} (Fig. 4), the RS value at the surface goes up from -285 MPa (for 3 impacts) to -160 MPa (for 12 impacts). Hence, this figure demonstrates that, for this case of high velocity, increasing the number of impacts (equivalent to coverage rate) has a negative effect on the RS distribution which can affect the fatigue performance of the treated component.

In order to predict the stabilized RS profiles, the pre-stressed peened target is submitted to a purely alternate ($R = -1$) cyclic tensile loading with an amplitude $\sigma_{xx}^{a} = 350$ MPa. The obtained results (Fig. 5) show that, for a high velocity (8 ms^{-1}), this cyclic mechanical loading results in a significant redistribution and evolution of the RS field. It can be noted that the relaxation causes a reduction of the induced CRS. In fact, the CRS at the surface are converted to tensile stresses after relaxation (i.e. the RS = $+100$ MPa for 12 impacts). Thus, this relaxation can affect the beneficial effects of the induced CRS which can lead to relevant degradation in the fatigue performance of peened parts.

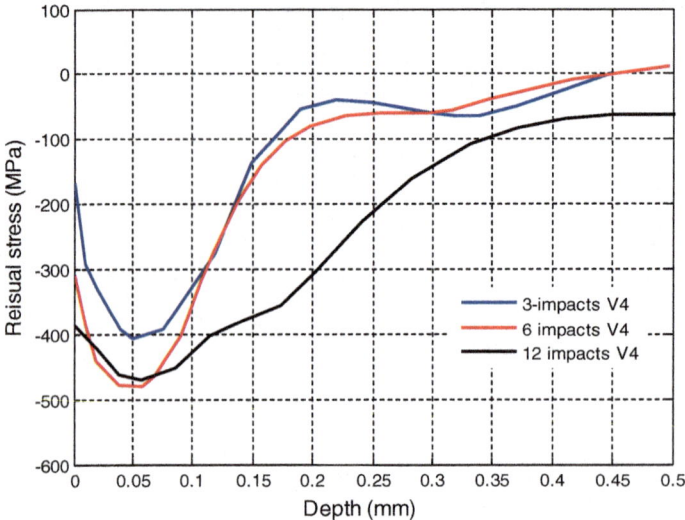

Fig. 3. Residual stress induced on the treated AISI 316L after repetitive impacts for a low velocity of impact (V = 4 m/s)

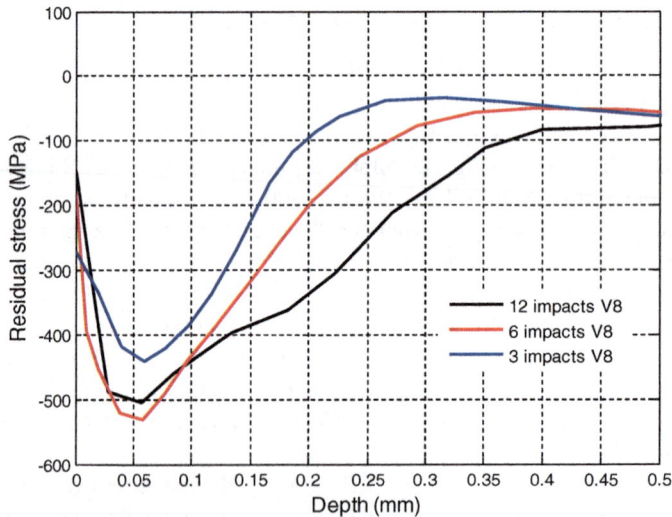

Fig. 4. Residual stress on the treated AISI 316L after repetitive impacts for a high velocity of impact (V = 8 m/s)

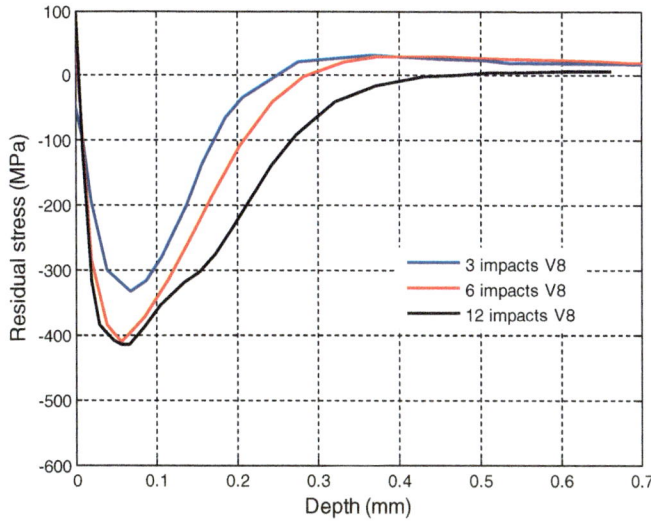

Fig. 5. Residual stress on the treated AISI 316L after repetitive impacts for a high velocity of impact (V = 8 m/s)

5.2 Effect of the Number and the Amplitude of Applied Cycles

Figure 6 depicts the influence of the applied cycles on the RS distribution. It proves that the most significant changes in the RS fields are achieved after the first applied-cycle, while the quasi-stabilized RS curves are obtained after few cycles ($\sim$ 25 cycles).

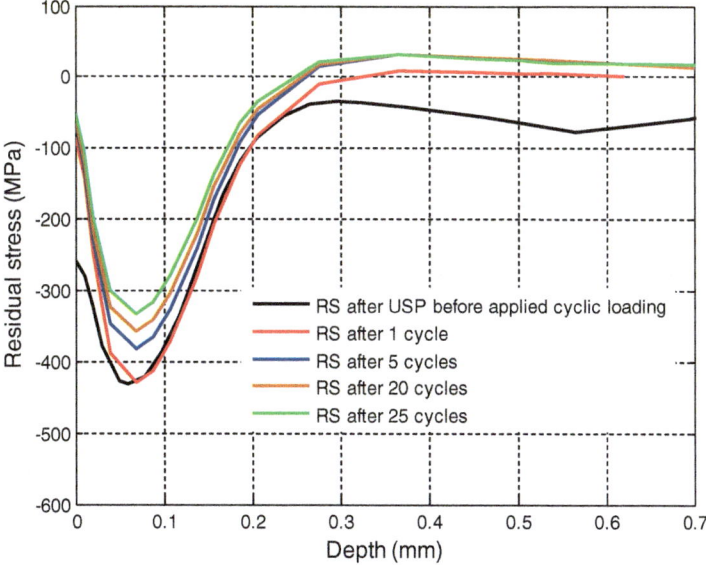

Fig. 6. The effect of the amplitude of the mechanical cyclic loading

To study the effect of the amplitude of the cyclic loading on the RS relaxation, different stress levels ($\frac{\sigma_{xx}^a}{\sigma_y} = 1.16$, $\frac{\sigma_{xx}^a}{\sigma_y} = 1.5$, and $\frac{\sigma_{xx}^a}{\sigma_y} = 1.6$) are applied to the treated parts. Figure 7 shows the evolution of the RS profiles, upon the peened target after 25 applied cycles of a purely-alternate tension (R = −1). It is noticeable that the amplitude of the cyclic loading affects significantly the RS relaxation, especially in the outer layers. Indeed, if the amplitude of the tensile loading is close the yield stress (σ_y); the RS relaxation is relatively negligible, especially, inside the compressed layers. Whereas, when the amplitude exceeds the yield stress, a relevant decrease in the induced CRS is noted for all affected layers. It can be deduced that the amplitude of cyclic loading plays an important role in the RS relaxation phenomena.

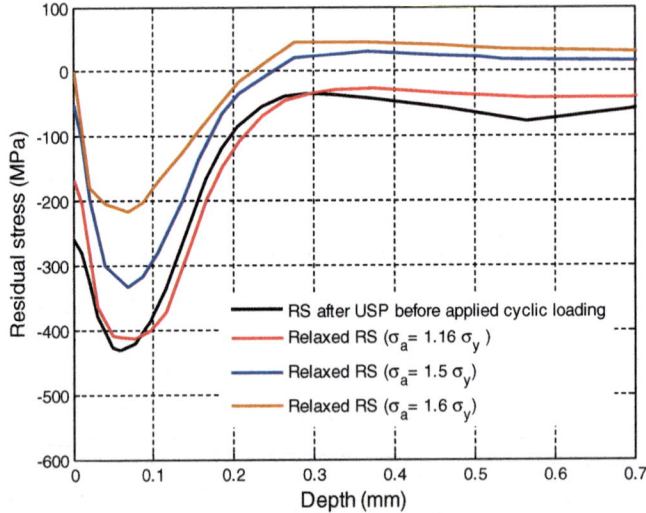

Fig. 7. The influence of the applied cycles on the RS distribution

6 Conclusion

A numerical model is conducted to simulate the USP process taking into account the cyclic work-hardening behavior of the treated material. A satisfactory correlation between the computed residual stresses and the experimental ones is observed, particularly in the first outer layers where the X-ray analyses are generally more precise than in the deeper layers. Using this proposed model, the CRS induced by a different number of shot impacts are predicted for two velocities. The obtained results show that increasing the number of impact for a low velocity (V = 4 m/s) introduce more CRS in the affected layers, which can improve the fatigue performance of the peened-part. However, for high velocity (V = 8 m/s) the benefit effect of increasing the number of impacts is annihilated. Accordingly, some precautions must be performed so as to avoid the unfavorable cases of over-peening in USP treatment. Our work consists also in modeling the RS relaxation for a purely alternate tensile loading with a load ratio $R_\sigma = -1$. It allows assessing the change of the initial CRS profiles which affect the fatigue performance of treated parts.

References

Chaise, T., Li, J., Nelias, D., Kubler, R., Taheri, S., Douchet, G., Robin, V., Gilles, P.: Modeling of multiple impacts for the prediction of distortions and residual stresses induced by ultrasonic shot peening (USP). J. Mater. Process. Technol. **212**, 2080–2090 (2012)

Rousseau, T., Hoc, T., Gilles, P., Nouguier-Lehon, C.: Effect of bead quantity in ultrasonic shot peening: surface analysis and numerical simulations. J. Mater. Process. Technol. **225**, 413–420 (2015)

Dalaei, K., Karlsson, B., Svensson, L.E.: Stability of residual stresses created by shot peening of pearlitic steel and their influence on fatigue lifetime. Mater. Sci. Eng. **A5282**, 1008–1015 (2011)

Zaroog, O.S., Aidy, A., Sahari, B.B., Zahari, R.: Modeling of residual stress relaxation of fatigue in 2024-T351 aluminium alloy. Int. J. Fatigue **33**, 279–285 (2011)

Lemaitre, J., Chaboche, J.L.: Mécanique des matériaux solides, 2nd edn. Dunod, Paris (2002). ISBN 210005662X

Laamouri, A., Sidhom, H., Braham, C.: Evaluation of residual stress relaxation and its effect on fatigue strength of AISI 316L stainless steel ground surfaces: experimental and numerical approaches. Int. J. Fatigue **48**, 109–121 (2013)

Li, J.: Simulation de Réparation par Soudage et Billage Ultrasonore d'un Alliage à Base Nickel. Ph.D. thesis LaMCoS Lyon (2011)

Finite Element Modeling of Shot Peening Process

Raoudha Seddik[1,2(✉)], Akram Atig[1,2], Rabii Ben Sghaier[1,3],
and Raouf Fathallah[1,2]

[1] UGPMM, ENIS, Route de Soukra Km 2.5, BP. 1173–3038, Sfax, Tunisia
Seddik.raoudha@yahoo.fr
[2] National Engineering School of Sousse, University of Sousse,
BP 264 Erriadh, 4023 Sousse, Tunisia
[3] Higher Institute of Applied Sciences and Technology of Sousse (ISSATS)
CitéTaffala (Ibn Khaldoun), University of Sousse, 4003 Sousse, Tunisia

Abstract. Shot Peening is common industrial cold-working process. It is widely used in several industrial fields particularly in automotive, aerospace and marine industries. This treatment is applied to enhance the fatigue performance of metallic components by: (i) retarding the crack growth due to the induced compressive residual stresses fields and (ii) inhibiting the crack initiation through the surface work-hardening. However, this process needs to be carefully controlled in order to avoid over-peening cases. The aim of the current study is to develop a dynamic and multi-impact shot peening process's model using the finite elements method. It is leading to predict the initial shot peening surface properties, which are classified, into three categories: (i) the outer layers compressive residual stresses, (ii) the induced plastic deformations and (iii) the superficial damage. To validate the proposed model, the obtained numerical results were compared with experimental ones analyzed by X-ray diffraction (XRD) for three materials the aeronautical-based Nickel super-alloy material Waspaloy and the AISI 316L stainless. The predictions are in good correlation and physically consistent with the experimental investigations. This proposed finite elements model is very interesting for engineering to predict the fatigue behavior of mechanical shot-peened components and to optimize the operating parameters of this process.

Keywords: Shot peening · Compressive residual stress
Surface work hardening · Superficial damage · Finite elements method

1 Introduction

Controlled shot peening is a cold surface treatment widely used in automotive and aerospace industries (Mylonas and Labeas 2011). It consists of bombarding metallic component surfaces, at relatively high velocities (20–120 ms^{-1}), with small spherical shots made, generally, of cast-steel, glass or ceramic (O'Hara 1984). Several studies (O'Hara 1984; Fathallah et al. 2004) show significant effects of shot peening on the fatigue behavior of treated components. The majority of experimental investigations

© Springer Nature Switzerland AG 2019
T. Fakhfakh et al. (Eds.): ICAV 2018, ACM 13, pp. 266–275, 2019.
https://doi.org/10.1007/978-3-319-94616-0_27

(Wang et al. 1998; Li et al. 1991) were focused, principally, on the prediction of compressive residual stresses fields. Furthermore, they pre-suppose that the compressive residual stresses are the key factor affecting the fatigue behavior of shot-penned metallic components. However, other works show clearly that shot peening surface modifications, such as: surface work-hardening, roughness and surface integrity, have also considerable influence on the fatigue performance of mechanical treated components (Ochi et al. 2001; Tekili 2002). Due to the difficulties and limitations of the experimental analysis and characterizations of shot peening surface modifications, a particular importance has been given to the numerical simulations of this mechanical process. Numerous finite elements models have been conducted to simulate the shot peening process. An initial simple model was performed by Al-Obaid (1990). It is based on three dimensional isoparametric finite elements. A quarter-symmetric shot peening model was presented by Meguid et al. (1999). It introduces contact elements to represent the physical contact between the shot and the target steel plate. This model has been exploited, in another work, to predict the equivalent stress, equivalent plastic strain and elastic strain as function of time (Meguid et al. 2002). In this study, importance was given to numerical convergence and to the validity of the compressive residual stresses fields. Frija et al. (2006) presented a three-dimensional finite element shot peening model leading to predict the compressive residual stresses fields, plastic strain profiles and, particularly, the superficial shot peening damage value. However, authors have applied the model for the case of isotropic hardening. The cyclic elastic-plastic hardening has not been taken into account. In the present work, we will develop a finite elements model by using the cyclic hardening law. In order to validate the proposed model, we are based on experimental results for three types of materials: the based-Nickel super-alloy Waspaloy and the AISI 316L stainless.

2 Finite Element Shot Peening Simulation Model

The general principle of the model (Fig. 1) is to simulate the impact of several shots (multi-impact model) on a structural element that can be extracted from the most critical region of the studied component. The modelling has been carried out using the finite element commercial code ABAQUS Explicit 6.10. In order to automatically generate several cases of simulations and/or parameters' optimization, a Python code has been developed and connected to Abaqus (Fig. 2). The friction between the shots and the treated surface has been characterized by the Coulomb friction model.

$$F_f = \mu F_n \tag{1}$$

Where F_f is the friction force, F_n is the normal force and μ is the friction coefficient.

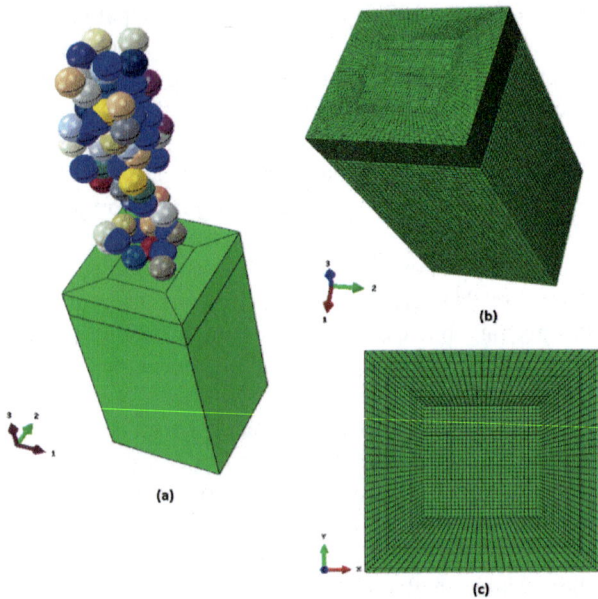

Fig. 1. Multi-shot-peening finite element model

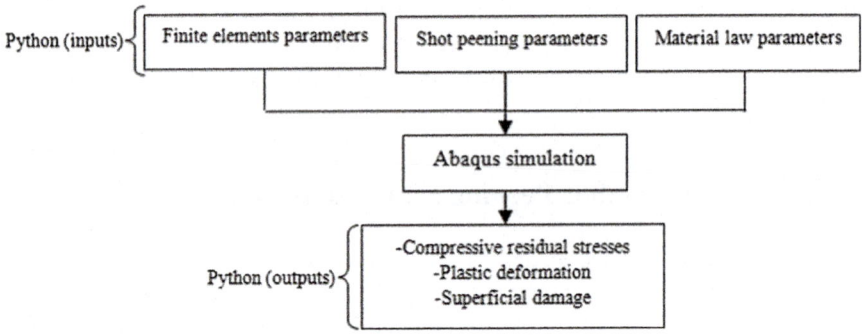

Fig. 2. Modeling steps

2.1   General Hypotheses

The assumptions adopted in the modeling of the shot peening process are:

- The shots are considered as rigid spheres of uniform radius.
- The diameter of the shot taken into account is the "nominal" diameter conventionally defined by the SAE J444 (2001)
- The mechanical response of the treated material is conforming to elastic-plastic behavior coupled with damage.
- The velocity of the shot is assumed to be constant during the impact.
- The angle of impact is considered equal to 90°.

2.2 Target Geometry, Boundary Conditions and Mesh

The target component has been modelled as a rectangular body with a width of 2 mm, a length of 2 mm and a height of 5 mm (Fig. 1). For the boundary condition, the bottom surface of the target has been fixed. The target was meshed by means of eight-node brick solid elements. In order to find the best compromise between the quality of the obtained results and the calculation time, the refinement of the mesh is located in the area of the contact shots/surface. Indeed, several calculations have been made to check the fineness of meshing required in the contact zone. The size of the smallest element is: 0.01 mm $\times$ 0.01 mm $\times$ 0.01. The shots are meshed with C3D4 elements.

2.3 Material Model of Shot Peening Process

To describe the shot peening cyclic loading, we adopt, in the present work, the combined isotropic-nonlinear kinematic hardening model (Chaboche 1977). It is expressed as follows:

$$f(\underline{\sigma}, \underline{X}, R) = J_2(\underline{\sigma} - \underline{X}) - R - \sigma_{y_0} \leq 0 \tag{2}$$

The nonlinear hardening tensor is defined by:

$$\underline{dX} = \frac{2}{3}C\underline{d\varepsilon}^p - \gamma\underline{X}dp \tag{3}$$

The isotropic hardening variable is defined by:

$$dR = b(Q - R)dp \tag{4}$$

The coefficients depending on the material are: the initial yield stress σ_{y_0}, two coefficients to represent the evolution of the isotropic hardening, b and Q, and two coefficients to represent the evolution of the kinematic hardening, C and γ.

In order to predict the shot peening superficial damage, Chaboche et al. (1977) three-dimensional ductile plastic model of damage is used:

$$D \cong \frac{D_c}{\varepsilon_R - \varepsilon_D}\left(p\left[\frac{2}{3}(1+\upsilon) + 3(1-2\upsilon)\left(\frac{\sigma_H}{\sigma_{eq_{VM}}}\right)^2\right] - \varepsilon_D\right) \tag{5}$$

Where υ is the Poisson's ratio, σ_H the hydrostatic stress of the applied stress tensor and σ_{eq} the Von Mises' equivalent stress. The three variables D_c, ε_R and ε_D are considered constants, where ε_D is the initial critical deformation for damage and ε_R the deformation at rupture for which the damage is equal to D_c. p is the cumulated plastic strain.

3 Application and Validation of the Proposed Model

The application and validation of the proposed finite element model was based on experimental results obtained on three types of materials: Waspaloy, AISI 316L and AISI 2205. The mechanical proprieties (Table 1) and the damage parameters (Table 2)

(Abdul-Latif 1996; Pedro et al. 2014; Laamouri et al. 2013) of the studied materials have been largely discussed in the open literature. Table 3 summarizes the used shot peening conditions for the studied cases (Pedro et al. 2014; Ahmed et al. 2015; Fathallah 1994).

Table 1. Mechanical properties (Abdul-Latif 1996; Pedro et al. 2014; Laamouri et al. 2013).

Material	$E(GPa)$	υ	$\sigma_{y_{0.2\%}}(MPa)$	$R_m\ (MPa)$	$A\,(\%)$	$C(MPa)$	γ	$Q(MPa)$	b
Waspaloy	210	0.3	900	1275	25	1185356	435	−100	145
AISI 2205	192	0.3	632	799	38	192777	575	−23	13
AISI 316L	196	0.29	220	600	80	30000	60	150	1

Table 2. Damage parameters.

Material	ε_R	ε_D	D_c
Waspaloy	0.6	0.02	0.8
AISI 2205	0.75	0.02	0.8
AISI 316L	0.8	0.02	0.5

Table 3. The used shot peening conditions (Pedro et al. 2014; Ahmed et al. 2015; Fathallah 1994).

Material	Angle (%)	Shot velocity (m/s)	Shot diameter (mm)	Peening coverage (%)	Friction coefficient
Waspaloy	90	52	0.6	100	0.2
AISI 2205	90	40	0.6	200	0.1
AISI 316L	90	40	0.8	100	0.1

3.1   Validation of the Proposed Model: Waspaloy

Figure 3 shows a comparison between the analyzed X-ray diffraction and the calculated compressive residual stress profiles obtained in-depth of the peened Waspaloy part (Fathallah 1994). It is observed that the difference between the depth of the compressed layers obtained using our finite element model (0.22 mm) and that obtained by the experimentation (0.25 mm) is very small.

Figures 4 and 5 present a qualitative comparison between the calculated in-depth Von Mises's plastic deformations profile induced by shot peening and the Full Width at Half Maximum (FWHM) of the X-ray diffraction peak profile (Fathallah 1994).

Figure 4 shows that the depth of the deformed layers is 0.22 mm, which is very close to the depth of the compressed layers (Fig. 3). The qualitative comparison between the calculated Von Mises' equivalent plastic deformation profiles and the FWHM of the X-ray diffraction peak profile shows that the depth of deformed layers is well predicted by the proposed finite element shot peening model.

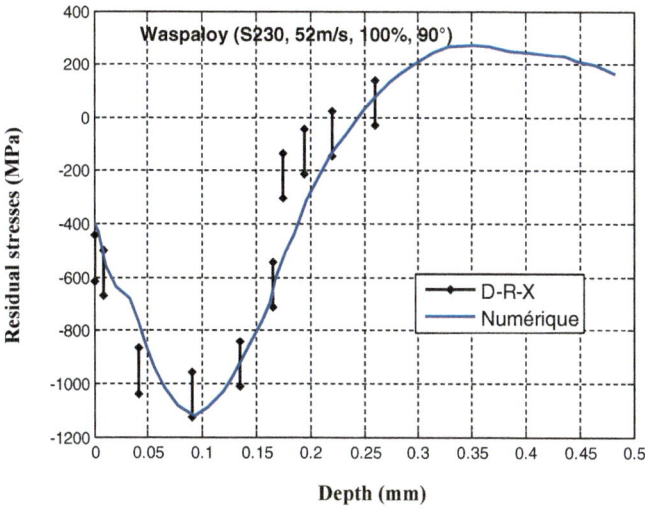

Fig. 3. Calculated and the analyzed X-ray residual-stress profiles in depth of shot-peened Waspaloy (Fathallah 1994)

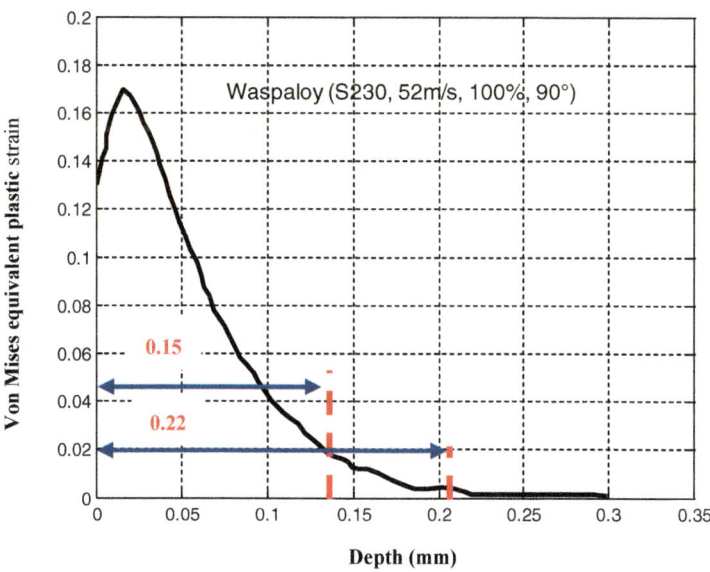

Fig. 4. Von Mises equivalent plastic strain profiles in depth of shot-peened Waspaloy

272   R. Seddik et al.

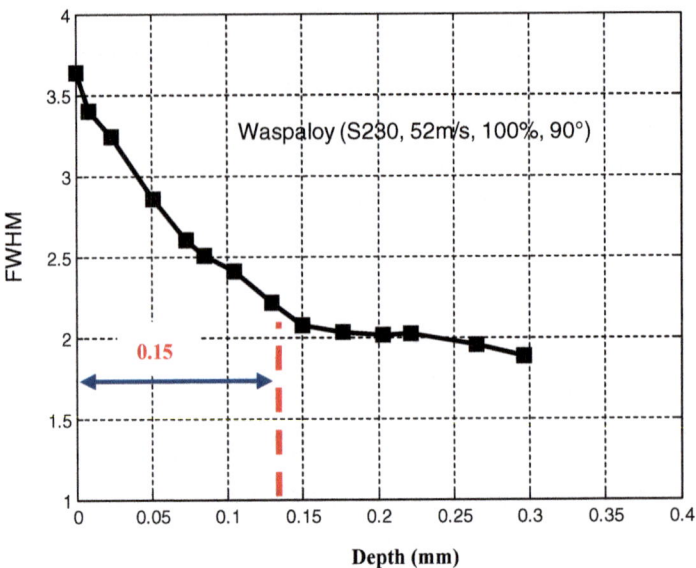

Fig. 5. Full width at half maximum of the X-ray diffraction peak profile (Fathallah 1994)

3.2 Validation of the Proposed Model: AISI 316 L

Figure 6 shows a comparison between the analyzed X-ray diffraction and the calculated compressive residual stress profiles obtained in-depth of the peened AISI 316L part. It is observed that the difference between the depth of the compressed layers obtained

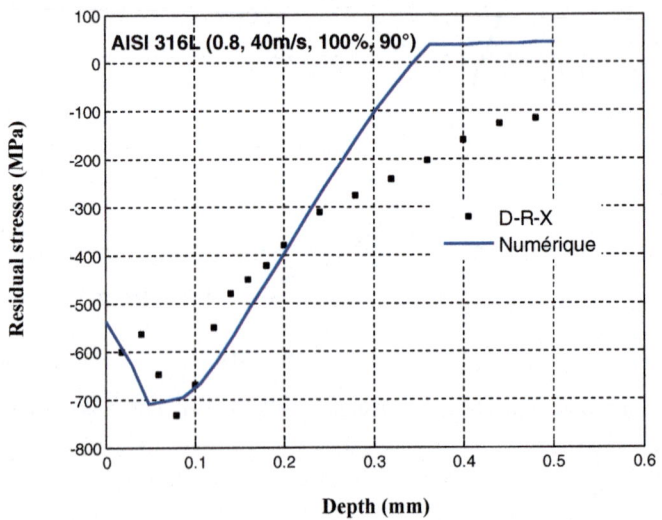

Fig. 6. Calculated and the analyzed X-ray residual-stress profiles in depth of shot-peened AISI 316 L (Ahmed et al. 2015)

using our finite element model (0.3 mm) and that obtained by the experimentation (0.5 mm) is important.

Figures 7 and 8 present a qualitative comparison between the calculated in-depth Von Mises's plastic deformations profile induced by shot peening and the FWHM of the X-ray diffraction peak profile (Ahmed et al. 2015).

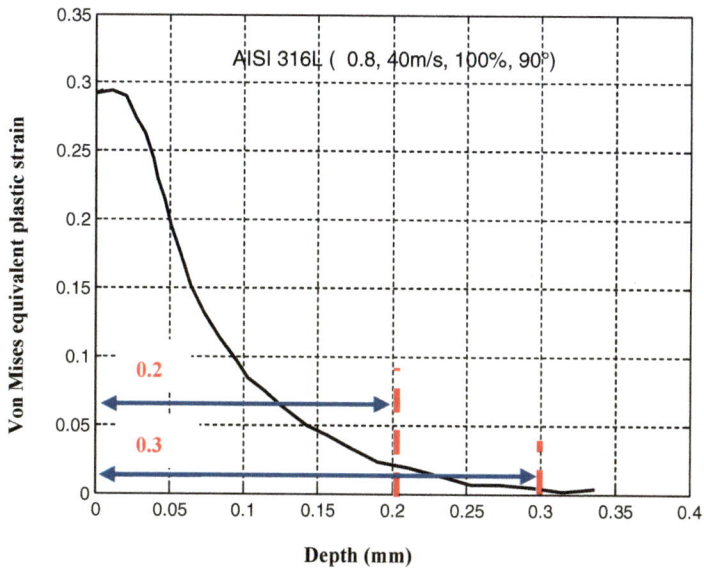

Fig. 7. Von Mises equivalent plastic strain profiles in depth of shot-peened AISI 36 L

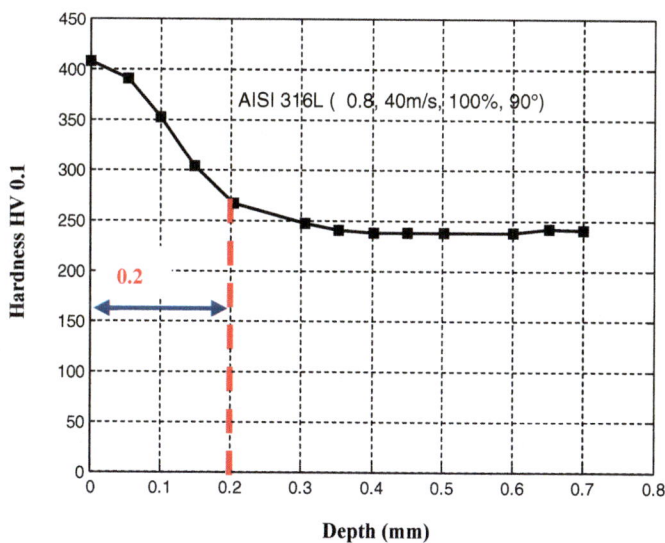

Fig. 8. Hardness profile AISI 316L (Ahmed et al. 2015)

Figure 7 shows that the depth of the deformed layers is 0.3 mm, which is very close to the depth of the compressed layers (Fig. 6). The qualitative comparison between the calculated Von Mises' equivalent plastic deformation profiles and the hardness profile shows that the depth deformed layers is well predicted by the proposed finite element shot peening model.

4   Discussion

The obtained results show the effect of the material (Figs. 2 and 5). For hard material Waspaloy a good correlation is observed between the residual stress profiles obtained by finite element calculations and those analyzed by X-ray diffraction. However, for soft material AISI 316L we note that the gap is very important. The gap between the experimental and numerical values can be explained by the uncertainties and the technical limitations of X-ray diffraction analysis and the control of shot peening treatment parameters. For the different studied materials, the depth of the compressed layers and the deformed ones are almost the same. This proves the validity of the proposed model. Figures 4, 5, 7 and 8 show a good qualitative correlation between the calculated equivalent plastic deformations and the FWHM.

5   Conclusion

An improvement 3D random dynamic model has been proposed to simulate the shot peening process via finite element method. Such improvement consists in including the repetitive random impacts of the shots and the cyclic work-hardening behavior coupled to the damage of the treated material. The compressive residual stress, the plastic strain and the damage variable in-depth of the affected layers can be predicted using our proposed model.

References

Mylonas, G.I., Labeas, G.: Numerical modelling of shot peening process and corresponding products residual stress, surface roughness and cold work prediction. Surf. Coat. Technol. **258**, 4480–4494 (2011)
O'Hara, P.: Developments in the shot peening process. Mater. Des. **5**(4), 161–166 (1984)
Fathallah, R., Laamouri, A., Sidhom, H.: High cycle fatigue behavior prediction of shot-peened parts. Int. J. Fatigue **26**, 1053–1067 (2004)
Wang, S., Li, Y., Yao, M., Wang, R.: Compressive residual stress introduced by shot peening. J. Mater. Process. Technol. **73**, 64–73 (1998)
Li, J.K., Zhang, R., Yao, M.: Experimental study on the compressive residual stress field introduced by shot-peening. In: Third International Conference on Residual Stresses (ICRS3), London, pp. 750–757 (1991)
Ochi, Y., Masaki, K., Matsumura, T., Sekino, T.: Effect of shot peening treatment on high cycle fatigue ductile cast iron. Int. J. Fatigue **23**, 441–448 (2001)

Tekili, S.: Enhancement of fatigue strength of SAE 9245 steel by shot peening. Mater. Lett. **57**, 604–608 (2002)

Al-Obaid, Y.F.: Three dimensional dynamic finite element analysis for shot peening. Mech. Comput. Struct. **3**, 681–689 (1990)

Meguid, S.A., Shagal, G., Stranar, J.C.: Finite element modelling of shot peening residual stresses. J. Mater. Process. Technol. **92–93**, 401–404 (1999)

Meguid, S.A., Shagal, G., Stranart, J.C.: 3D FE analysis of peening of strain-rate sensitive materials using multiple impingement model. Int. J. Impact Eng. **27**, 119–134 (2002)

Frija, M., Hassine, T., Fathallah, R., Bouraoui, C., Dogui, A.: FEM modelling of shot peening process: Prediction of the compressive residual stresses, the plastic deformations and the surface integrity. Mater. Sci. Eng. **426**, 173–180 (2006)

SAE J441: Cut Wire Shot. Society of Automotive Engineers, Warrendale (2001)

Chaboche, J.-L.: Sur l'utilisation des variables d'état interne pour la description de la viscoplasticité cyclique avec endommagement. In: Problèmes Non Linéaires de Mécanique, Symposium Franco-Polonais de Rhéologie et Mécanique, pp. 137–159 (1977)

Laamouri, A., Sidhom, H., Braham, C.: Evaluation of residual stress relaxation and its effect on fatigue strength of AISI 316L stainless steel ground surfaces: experimental and numerical approaches. Int. J. Fatigue **48**, 109–121 (2013)

Pedro, S., Rodríguez, C., Peñuelas, I., García, T.E., Belzunce, F.J.: Influence of the target material constitutive model on the numerical simulation of a shot peening process. Surf. Coat. Technol. **258**, 822–831 (2014)

Abdul-Latif, A.: Constitutive equations for cyclic plasticity of Waspaloy. Int. J. Plast. **12**, 967–985 (1996)

Ahmed, A.A., Mhaede, M., Basha, M., Wollmann, M., Wagner, L.: The effect of shot peening parameters and hydroxyapatite coating on surface properties and corrosion behavior of medical grade AISI 316L stainless steel. Surf. Coat. Technol. **280**, 347–358 (2015)

Fathallah, R.: Modélisation du Procédé de Grenaillage: Incidence des Billes et Taux de Recouvrement. Ph.D. thesis ENSAM Paris (1994)

Predicting the A356-T6 Cast Aluminum Alloy's High-Cycle Fatigue Life with Finite Elements

Amal Ben Ahmed$^{(\boxtimes)}$, Mohamed Iben Houria, and Raouf Fathallah

unité de production mécanique et matériaux, universié de Sousse,
Sousse, Tunisia
amalbenahmed@gmail.com, benhouria007@yahoo.fr,
raouf.fathallah@gmail.com

Abstract. This attempt proposes an engineering framework to predict the AL-Si-Mg casting alloy's High Cycle Fatigue (HCF) response considering the microstructural heterogeneities (Secondary Dendrite Arm Spacing (SDAS)) and its correlation with the casting defects effect. The developed approach is based on the evaluation of the highly stressed volume caused by local porosities and defined as the Affected Area (AA), using Finite Element (FE) analysis. Therefore, a 3D Representative Elementary Volume (REV) describing the defective material, was embedded to evaluate the cast aluminum alloy's High Cycle Fatigue behavior under various load conditions. Work hardening due to cyclic loading is considered by applying the Lemaitre-Chaboche model. The Kitagawa-Takahashi Diagrams were simulated, using the Affected Area Method, under fully reserved tension and torsion loadings for different SDAS values. The generated diagrams were compared to experimental data carried out on cast aluminium alloy A356 with T6 post heat-treatment with different microstructure (39–72 μm). The results show clearly that the proposed approach provides a good estimation of the A356-T6 fatigue limit and exhibits good ability in simulating the Kitagawa-Takahashi Diagrams for fine and coarse microstructures. The developed framework is practical tool able to generate the Kitagawa diagrams for fine and coarse microstructures, at different fatigue loads.

Keywords: High Cycle Fatigue · A356-T6 · Kitagawa diagrams
Secondary Dendrite Arm Spacing (SDAS)

1 Introduction

The A356-T6 is a classic Al-Si casting alloy that is widely employed in aerospace and automobile fields due to its low density, good process ability and high strength. The A356-T6 mechanical components, such as engine blocks and engine heads, are generally subjected to cyclic fatigue loads.

In the literature, there is a huge amount of data [1–6] proving that the A356-T6 HCF behavior is mainly influenced by microstructural heterogeneities characterized by the Secondary Dendrite Arm Spacing (SDAS) and local porosities.

Experimental investigations [3–5] have confirmed that: (i) for defect-free A356 alloy, the fatigue cracks initiate from the SDAS and the higher fatigue limit is obtained

© Springer Nature Switzerland AG 2019
T. Fakhfakh et al. (Eds.): ICAV 2018, ACM 13, pp. 276–283, 2019.
https://doi.org/10.1007/978-3-319-94616-0_28

for the finer SDAS value and (ii) for defective A356-T6 alloy, fatigue cracks initiate from surface porosities and the fatigue limit is directly affected by the defect size that is generally defined by the Murakami parameter "$\sqrt{area}$" [7]. It has been also shown that the Dang Van criterion and the Linear Elastic Fracture Mechanics (LEFM) are enable to predict correctly the fatigue behavior of casting Al-Si alloy [2, 3].

Therefore, this work aims to develop a predictive approach able to determine the A356-T6 fatigue limit by considering the SDAS and the porosities effects. The goal was to evaluate the local stress concentration caused by the defect by using the affected surface methodology [8]. The proposed approach was used to generate the A356-t6 Kitagawa diagrams for different microstructures and load conditions. A comparison between the generated diagrams and the experimental Data [1] for fine and coarse SDAS values were performed.

2   Modelling Frame Work

2.1   Short Review of the Affected Area Methodology

The Affected Area (AA) method [8] is proposed to evaluate the stress distribution near the defect and to quantify its impact on the fatigue response under various cyclic loads. The AA was defined as the as the part of the High Loaded Plane where the considered fatigue criterion is violated ($\sigma_{eqM} \geq \beta$) (Fig. 1).

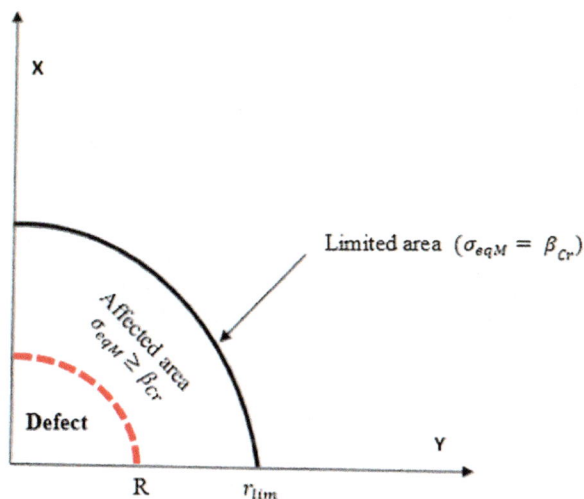

Fig. 1. Affected area

Based on stress analysis and numerical simulations, the authors [8] have showed that for all the defect sizes, when the applied load (σ_a) is equal to the fatigue limit (σ_D) the same affected area value is obtained. In this case ($\sigma_a = \sigma_D$), the affected area is defined as the Affected Area Limit (AAL) and the authors have proposed the material parameter (K_l) that links between the AAL and the fatigue limit (1):

$$K_l = \sqrt{AAL} \times \sigma_D \qquad (1)$$

2.2   Finite Element Analysis

In this study, a 3D Finite Element (FE) simulations using ABAQUS software were carried out to determine the fatigue response of A356-T6 aluminum alloy. The RVE model used to determine the stress distribution in the vicinity of the defect is a cube containing hemi-spherical defect. Due to the symmetry of the problem only ¼ of the numerical specimen is considered. Boundary conditions and symmetries are implemented as shown in Fig. 2.

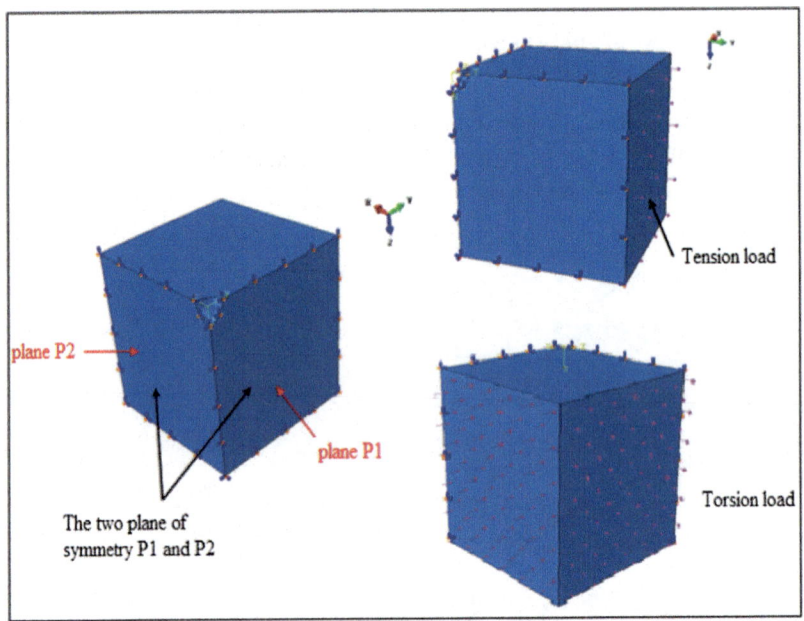

Fig. 2. FE model: load and boundary conditions: (a) tension loading (b) torsion loading

The nonlinear isotropic/kinematic hardening model is used to evaluate the defective A356-T6 fatigue behavior. This advanced model is able to simulate the cyclic plastic response (the Baushinger effect and the mean stress relaxation). The material parameters for the Lemaitre-Chaboche model [9] are sum up in Table 1.

Table 1. The A356-T6 cyclic parameters [10].

Material	E (GPa)	v	R0 (MPa)	Q	b	C	D
A356-T6	66	0.33	200	30	10	58000	680

2.3   Crossland Equivalent Stress Interpolation Through FE Analysis

In this study, the Crossland criterion (σ_{eq}^{Cr}) [11] is chosen to evaluate the A356-T6 fatigue stress distribution around the defect in the HLP.

In previous works [5], EF analysis of the σ_{eq}^{Cr} have showed that, for a given point in the High Loaded Plane, the fatigue response depends on (i) the defect size, (ii) the applied load and (iii) and its position. In fact, the simulated σ_{eq}^{Cr} remains constant for an arc having the same center as the defect. More details are provided in [5]. As it is reported in [5], σ_{eq}^{Cr} can be interpolated in the case of fully reserved tension and torsion loads as follow:

$$\sigma_{eq}^{Cr} = \frac{\tau_{-1}}{\sigma_{-1}}\sigma_a\left(\frac{1}{\left(\frac{r}{R}\right)^4}\right) + 1 \qquad \text{Fully reserved Tension} \qquad (2)$$

$$\sigma_{eq}^{Cr} = \frac{\tau_{-1}}{\sigma_{-1}}\sigma_a\left(\frac{1}{\left(\frac{r}{R}\right)^5}\right) + 1 \qquad \text{Fully reserved Torsion} \qquad (3)$$

Where:

R: the defect radius

r: Distance from the defect center to a considered point in the HLP.

σ_{-1} and τ_{-1} are respectively fatigue limit under fully reserved tension and torsion loading of defect free material.

2.4   Simulation of the Kitagawa Diagrams Using the Affected Area Methodology

The affected area describes the fatigue resistance behavior of cast material with pre-existing defects. As it shown in Fig. 1, the affected area may be calculated as follow:

$$\textit{Affected Area} = \frac{\Pi}{4}\left(r_{\text{lim}}^2 - R^2\right) \qquad (4)$$

Using the Murakami method [7], the defect size can be expressed as following:

$$\sqrt{area} = \sqrt{\frac{\Pi R^2}{2}} \qquad (5)$$

Then, the AA's expression becomes:

$$\textit{Affected Area} = \frac{1}{2}\left(\varepsilon^2 - 1\right)\left(\sqrt{area}\right)^2 \qquad (6)$$

Where

$$\varepsilon = \frac{r_{\lim}}{R} \tag{7}$$

From Eqs. 2 and 3, ε may be written as:

$$\varepsilon = \left(\frac{1}{\frac{\sigma_{-1}}{\sigma_a} - 1}\right)^{0.25} \quad \text{Fully reserved tension} \tag{8}$$

$$\varepsilon = \left(\frac{1}{\frac{\sigma_{-1}}{\sigma_a} - 1}\right)^{0.2} \quad \text{Fully reserved torsion} \tag{9}$$

Substituting the Eqs. 8 and 9 into the Affected Area Expression (Eq. 6), the following expressions are obtained:

$$Affected\,Area = \begin{cases} 0.5\left(\left(\frac{1}{\frac{\sigma_{-1}}{\sigma_a}-1}\right)^{0.25}\right)^2 \left(\sqrt{area}\right)^2 & \textbf{Tension} & (10)\\[4mm] 0.5\left(\left(\frac{1}{\frac{\sigma_{-1}}{\sigma_a}-1}\right)^{0.2}\right)^2 \left(\sqrt{area}\right)^2 & \textbf{Torsion} & (11) \end{cases}$$

In previous work [5], it has been proved that the fatigue limit under fully reserved tension loading of defect free material (σ_{-1}) may be expressed as a function of the SDAS (λ_2) as follow:

$$\sigma_{-1}(\lambda_2) = \frac{3\beta_0 \exp\left(\frac{-\lambda_2}{\lambda_0}\right)}{\alpha_0 \exp\left(\frac{-\lambda_2}{\lambda_0}\right) + \sqrt{3}} \tag{12}$$

Finally, substituting (12) into (10) and (11) respectively, new expressions of the Affected Area depending on the defect size ($\sqrt{area}$) and the microstructural parameter (λ_2) are obtained. In the following, these new expressions will be calculated for different SDAS values and defect sizes. Then, they will be compared to the Affected Area Limit given by (1). The fatigue limit will be given when the AA and the AAL are equals.

The identification of the K_1 parameters was made with experimental results [1] performed under alternate tension and torsion loadings. The obtained results as well as the experimental data used for the identification are illustrated in Table 2.

The improved Affected Area expressions have been employed to generate the A356-T6 Kitagawa diagrams for alternate tension and torsion loadings by accounting for both SDAS and defect size effects.

Table 2. Experimental results and identified K_l [1]

Load	SDAS (μm)	$\sqrt{area}$ (μm)	σ_D (MPa)	K_l (MPa μm)
Alternate tension	39	688	75	60 500
	72	900	57	92 200
Alternate torsion	39	380	88	74 450
	72	730	44	121 350

3 Comparison Between Simulations and Experimental Data

In this section, the A356-T6 Kitagawa diagrams will be predicted, for the upper and lower SDAS values (39 and 72 μm) used for the experimental tests [1], under tension and torsion fatigue loadings. The generated diagrams are plotted and compared with experimental results in Figs. 3 and 4. The obtained results lead to the conclusion that the improved Affected Area method exhibits good ability in simulating the A356-T6 fatigue limit at different load conditions. Consequently, it seems that the developed approach is able to describe adequately the A356-T6 fatigue behavior.

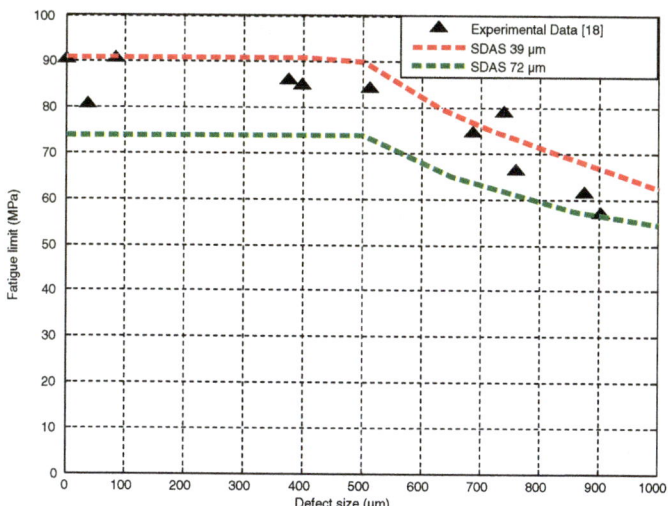

Fig. 3. Predicted Kitagawa diagrams under fully reserved tension

The obtained results show the following observations:

- From Figs. 3 and 4, it worth noticing that the predicted Kitagawa diagrams are constituted bye two zones: in the first zone (small defects), there is no considerate impact of the defect and the SDAS has the major role in controlling the Al 356-T6 fatigue limit. In the second zone (big defects), the fatigue limit is strongly affected by both the SDAS and the presence of defects.

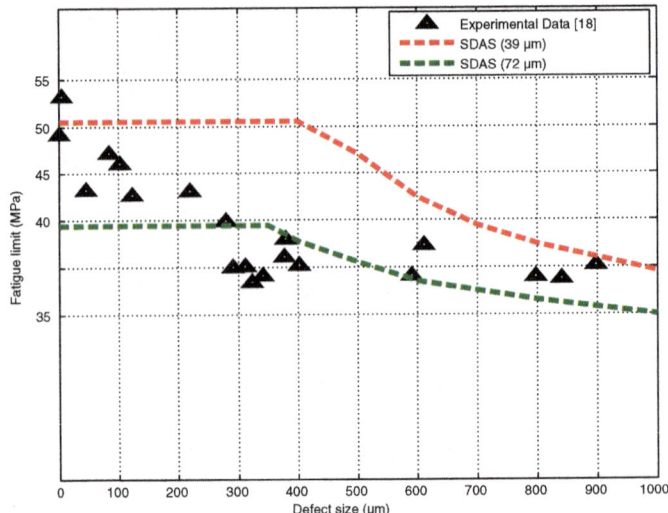

Fig. 4. Predicted Kitagawa diagrams under fully reserved torsion

- When the SDAS increases, the fatigue resistance decreases notably: the highest fatigue limits are obtained for finest microstructure (SDAS = 39 μm) and *vice versa*. However, in the presence of big defects, the simulated Kitagawa diagrams corresponding to the fine and the coarse microstructures converge. This result indicates that the SDAS effect decreases with the increase of the defect size.
- For torsion loading results (Fig. 4), it is obvious that the SDAS has the most detrimental impact on fatigue response, especially for small defects. In fact, in this zone, the difference between the two fatigue limits obtained for fine and coarse microstructure exceeds 40 MPa. Even for big defect sizes, the SDAS still dominate despite of its decreasing effect.

4   Conclusions

This paper proposes a predictive approach to simulate the Kitagawa diagrams for defective A356-T6 cast alloy taking into consideration the microstructure heterogeneities. From this study, it can be concluded that:

(i) The High Cycle Fatigue behavior of Al-Si alloy is dominated by microstructure heterogeneities characterized by DAS/SDAS and cast defects.

(ii) In this work, the affected area approach was modified by introducing a microstructure parameter (λ_2) that describe the SDAS impact, in order to generate the A356-T6 Kitagawa diagrams for different SDAS values.

(iii) The suggested modelling frame work represents an easy way to evaluate Al fatigue behavior with respect to the microstructure, mean stress and cast defect impacts. It gives appreciable results even for small defect sizes.

Acknowledgements. This work is partially supported by Sousse University. The authors also gratefully acknowledge the helpful comments and suggestions of the reviewers, which have improved the presentation.

References

1. Houria, M.I., Nadot, Y., Fathallah, R., Roy, M., Maijer, D.M.: Influence of casting defect and SDAS on the multiaxial fatigue behaviour of A356-T6 alloy including mean stress effect. Int. J. Fatigue **80**, 90–102 (2015)
2. Roy, M.J., Nadot, Y., Nadot-Martin, C., Bardin, P.G., Maijer, D.M.: Multiaxial Kitagawa analysis of A356-T6. Int. J. Fatigue **33**(6), 823–832 (2011)
3. Koutiri, I., Bellett, D., Morel, F., Augustins, L., Adrien, J.: High cycle fatigue damage mechanisms in cast aluminium subject to complex loads. Int. J. Fatigue **47**, 44–57 (2013)
4. Ahmed, A.B., Nasr, A., Bahloul, A., Fathallah, R.: The impact of defect morphology, defect size, and SDAS on the HCF response of A356-T6 alloy. Int. J. Adv. Manuf. Technol. **92**(1–4), 1113–1125 (2017)
5. Ahmed, A.B., Nasr, A., Fathallah, R.: Probabilistic high cycle fatigue behavior prediction of A356-T6 alloy considering the SDAS dispersion. Int. J. Adv. Manuf. Technol. **90**(9–12), 3275–3288 (2017)
6. Wang, Q.G., Praud, M., Needleman, A., Kim, K.S., Griffiths, J.R., Davidson, C.J., Cáceres, C.H., Benzerga, A.A.: Size effects in aluminium alloy castings. Acta Mater. **58**(8), 3006–3013 (2010)
7. Murakami, Y.: Metal Fatigue: Effects of Small Defects and Nonmetallic Inclusions. Elsevier, Amsterdam (2002)
8. Wannes, H., Nasr, A., Bouraoui, C.: New fatigue limit assessment approach of defective material under fully reversed tension and torsion loading. Mech. Ind. **17**(3), 310 (2016)
9. Chaboche, J.L.: Continuous damage mechanics—a tool to describe phenomena before crack initiation. Nucl. Eng. Des. **64**(2), 233–247 (1981)
10. Le Pen, E., Baptiste, D.: Prediction of the fatigue-damaged behaviour of Al/Al2O3 composites by a micro-macro approach. Compos. Sci. Technol. **61**(15), 2317–2326 (2001)
11. Crossland, B.: Effect of large hydrostatic pressures on the torsional fatigue strength of an alloy steel. In: Proceeding of International Conference on Fatigue of Metals, vol. 138. Institution of Mechanical Engineers London (1956)

Tensile Fatigue Behavior of Carbon-Flax/Epoxy Hybrid Composites

Mariem Ben Ameur[1,2(✉)], Abderrahim El Mahi[1], Jean-Luc Rebiere[1],
Moez Beyaoui[2], Moez Abdennadher[2], and Mohamed Haddar[2]

[1] Laboratoire d'acoustique de l'université du Maine (LAUM UMR CNRS
6613), Le Mans Université, Av. O. Messiaen, 72085 Le Mans Cedex 9, France
{Mariem.Ben_Ameur.Etu,abderrahim.elmahi,
jean-luc.rebiere}@Univ-lemans.fr
[2] Laboratoire de recherche de Mécanique, Modélisation et Production (LA2MP),
Département Génie Mécanique, Ecole Nationale d'Ingénieurs de Sfax,
Route Soukra, 3038 Sfax, Tunisia
moez.beyaoui@yahoo.fr,moezabd@yahoo.fr,
Mohamed.haddar@enis.rnu.tn

Abstract. Hybridation of carbon fiber composites with flax fiber offer interesting bio-degradability, respect of the environment, reduced cost and important
dynamic properties. The purpose of this work is to study the effect of hybridation on the mechanical fatigue behavior of unidirectional carbon-flax hybrid
composites. Static and fatigue tensile tests were realized for different laminates
made of carbon fibers and carbon-flax hybrid fibers with an epoxy resin. The
carbon laminates and two different staking sequences of hybrid laminates were
manufactured by hand lay-up process. Monotonic tensile tests were realized to
identify the mechanical properties of composites and the ultimate loading. Then,
load-controlled tensile fatigue tests were conducted on standard specimens with
applied load ratio R_F of 0.1. Specimens were subjected to different applied
fatigue load level until the failure (60%, 65%, 75% and 85%). Damage was
observed early after a few loading cycles. The decrease in the Young's modulus
was depending on the ratio of fibers on the composites. Overall, the stiffness
decreases by showing three stages for all studied samples. It has been found that
the stress-number of cycle S-N curves show that carbon laminates have higher
fatigue endurance than hybrid composites.

Keywords: Flax fiber · Carbon fiber · Hybrid composites · Static behavior
Fatigue behavior

1 Introduction

Laminated composite materials reinforced with conventional fibers, such as Kevlar,
Glass and Carbon are extensively used in industrial applications. However, the use of
these synthetic fibers raises many problems for health and environment. Over recent
years, the use of agro-based composites become increasingly higher because of their
biodegradability and their eco-friendly issues (Stamboulis et al. 2001).

© Springer Nature Switzerland AG 2019
T. Fakhfakh et al. (Eds.): ICAV 2018, ACM 13, pp. 284–291, 2019.
https://doi.org/10.1007/978-3-319-94616-0_29

The most interesting plant fibers are the flax fibers because they offer a good specific mechanical property (Wambua et al. 2003). Flax fiber reinforced composites have been investigated by many researchers (Faruk et al. 2012). Yan et al. (2014), Liang et al. (2014), Monti et al. (2016), Monti et al. (2018) and Haggui et al. (2018) studied the mechanical behavior of flax fiber polymers. Monti et al. (2017) and Duc et al. (2014) investigated the damping and mechanical properties of flax, glass and carbon laminates. They demonstrated that flax fiber reinforced composites present a higher performance for damping properties but lower performance in mechanical properties than the glass and carbon laminates. Le Guen et al. (2016) evaluated the relationship between the damping and the modulus for carbon-flax hybrid composites. They found that damping properties was increased by increasing the flax fiber content, but the mechanical properties were decreased.

However, few or no studies have looked on the fatigue behavior of flax/carbon hybrid composites. In this context, this work consists of studying the mechanical behavior of carbon/epoxy composites and hybrid carbon-flax/epoxy composites. Experimental tensile tests were carried out to characterize the different stacking sequences. Then tensile fatigue tests are also realized to follow the evolution of the mechanical properties during the tests. The stiffness evolution, the hysteresis loops and the fatigue life were studied. Moreover, the effect of hybridation on the mechanical properties of the carbon/flax composite materials is investigated.

2 Experimental Methods

2.1 Materials

The materials under study are carbon laminate and hybrid laminates made of unidirectional carbon fabric and unidirectional flax tape supported in epoxy-based SR 1500 resin with SD 2505 hardener. The weights of the unidirectional carbon and flax fibers were 300 g/m^2 and 200 g/m^2, respectively. The stacking sequence of laminates consists of 6 layers all oriented on the 0° direction of fibers are shown in Table 1. The composite plates (300 × 300 mm^2) were manufactured using a hand lay-up process. They were cured at room temperature (20 °C) at a pressure of 50 kPa using vacuum molding process for 7 h.

Table 1. Studied materials

Laminates	Stacking sequences	Thickness (mm)
$[C_3]_s$	[C/C/C/C/C/C]	2
$[F/C_2]_s$	[F/C/C/C/C/F]	2.5
$[F_2/C]_s$	[F/F/C/C/F/F]	2.9

Rectangularly test specimens with a length and width of 200 mm and 15 mm were cut from the laminated plates with a high speed of diamond saw. In order to avoid moisture absorption, no lubrication fluid was used while cutting the specimens. After cutting, the edges were slightly polished with fine sandpaper.

2.2   Testing Procedures

The experimental uniaxial loading was performed on a standard hydraulic testing machine Instron-8801 with a capacity of 100 kN. The load was determined by the load cell and the displacement in the tensile direction was measured using an extensometer. The machine is interfaced with a computer for test control and data acquisition. All types of laminated materials were subjected to static and fatigue tensile tests according to ASTM D3039/D3039 M standard test method. For each configuration, three specimens were tested to check the validity of the results. The monotonic tensile loading was conducted in order to determine the ultimate tensile load F_u at a constant displacement rate of 1 mm/min (Fig. 1).

a)   b)

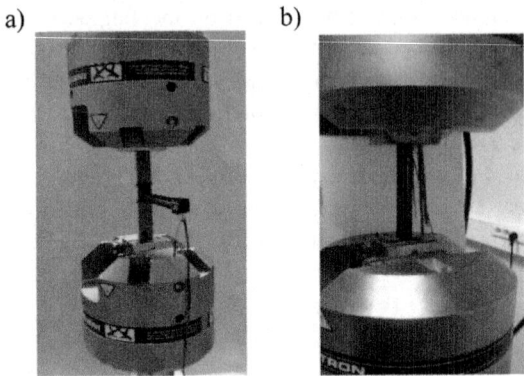

Fig. 1. Experimental setup: (a) Specimen under static tests and (b) Specimen under fatigue tests

The fatigue tests were carried out using a sinusoidal waveform at a constant frequency rate of 10 Hz for all tests. Specimens were tested under load control with various maximum load level. The applied load level $r_F = F_{max}/F_u$ (ratio between the maximum applied load and the ultimate tensile load) was varied from 60% to 85% (60%, 65%, 75% and 85%). The applied load ratio $R_F = F_{min}/F_{max}$ (ratio between the minimum and maximum applied load) was maintained constant at $R_F = 0.1$. All the fatigue tests were tested until the failure of specimens.

3   Results and Discussion

3.1   Static Tests

Typical stress-strain curves derived from experimental tests for carbon fiber laminates and carbon-flax hybrid fiber laminates are compared in Fig. 2. It can be seen that when the percentage of carbon increase the laminates exhibit better performance. Mechanical properties, e.g. Young's modulus, maximum strain, maximum stress as well as ultimate tensile load are presented in Table 2. The Young modulus of $[F_2/C]_s$ and $[F/C_2]_s$ laminates are respectively 59% and 35% lower than carbon fiber laminates.

The ultimate tensile load for $[F_2/C]_s$ and $[F/C_2]_s$ laminates are respectively 60% and 30% lower. This observation is due to higher intrinsic mechanical properties of the carbon fiber compared to flax fiber. In the next part, the performance of these stacking sequences on dynamic is studied.

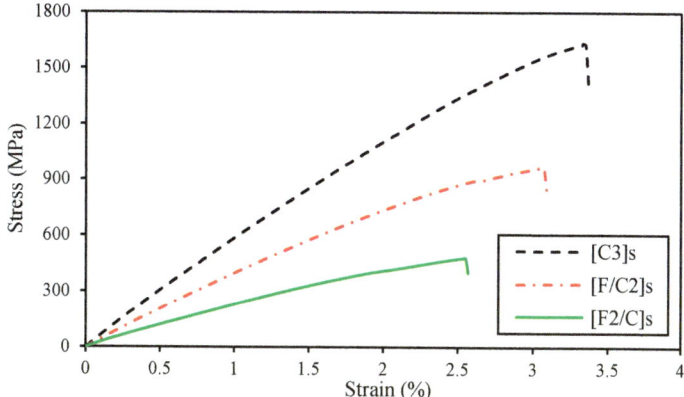

Fig. 2. Static response in tensile loading

Table 2. Mechanical properties of carbon and hybrid fiber composites

Laminate	E (GPa)	Maximum strain (%)	Maximum stress (MPa)	Ultimate load (kN)
$[C_3]_s$	85	3.3	1585	54
$[F/C_2]_s$	55	3.1	960	38
$[F_2/C]_s$	35	2.5	460	22

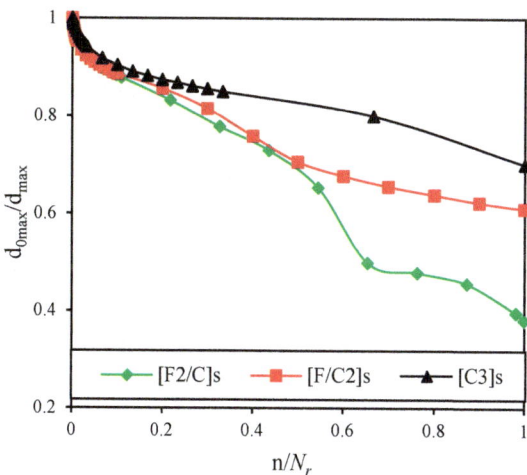

Fig. 3. Typical evolution of normalized displacement (d0max/dmax) as function of life ratio (n/N_r) at $r_F = 0.6$.

3.2   Fatigue Tests

3.2.1   Stiffness Degradation

The evolution of the specimen's stiffness gives information on the material's damage propagation. During fatigue tests, the increase in the maximum displacement d_{max} in each cycle was recorded. The normalized displacement ratio of (d_{0max}/d_{max}) was plotted as a function of the specimen's life ratio (n/N_r) for the different stacking sequences (Figs. 3 and 4), where d_{0max} is the value of maximum displacement at early cycle.

For all specimen types, the normalized displacement decreases with the life ratio. The maximum loss was of 30% and 40% for $[C_3]_s$ and $[F/C_2]_s$ respectively, but reached 60% for $[F_2/C]_s$ (Fig. 3). The stiffness of all specimens decreases in three stages. This behavior is similarly to the general behavior of composite materials under fatigue tests (Case and Reifsnider 2003). The initial stage is with steep stiffness reduction which involve micro-cracks in matrix. Followed by the intermediate second stage with slow decrease which involve the propagation of the microscopic damage (matrix cracking, fiber-matrix debonding and delamination between plies. And final stage with abrupt

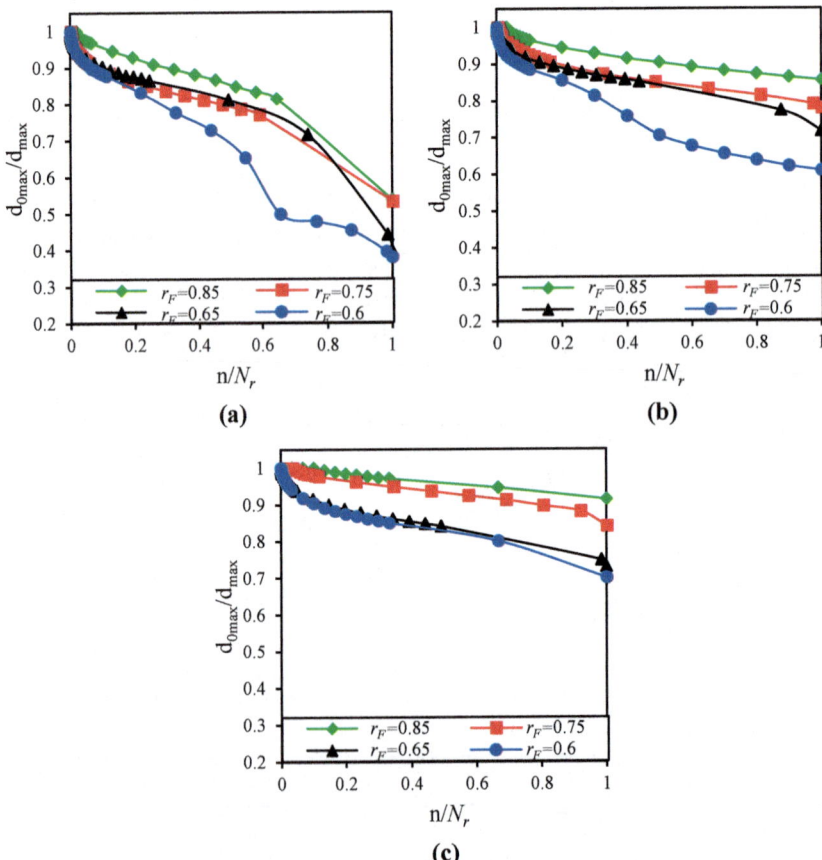

Fig. 4. Evolution of normalized displacement (d0max/dmax) as function of life ratio (n/N_r): (a) $[F_2/C]_s$, (b) $[F/C_2]_s$ and (c) $[C_3]_s$.

stiffness degradation followed by the specimen's failure that the most dominant is fiber breakage.

On the other hand, the stiffness degradation depends on the loading conditions (Fig. 4a–c). In fact, we clearly observe that the stiffness degradation decreases with the increase of the applied load level as function of the specimen's life ratio (n/N_r).

3.2.2   Hysteresis Curves

During fatigue tests, 100 experimental data points are recorded for each cycle. The hysteresis loops are obtained from the load-displacement curves. Typical load-displacement hysteresis loops at an early cycle (n = 1) and at a lately cycle (n = 3000) for all types of samples at the applied load level $r_F = 0.6$ are plotted in Fig. 5a–c. For all laminates, the behaviors are similar whereas the peak loads on hysteresis curves are different.

For any given type of specimen, the hysteresis loops move towards higher strains at constant stress level. Displacement corresponding to the minimum and maximum loading of loops increase with the number of cycle. We also clearly observe from these

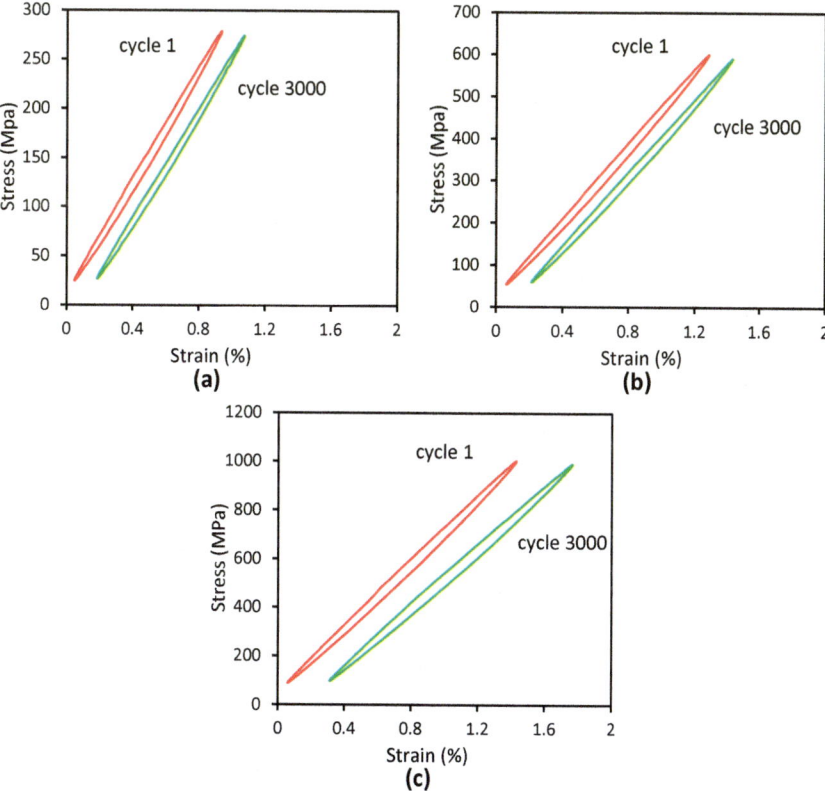

Fig. 5. Hysteresis loops under load controlled fatigue tests at $r_F = 0.6$ for: (a) $[F_2/C]_s$, (b) $[F/C_2]_s$ and (c) $[C_3]_s$.

hysteresis loops that the strain increase with the increase of the carbon fiber content which is in accordance with the maximum static strain of samples (Table 2). Finally, the area of the hysteresis loops decreases when the number of cycle increase.

3.2.3   Fatigue Life (S-N Curves)

The average numbers of cycle to failure N_r^{avg} which express the fatigue life for the different specimens are listed in Table 3. Also, the standard deviations for all specimens are given. The maximum loading stress versus number of cycles are plotted in Fig. 6. Wohler law (Eq. 1) was used to predict the specimen's life under tensile fatigue with imposed loading (Koricho et al. 2014).

$$\sigma = A - B\ln(N_r), \tag{1}$$

where σ is the maximum loading stress (MPa), A and B are constants depend on the type of material and N_r correspond to number of cycle at failure. The regression coefficient R^2 of the median Wohler's curve is closed to 1, indicating that the linear relation used fits well with the experimental data.

The presented curves reveal that specimens with higher carbon fiber volume fraction exhibit higher resistance to fatigue loading. This result is in accordance with the higher ultimate static strength of carbon fibers.

Table 3. Average fatigue life N_r^{avg} (standard deviation) of the studied stacking sequences

Laminate	$r_F = 0.85$	$r_F = 0.75$	$r_F = 0.65$	$r_F = 0.6$
$[C_3]_s$	20 (7)	93 (7)	2670 (1240)	4094 (790)
$[F/C_2]_s$	185 (110)	686 (120)	2712 (390)	8258 (1160)
$[F_2/C]_s$	112 (40)	2072 (380)	3860 (570)	7656 (2305)

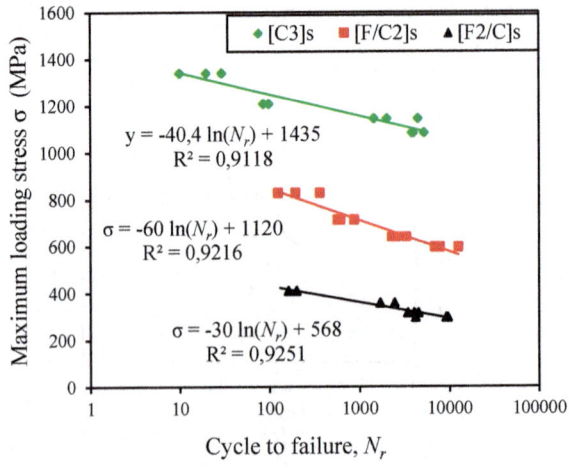

Fig. 6. Wohler curves for all types of samples

4   Conclusion

Experimental investigations of the fatigue behavior were conducted on carbon laminates $[C_3]_s$ and flax-carbon hybrid laminates $[F/C_2]_s$ and $[F_2/C]_s$. Specimens were submitted to static and fatigue tensile tests with load control. At first, the characteristics at failure of the studied composite specimens were determined from the static tensile tests. The stiffness evolution, the hysteresis loops and the S-N curves were studied for samples subjected to fatigue loading. The damage propagation was studied using the stiffness degradation method. For each laminate, the stiffness degradation depends on the applied load level. Resulting S-N curves of $[C_3]_s$ specimens show higher fatigue resistance compared to $[F/C_2]_s$ and $[F_2/C]_s$. It was also shown that the fatigue properties increase with the increase of carbon fiber content.

References

Case, S.W., Reifsnider, K.L.: Fatigue of composite materials. In: Comprehensive structural integrity, pp. 405–410 (2003). (Chap. 4.16)

Duc, F., Bourban, P.E., Manson, J.A.E.: Damping of thermoset and thermoplastic flax fibre composites. Compos. Part A **64**, 115–123 (2014)

Faruk, O., Bledzki, A.K., Fink, H.P., Sain, M.: Biocomposites reinforced with natural fibers: 2000–2010. Prog. Polym. Sci. **37**, 1552–1596 (2012)

Haggui, M., El Mahi, A., Jendli, Z., Akrout, A., Haddar, M.: Static and fatigue characterization of flax fiber reinforced thermoplastic composites by acoustic emission. Appl. Acoust. (2018)

Koricho, E.G., Belingardi, G., Beyene, A.T.: Bending fatigue behavior of twill fabric E-glass/epoxy composite. Compos. Struct. **111**, 169–178 (2014)

Le Guen, M.J., Newman, R.H., Fernyhough, A., Emms, G.W., Staiger, M.P.: The damping-modulus relationship in flax-carbon fibre hybrid composites. Compos. Part B **89**, 27–33 (2016)

Liang, S., Gning, P.B., Guillaumat, L.: Properties evolution of flax/epoxy composites under fatigue loading. Int. J. Fatigue **63**, 36–45 (2014)

Monti, A., El Mahi, A., Jendli, Z., Guillaumat, L.: Mechanical behaviour and damage mechanisms analysis of a flax-fibre reinforced composite by acoustic emission. Compos. Part A **90**, 100 (2016)

Monti, A., El Mahi, A., Jendli, Z., Guillaumat, L.: Experimental and finite elements analysis of the vibration behaviour of a bio-based composite sandwich beam. Compos. Part B **110**, 466–475 (2017)

Monti, A., EL Mahi, A., Jendli, Z., Guillaumat, L.: Quasi-static and fatigue properties of a balsa cored sandwich structure with thermoplastic skins reinforced by flax fibres. J. Sandw. Struct. Mater. 1–24 (2018)

Stamboulis, A., Baillie, C.A., Peijis, T.: Effects of environmental conditions on mechanical and physical properties of flax fibres. Compos. Part A: Appl. Sci. Manuf. **32**, 1105–1115 (2001)

Yan, L., Chouw, N., Jayaraman, K.: Flax fibre and its composites—a review. Compos. Part B **56**, 296–317 (2014)

Wambua, P., Ivens, J., Verpoest, I.: Natural fibres: can they replace glass in fibre reinforced plastics? Compos. Sci. Technol. **63**, 1259–1264 (2003)

Effect of Injection Direction in Elaboration of Polypropylene Reinforced with Olive Wood Flour on Ultrasonic and Morphological Properties

Nesrine Bouhamed[1,2], Slim Souissi[1(✉)], Pierre Marechal[2], Mohamed Benamar[1], and Olivier Lenoir[2]

[1] Laboratory of Electromechanical Systems/ENIS, University of Sfax, Sfax, Tunisia
bouhamednesrine4@gmail.com, slim.souissi@ymail.com
[2] Laboratory Waves and Complex Environments, University of Le Havre, Le Havre, France

Abstract. Although the anisotropy of wood fibers is reasonably well established, the anisotropy of injection molded wood fiber composites is not well understood. For this, fiber distribution is an important parameter in determining the properties of the composite. This work investigates the application of ultrasonic testing in evaluating natural fiber thermoplastic composites reinforced with olive wood flour (OWF). The characterization of sound propagation speed in the composite is intended to be a tool for evaluating the bio-composite namely fiber distribution and the effects of the direction of injection during the elaboration of the composite. The quality of fiber distribution homogeneity can be assessed by mapping the returning signals of the emitted longitudinal ultrasonic wave. This study presents the measured sound speeds for a composite system of OWF and polypropylene (PP) using immersion measurements. It is known that the longitudinal wave velocity is a function of the material property, which in turn is a function of fiber content and adhesion efficiency. Therefore, the aim of this work is to study the feasibility of using the ultrasonic longitudinal sound wave and the time of flight TOF instead of the morphological analysis with the scanning electron microscope, which is much more expensive and complicate.

Keywords: Ultrasonic properties · Bio-composite · Injection direction Time of flight

1 Introduction

Faced with the growing interest in the correct use of natural fibers, in parallel with glass and carbon fibers because of their low cost, high specific modulus, light weight, high availability and biodegradability [1, 2], the use of natural renewable lignocellulosic materials such as reinforcing fillers in thermoplastic or thermosetting polymers has recently increased in order to preserve environmental resources while improving economic activities. The industrial use of wood-plastic composite (WPC) cross for

© Springer Nature Switzerland AG 2019
T. Fakhfakh et al. (Eds.): ICAV 2018, ACM 13, pp. 292–299, 2019.
https://doi.org/10.1007/978-3-319-94616-0_30

several years in many fields like engineering and technology [3], these composites predominate in outdoor deck applications, the automotive field [4], the aerospace industry and others. Concerned thermoplastic polymers reinforced with wood fibers or flour the most widespread matrices are polypropylene (PP) [5] polyethylene (PE) [6], and polyvinyl chloride (PVC) [7, 8], because of their low cost. Furthermore, the processing temperature of these thermoplastics is less than 220 °C, which makes it possible to avoid the degradation of cellulosic fibers. This work will focus on olive wood flour reinforced with polypropylene.

Although several studies have reported the possibility of using wood flour as thermoplastic polymer filler, but no thorough investigation into the incorporation of OWF into a polypropylene (PP) matrix has been conducted. Only a few research studies reported in the literature have addressed the use of olive stones as filler. The use of olive nut flour as a filler for the PP matrix was considered by Amar et al. [1]. Composites based on olive stones and PVC as a matrix and the study of their mechanical and thermal stability were studied by [2, 9].

The applications of the non-destructive control (NDC) for the mechanical characterization materials are more and more numerous [10], and return the experimental approaches destructive mechanical tests.

In particular, the ultrasonic techniques, based on the analysis of the distribution of the elastic waves through a sample, are very used seen the direct connection between the characteristics of the elastic answer of the material and the characteristics of the distribution of the waves [11]. This property can be usefully exploited for the mechanical characterization of innovative and complex materials, as, for example, composites and/or laminated materials, materials anisotropes, biological materials, etc. [12, 13].

There are many advantages of the ultrasonic tests non-destructive: the freedom of choice of the geometry of samples, the possibility of obtaining a high level of precision, the quantity of information, the speed and the low(weak) price(prize) of the experiences (experiments).

We know that the longitudinal sound speed is a function of the material property, which, in his/her turns, is a function of the content in fibers and the efficiency of adhesion. Consequently, in this study, we prepared composites with polypropylene strengthened by fibers of wooden flour of olive tree by studying the feasibility to use sound waves to obtain the mappings of composites to realize one morphological and thermal analysis compare the results obtained with those realized by the microscope with electronic sweeping and tries thermal.

2 Materials and Methods

2.1 Raw Materials

Polypropylene (PP) used in this study is a standardized homopolymer PP H9069. This polymer is a simple flow grade with a melt flow index of 25 g/10 min (230 _C, 2.16 kg) according to ISO 1133 standard in order to facilitate the adhesion of the fibers in the composite material. The tensile strength of this polymer, measured by ISO 527-2, is 23 MPa, the tensil modulus is 1.6 GPa and the hardness Rockweil is 95 under ISO 2039-2.

The olive wood was from the region of Sfax in the center of Tunisia. It was recovered from the waste wood obtained from the craftwork through an aspirator, it was directly oven dried at 105 °C for 24 h to reduce its moisture content. It was then sieved and stored in plastic bags to protect it against moisture.

2.2   Elaboration of Composites

To eliminate all the absorbed humidity and avoid the formation of agglomerates, the OWF was pre-dried in 105 °C during 24 h before the extrusion. The polymer PP and the OWF were mixed in an extruder mini (DSM Xplore Netherlands) in screws coupled in co-rotation parallel with three zones of heating. The rotation speed of the screw was settled in 100 rpm and the exit (release) of the material was 200 g/h, the diameter of the screw was 10 mm. First, the OWF and the PP were manually mixed and placed in the hopper of extruder. The compound materials crossed the various zones and were extruded through the cylindrical matrix (diameter 1 mm). Then, the extruded were cooled then cut. The granules obtained subsequently underwent injection in order to obtain panels of dimensions l = 100 mm, L = 120 mm and e = 4 mm.

2.3   Ultrasonic Testing

To make the images of C-scan maps, focusing transducers of 0.5″ in diameter are used. The water in the tank is maintained at a constant and homogeneous temperature by a double tank system (Fig. 1). The transducer is positioned 43 mm above the plate and a precise X-Y scan is performed (Fig. 2). The equipment used mainly comprises an X-YZ rack completed by a control device (MISTRAS) and a signal acquisition and processing software (UTwin from Eurosonics).

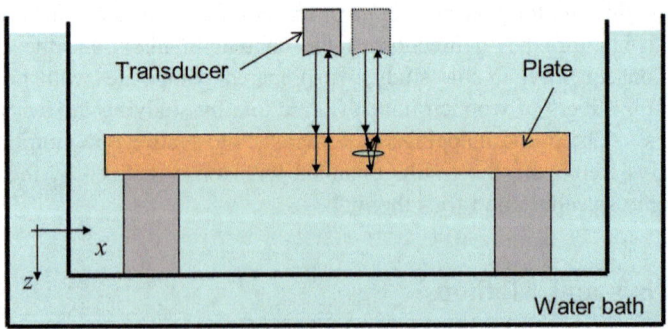

Fig. 1. Explanatory diagram of the equipment used for the ultrasonic measurement

Fig. 2. Immersion control of the plate using the ultrasonic method

2.4 Acquisition of Data

The mappings obtained are used to inform about the surface state as well as the homogeneity of the samples, it was also possible to interpret echoes based on these maps as shown in Fig. 3, the echoes obtained allows us to evaluate the properties composite by determining its longitudinal velocities (C_L) and its attenuation (Att_L).

Signals were processed by Matlab software using two estimation methods: the time of flight and the spectral method. The calculation of the speed of propagation of the wave in a material by flight time measurements is made from two echoes.

By time of flight and amplitude ratio measurements (Fig. 3b), properties such as speed and attenuation in the time domain can be evaluated from two echoes.

The propagation longitudinal speed C_L given by the Eq. (1) and the ultrasonic attenuation att_L by Eq. (2) are deduced from the transfer function:

$$C_L = \frac{2e_p}{\Delta t} \tag{1}$$

$$att_L = \frac{1}{2e_p} \ln\left(\frac{A_2}{A_1}\right) \tag{2}$$

Where:

$$\Delta t = T_2 - T_1$$

- T_1: is the projection on the time axis of the maximum of first signal
- T_2: is the projection on the time axis of the maximum of second signal
- A_1: is the amplitude of the first emitted signal

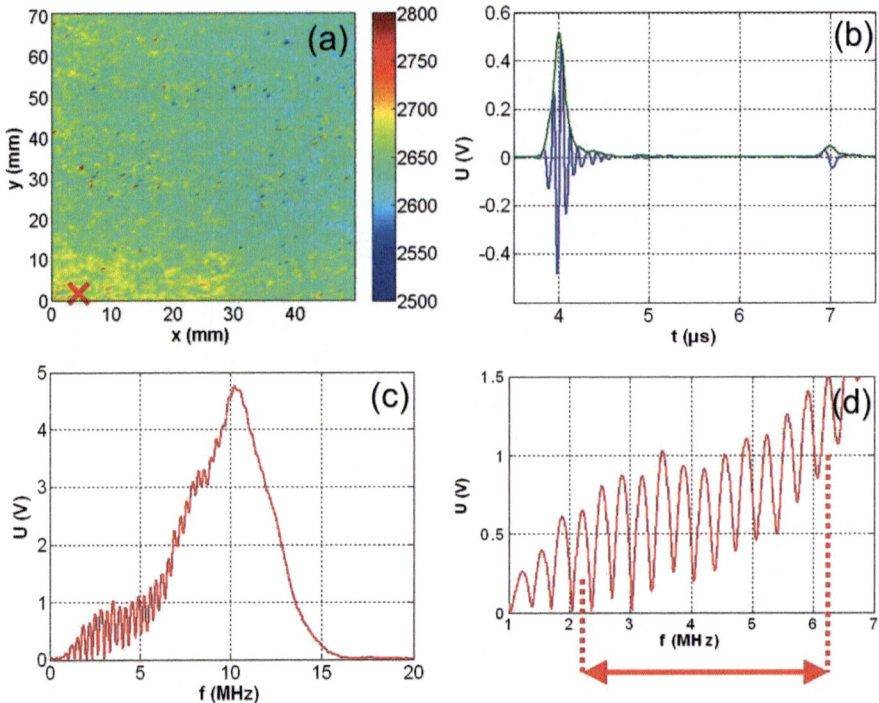

Fig. 3. Methods for calculating C_L in a marked point of the cartography (a) by the calculation of the time of flight (b) and the spectral method (c, d)

- A_2: is the amplitude of the second emitted signal
- e_p: is the thickness of the measuring point

The spectral method is more complicated, it consists in expressing the temporal signal (Fig. 3c–d) according to the spectral ratio of two echoes indeed, by applying the Fourier Transform (FT) via the Matlab signal processing software, we transform the real time signal s(t) in a complex frequency signal called spectrum $\underline{S}(f)$. The spectral method consists in expressing these properties according to the spectral ratio of two echoes, in the case where frequency evolutions of speed and attenuation can be observed. This method allows finding the ultrasonic dispersion properties with the frequency. The spectral method allows us to calculate the longitudinal speed propagation C_L in the plate from the Eq. (3):

$$C_L = 2e_p \cdot \Delta f \tag{3}$$

2.5   Scanning Electron Microscopy (SEM)

Scanning electron microscopy (SEM) was performed with a JEOL JSM-5400 Micro-scope operate in gat 15 kV. Prior to analysis, all samples were coated with a layer of gold to avoid sample charging under the electron beam.

3   Results and Discussion

Following the observation of the ultrasonic speed mapping (Fig. 4), we noticed a color difference from left to right depending on the injection direction of the bio-composite plate. The distribution of colors varies from red (2750 m/s) at the beginning of injection followed by a great heterogeneity of colors (blue, green, yellow from 20 mm from a variation of ultrasonic speed. From the 90 mm length of the panel the colors going more and more towards low speeds (dominance of the blue color).

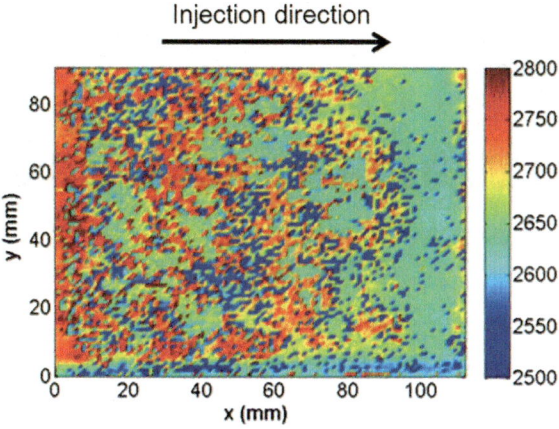

Fig. 4. C-scan maps of C_L of the PP /20% OWF plate.

This large variation between the two ends of the bio-composite panel according to the direction of injection and due to the poor distribution of the olive wood flour. The accumulation of red at the beginning of the panel shows the agglomeration of the OWF on this part.

This is confirmed after cutting and measuring the density of the noted samples (P1 and P2) as shown in Fig. 5, we observe a decrease in the density value from 0.963 (part P1) to 0.931 (part P2). This explains that the density of the plate decreases according to the direction of injection. This variation in density is accompanied by the presence of visually remarkable porosity on the cutting face (Fig. 5), this porosity increases with the direction of injection of the panel.

To be able to determine the adhesion rate between the fibers and the matrix on the surface of the plates according to their directions of injection, we could cut a piece of the lower part of the bio-composite plate as shown in the Fig. 5. We then made

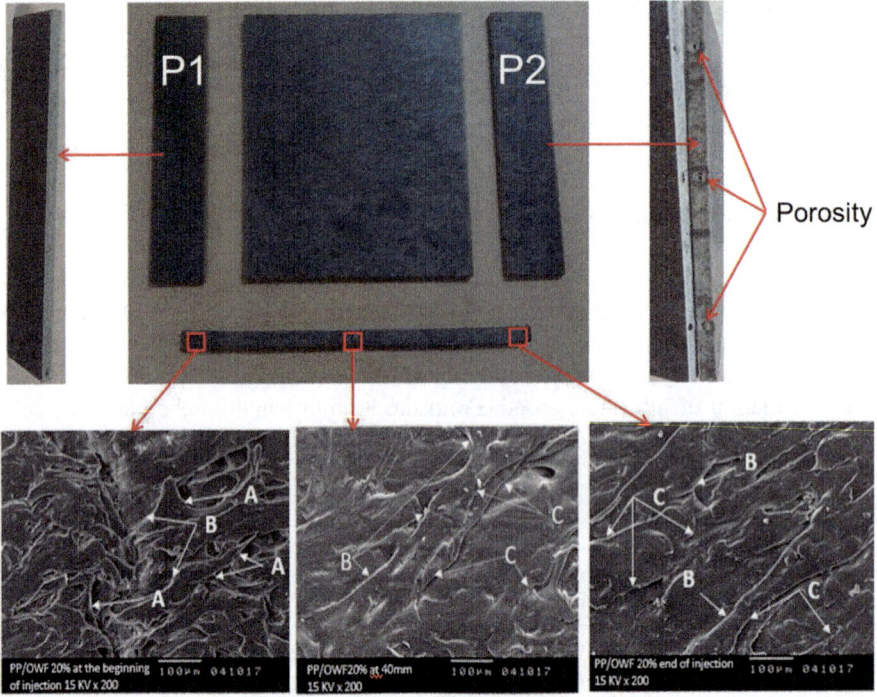

Fig. 5. Scanning electron micrographs at 200x enlargement at the surface of the samples made with 20% of fibers (legend: A = no close contact/good wetting, B = macro-fibrils, C = close contact/good wetting)

observations with the SEM on three different zones (begin of injection, sample medium and end of injection), the observations obtained show a difference in adhesion between the OWF and the PP according to the inspect area.

SEM micrographs showed variations in wetting at the fiber-matrix interface among the different sample area (Fig. 5). For example, at the beginning of the plate (first injected zone) fibers are not in close contact with PP (noted A) whereas at 40 and 120 mm length of the plate (middle and end of injection) we find same fibers completely wetted with PP (noted C), other fibers are not completely wet-ted either. SEM micrographs also showed variations in mechanical adhesion and interlocking at the fiber-matrix interface among the different sample area investigated. The three areas observe have macro-fibrils at the surface interlocking with the polymer matrix, thus increasing fiber reinforcement (noted B). The wetting and interlocking phenomena suggest a superior stress transfer in the case of fibers in the end-injected zone. It explained the better performance of this area among the three zone observe in the present study.

4   Conclusion

The direction of injection during the manufacture of bio-composite plates has a huge influence on the characteristics of the latter such as the density, the porosity and the adhesion ratio between fiber and matrix. It can be concluded therefore that moving away from the injection nozzle the ultrasonic speed decreases, consequently, the density of the composite decreases because of the agglomeration of the charge at the beginning of the plate. We also noticed the formation of pores at the end of the plate (end of injection). Regarding the adhesion between fiber and matrix it much better at the end of injection because of the lack of charge rate in this area.

References

1. Amar, B., Salem, K., Hocine, D., Chadia, I., Juan, M.J.: Study and characterization of composites materials based on polypropylene loaded with olive husk flour. J. Appl. Polym. Sci. **122**(2), 1382–1394 (2011)
2. Djidjelli, H., Benachour, D., Boukerrou, A., Zefouni, O., Martinez-Véga, J., Farenc, J., Kaci, M.: Thermal, dielectric and mechanical study of poly(vinyl chloride)/olive pomace composites. Express Polym. Lett. **1**, 846–852 (2007)
3. Bouafif, H., Koubaa, A., Perré, P., Cloutier, A.: Effects of fiber characteristics on the physical and mechanical properties of wood plastic composites. Compos. A Appl. Sci. Manuf. **40**(12), 1975–1981 (2009)
4. Panthapulakkal, S., Sain, M.: Injection-molded short hemp fiber/glass fiber-reinforced polypropylene hybrid composites—mechanical, water absorption and thermal properties. J. Appl. Polym. Sci. **103**(4), 2432–2441 (2007)
5. Raj, R.G., Kokta, B.V., Daneault, C.: Wood flour as a low-cost reinforcing filler for polyethylene: studies on mechanical properties. J. Mater. Sci. **25**(3), 1851–1855 (1990)
6. Harper, D., Wolcott, M.: Interaction between coupling agent and lubricants in wood–polypropylene composites. Compos. A Appl. Sci. Manuf. **35**(3), 385–394 (2004)
7. Keener, T.J., Stuart, R.K., Brown, T.K.: Maleated coupling agents for natural fibre composites. Compos. A Appl. Sci. Manuf. **35**(3), 357–362 (2004)
8. Jiang, H., Kamdem, D.P.: Characterization of the surface and the interphase of PVC–copper amine-treated wood composites. Appl. Surf. Sci. **256**(14), 4559–4563 (2010)
9. Naghmouchi, I., Espinach, F.X., Mutjé, P., Boufi, S.: Polypropylene composites based on lignocellulosic fillers: how the filler morphology affects the composite properties. Mater. Des. 1980–2015 **65**, 454–461 (2015)
10. Kromine, A.K., Fomitchov, P.A., Krishnaswamy, S., Achenbach, J.D.: Laser ultrasonic detection of surface breaking discontinuities: scanning laser source technique. Mater. Eval. **58**, 173 (2000)
11. Castellano, A., Foti, P., Fraddosio, A., Marzano, S., Piccioni, M.D.: Mechanical characterization of CFRP composites by ultrasonic immersion tests: experimental and numerical approaches. Compos. B Eng. **66**, 299–310 (2014)
12. Every, A.G., Sachse, W.: Determination of the elastic constants of anisotropic solids from acoustic-wave group-velocity measurements. Phys. Rev. B **42**(13), 8196 (1990)
13. El-Sabbagh, A., Steuernagel, L., Ziegmann, G.: Characterisation of flax polypropylene composites using ultrasonic longitudinal sound wave technique. Compos. B Eng. **45**(1), 1164–1172 (2013)

Comparison Between Ultrasonic and Mechanical Young's Modulus of a Bio-composite Reinforced with Olive Wood Floor

Slim Souissi[1(✉)], Karim Mezghanni[1,2], Nesrine Bouhamed[1,2],
Pierre Marechal[2], Mohamed Benamar[1], and Olivier Lenoir[2]

[1] Laboratory of Electromechanical Systems, LASEM/ENIS, University of Sfax,
Sfax, Tunisia
Slim.souissi@ymail.com
[2] Laboratory Waves and Complex Environments LOMC,
University of Le Havre, Le Havre, France

Abstract. Ultrasonic testing is a technique frequently used in the field of nondestructive evaluation given the fact that ultrasonic waves are directly related to the mechanical behavior of materials. It is for this reason that mechanical waves are often involved in solid material testing and the evaluation of their mechanical properties. As such, ultrasonic velocity is often used to identify so-called healthy concrete in comparison to deteriorated concrete. The objective of the present study is to determine Young's modulus of a bio composite using two methods: ultrasonic and mechanical methods. For this, a bio-composite based on polypropylene (PP) as a matrix and the olive wood flour (OWF) as a reinforcement was elaborated with extrusion using a twin extruder following by the injection in the form of 4 mm thick plate for ultrasonic control and standardized specimens for tensile testing. The longitudinal and transversal velocity of propagation of the wave in the plates is measured with the technique of immersion in water using transducer at 5 MHz center frequency in order to determinate the ultrasonic Young's modulus. Results show that the ultrasonic Young's modulus of the studied bio-composite is different than that mechanical Young's modulus. The causes of this difference will be studied.

Keywords: Ultrasonic Young's modulus · Mechanical Young's modulus
Bio-composite · Polypropylene · Olive wood floor

1 Introduction

Increasing attention to environmental protection from industrial pollution has raised interest in biomaterials, and in particular towards bio-composites, materials obtained usually by reinforcing matrices by means of natural fibers [1]. In this context, Natural fibers have recently become attractive for researchers, engineers and scientists as an alternative component for composite materials. Due to their low cost, fairly

© Springer Nature Switzerland AG 2019
T. Fakhfakh et al. (Eds.): ICAV 2018, ACM 13, pp. 300–309, 2019.
https://doi.org/10.1007/978-3-319-94616-0_31

good mechanical properties, high specific strength, non-abrasive, eco-friendly and bio-degradability characteristics, they are exploited as alternative for the conventional fiber, such as glass, aramid and carbon [2, 3]. To evaluate properties of these kinds of materials, ultrasonic non-destructive testing techniques and mechanical testing have proven to be effective and they were used to measure the Young's modulus. In fact, an important parameter to characterize the stiffness of material is the Young's modulus and its accurate determination is required in many fields ranging from medicine to structural mechanics [4].

In this paper, a bio-composite made of PP and OWF is elaborated, the longitudinal and transversal velocity of the composite are experimentally determined, the ultrasonic Young's modulus will be obtained through these velocities, then standardized specimens will be submitted to tensile tests through a tensile machine in order to determine the Young's modulus.

2 Materials and Methods

2.1 Materials

The PP used is a fluid injection-molding grade with a melt flow index of 25 g/10 min to facilitate the dispersion and process ability of the composite material. It is characterized by its low density, high inertia to chemical attack, high resistance to shocks and with temperatures of use higher than 100 °C.

OWF used as a reinforcement for PP composites. It is obtained by vacuuming woodwork waste from craft objects made by olive wood. The OWF used has an average grain diameter of 110 μm.

2.2 Methods

The composite plates are elaborated going through three steps:

The mixture used consists of two components: olive wood flour (OWF) and polypropylene (PP) in the form of granules. Three rate of reinforcement used in our mixtures are prepared: 10%, 20%, and 30%, PP plate using 0% reinforcement will also be elaborated.

The mixtures are extruded using a twin extruder containing three heating zones, the temperature used is 180 °C, the rotational speed is 100 towers per minute and the feed rate is 200 g/h.

This gives PP granules mixed with OWF.

After the preparation of the granules, they are putted in an injection machine in order to be injected in the form of plates having 4 mm of thickness and standard test pieces (ISO ½).

3 Ultrasonic Young's Modulus E$_{\text{ultra}}$

The ultrasonic Young's modulus is determined from the following expression

$$E = \frac{\mu(3\lambda + 2\mu)}{\lambda + \mu} \tag{1}$$

$$\lambda = \rho C_L^2 - 2\mu : \quad \text{first Lame coefficient} \tag{2}$$

$$\mu = \rho C_T^2 : \quad \text{second Lame coefficient} \tag{3}$$

To determinate the ultrasonic Young's modulus of plates, it is necessary to determine first the longitudinal and transversal velocity and the density of every plate. We will study now the case of a 10% reinforcement plate.

3.1 Longitudinal Velocity

The longitudinal velocity of propagation of the wave in the plate is determined with the technique of immersion in water at normal incidence. The transducer at 5 MHz center frequency and the plate are placed in the water bath and maintained with a support (Fig. 1).

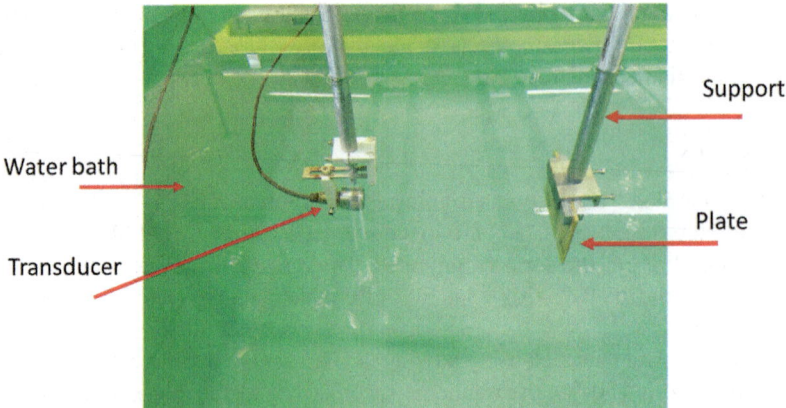

Fig. 1. Immersion control at normal incidence.

The signals obtained after the measurement made by the transducer, which inform about the longitudinal velocity, were processed using MATLAB software according to two estimation methods: the time of flight and the spectral method.

<u>Time of Flight Method</u>
The time of flight is the course time of a wave. This method consists in evaluating properties in the temporal domain from two echoes by measurements of time of flight and amplitude ratio as it's shown in Fig. 2.

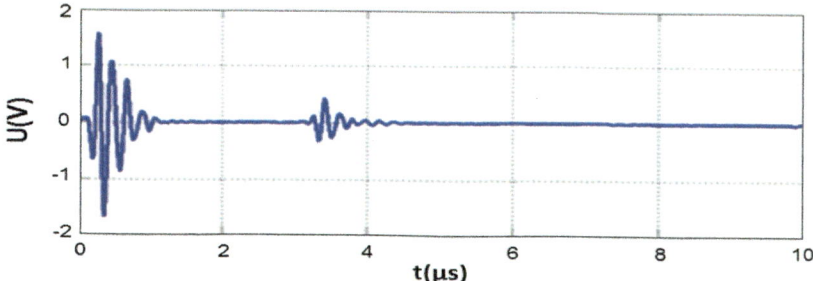

Fig. 2. Temporal signal of 10% reinforcement plate.

The longitudinal velocity in the plate is calculated through the following equation [5]:

$$C_L = \frac{2e_p}{\Delta t_L} \tag{4}$$

e_p is the plate thickness, $\Delta t_L = t_{2L} - t_{1L}$, with t_{1L} is the projection on the time axis of the maximum of the first echo $s_1(t)$, et t_{2L} is the projection on the time axis of the maximum of the second echo $s_2(t)$ (Fig. 3).

$$e_p = 4 \, mm.$$

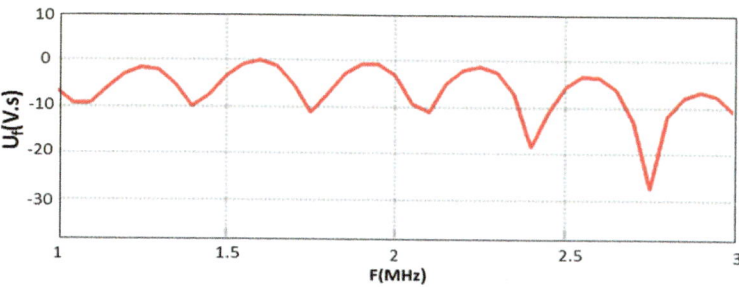

Fig. 3. Spectrum of 10% reinforcement plate.

The time between the two echoes is: $\Delta t = 3.4 - 0.4 = 3 \; \mu s. C_L = \frac{2e_p}{\Delta t_L} = \frac{2 \times 4.10^{-3}}{3.10^{-6}} = 2664 \, m/s.$

Spectral Method

The spectral method consists in expressing the temporal signal as a function of the spectral ratio of two echoes, in the case where frequency evolutions of speed and can be observed (Fig. 4).

By making the Fourier Transform of the signal $s(t)$ with Matlab, we obtain the spectrum $\underline{S}(f)$ whose module is shown in Fig. 3.

Fig. 4. Immersion in water control at oblique incidence.

The spectral method allows to calculate the longitudinal velocity C_L [6] in the plate through the following expression:

$$C_L = 2e_p \cdot \Delta f \tag{5}$$

Δf: interval between two consecutive minimums of frequencies.

According to the following spectrum $\Delta f = 0.333$ MHz $= 333$ kHz.

$$C_L = 2e_p \cdot \Delta f = 2 \times 4 \times 333 = 2664 \, \text{m/s}$$

These steps are repeated to calculate the longitudinal velocity of all plates.

3.2   Transversal Velocity

The transverse velocity of wave propagation in plates is determined through immersion in water method at oblique incidence. The experimental device is constituted of a water bath in which the plate and two 5 MHz center frequency transducers are placed. The first transducer acts as a transmitter, the second is a receiver. The angle made by the two transducers is set according to the Snell–Descartes law.

After the recovery of data and their analysis through Matlab software, we obtain the following spectrum:

The transverse velocity of 10% reinforcement plate is determined from the spectrum in Fig. 5.

The transverse velocity is obtained through the following expression [5]:

$$C_T = \frac{2e_p \cdot \Delta f}{\sqrt{1 + \left(2e_p \cdot \Delta f \frac{\sin\theta}{C_E}\right)^2}} \tag{6}$$

With C_E: water velocity $= 1470$ m/s.

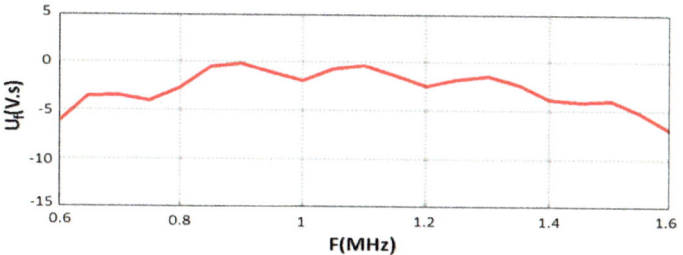

Fig. 5. Spectrum of 10% reinforcement plate.

From this spectrum, $\Delta f = 200$ kHz

$$C_T = \frac{2e_p.\Delta f}{\sqrt{1 + \left(2e_p \cdot \Delta f \frac{\sin\theta}{C_E}\right)^2}} = \frac{2 \times 4 \times 200}{\sqrt{1 + \left(2 \times 4 \times 200 \times \frac{\sin 40}{1470}\right)^2}} = 1272 \text{ m/s}$$

All previous steps are repeated for other plates at different percentages of reinforcement.

Table 1. C_L and C_T values.

% OWF	0	10	20	30
CL		2600	2664	2750	2816
CT		1230	1272	1311	1322

Table 1 presents the values of longitudinal and transverse plates velocities at different reinforcement percentage.

3.3 Density

The density was calculated practically by measuring the mass and volume of each plate with (Table 2):

$$\rho = \frac{m}{v} \tag{7}$$

Table 2. Densities values.

% OWF	0	10	20	30
ρ (g/cm3)	0.905	0.928	0.953	0.979

From the values of C_L C_T and ρ, the first and second coefficient of Lame are determined (Table 3)

The values of the ultrasonic Young's modulus are shown in Table 4.

Table 3. λ and μ values.

% OWF	0	10	20	30
λ (GPa)	3.37	3.41	3.56	3.82
μ (GPa)	1.36	1.57	1.63	1.71

Table 4. Ultrasonic Young's modulus values.

% OWF	0	10	20	30
E_{Ultra} (GPa)	3.71	4.05	4.39	4.60

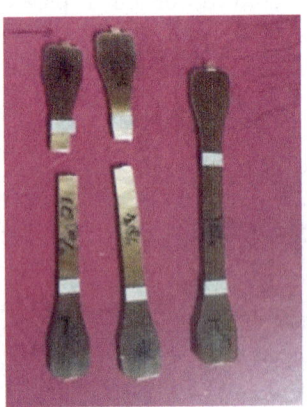

Fig. 6. Specimens before and after rupture.

4 Mechanical Young's Modulus E_{mec}

To determine the mechanical Young's modulus of the bio-composite, standard test specimens (ISO ½) were developed (Fig. 6) with the same material and the same portions of reinforcement of the composite plates (Table 5).

These specimens were subjected to tensile tests through a tensile machine (Fig. 7).

From the curves presenting the stress as a function of the deformation $\sigma = f(\varepsilon)$ obtained, the mechanical Young's modulus Emec is determined

Table 5. Mechanical Young's modulus values.

% OWF	0	10	20	30
E_{mec}(GPa)	2.02	2.33	2.83	3.36

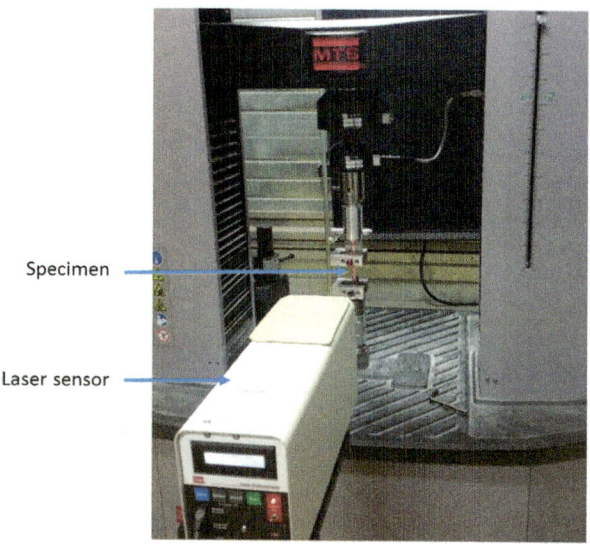

Fig. 7. Traction machine.

Table 6. Comparison between E_{ultra} and E_{mec} of plates.

% OWF	Eultra (GPa)	Emec (GPa)	Ratio
0	3.71	2.02	1.83
10	4.05	2.33	1.73
20	4.39	2.83	1.55
30	4.60	3.36	1.36

5 Comparison Between E_{ultra} et E_{mec}

Table 6 presents a comparison between the Young's modulus obtained by ultrasonic method and those obtained by mechanical method of our bio-composite.

Figure 8 presents a comparison between the Young's modulus obtained by ultra-sonic method and those obtained by mechanical method.

From the results found, we notice that the ultrasonic Young's modulus is more important than the mechanical Young's modulus with a ratio that varies between 1.36 and 1.83 (Fig. 9).

These results are expected and can be explained by the fact that we do not work in the same field of deformation for the two methods:

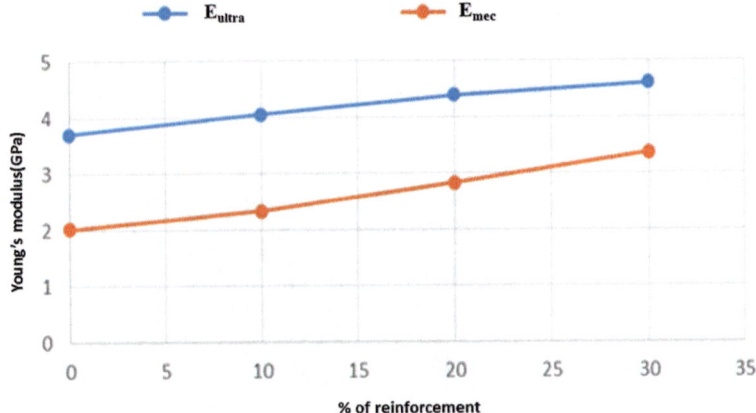

Fig. 8. Variation of E_{ultra} and E_{mec}.

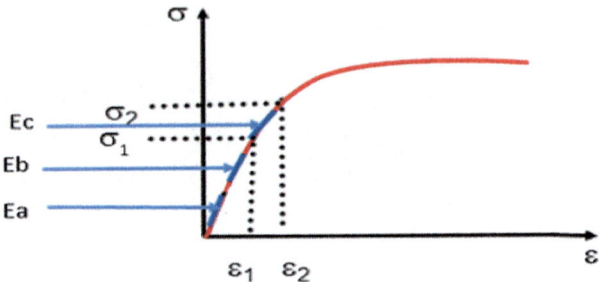

Fig. 9. Deformation domains.

According to ISO 527-1 [7], for the mechanical method, we work in the field 'Ec' for which:

- $\sigma1$ is the stress in MPa measured at a strain $\varepsilon1 = 0.05\%$.
- $\sigma2$ is the stress in MPa measured at a strain $\varepsilon2 = 0.25\%$.

For the ultrasounic method, one works in the field 'Ea', whose strain domain is between $\varepsilon A1 = 0\%$ et $\varepsilon A2 < \varepsilon1 = 0.05\%$.

For this deformation domain, the slope of E is greater than that of the 'EC' domain, which explains this difference in values between the two ultrasonic and mechanical methods with Eultra > Emec.

These results can be explained by the fact that the deformation domain is not the same [7, 8]. In addition, the deformation rates are not the same for both methods [9, 10]. To this it can also be added that although the compositions and methods of preparation are strictly the same for the test specimens and the plates, the geometries, the thicknesses and the stress sections are different for the ultrasonic and mechanical tests.

6 Conclusion

In this work, we have elaborated and characterized a bio-composite material based on PP as polymer matrix and OWF as reinforcement. The ultrasonic and mechanical Young's modulus was studied according to the rate of reinforcement.

The ultrasonic Young's modulus is more important than the mechanical Young's modulus, this is due to the difference of deformation domain, we also note that when the percentage of reinforcement increases, the Young's modulus increases also, hence a better rigidity of the plates.

References

1. Pantano, A., Zuccarello, B.: Numerical model for characterization of biocomposites reinforced by sisal fibers. Procedia Struct. Integr. **8**, 517–525 (2018)
2. Chikhi, M.: Young's modulus and thermophysical performances of bio-sourced materials based on date palm fibers. Energy Build. **129**, 589–597 (2016)
3. Porras, A., Maranon, A., Ashcroft, I.A.: Optimal tensile properties of a Manicaria-based biocomposite by the Taguchi method. Compos. Struct. **140**, 692–701 (2016)
4. Ratassepp, M., Rao, J., Fan, Z.: Quantitative imaging of Young's modulus in plates using guided wave tomography. NDT E Int. **94**, 22–30 (2018)
5. Laperre, J., Thys, W., Lenoir, O., Izbicki, J.L.: Experimental determination of the transversal wave velocity in plates. J. Acoust. **5**, 161–170 (1992)
6. Ghodhbani, N., Marechal, P., Duflo, P.: Ultrasonic broadband characterization of a viscous liquid: methods and perturbation factors. Ph.D. thesis, Université du Havre (2014)
7. ISO 527-1&2: Plastics-determination of tensile properties. International Standard Organization (2012)
8. Jones, D.R.H., Ashby, M.: Engineering Materials: An Introduction to Properties, Applications and Design. Butterworth-Heinemann, London (2012)
9. Yang, B.J., Kim, B.R., Lee, H.K.: Predictions of viscoelastic strain rate dependent behavior of fiber reinforced polymeric composites. Compos. Struct. **94**, 1420–1429 (2012)
10. Notta-Cuvier, D., Nciri, M., Lauro, F., Delille, R., Chaari, F., Robache, F., Haugou, G., Maalej, Y.: Coupled influence of strain rate and heterogeneous fibre orientation on the mechanical behaviour of short-glass fibre reinforced polypropylene. Mech. Mater. **100**, 186–197 (2016)

transformation on microstructure, hardness and residual stresses in the different regions of the weld. There have been few studies on the influence of void on the mechanical response of welded HSLA steel as that was presented by Xue et al. (2003).

However, the understanding of the effect of the local microstructure around porosity defect with the residual stress in the weld is still insufficient.

In this paper, the microstructure and the hardness in the weld and around the porosity have been studied. The contour method was used to determine the longitudinal residual stress in the butt-welded joint.

2   Materials and Experimental Methods

In this work, the base material used for the butt-welded plate is the S500MC HSLA steel. The mechanical properties of the base metal (BM) and the weld metal (WM) after welding are listed in Table 1. Butt-welded joint of the HSLA steel plates is carried out, in our laboratory, using conventional manual metal arc welding technique. To complete the joining, two weld passes were performed by means of a MIG-welding process.

Table 1. Mechanical properties of the base metal and the weld metal.

	Tensile strength, σ (MPa)	Yield strength, σ_y (MPa)	Elongation, ε (%)
Base metal	690	520	19
Weld metal	633	539	11

The current and the voltage of the welding are 128 A and 17 V respectively. As shown in Fig. 1, arc welding was used for producing the joint of HSLA steel plates of dimension $130 \times 100 \times 10$ mm^3. The bevel angle of the joint was 30° on either side. The joint preparation consisted of a 2 mm root gap with a 2 mm deep root face. Along the paper, x, y and z directions designate longitudinal, transverse and normal directions, respectively, as shown in Fig. 1a.

For the microstructural examinations, the welded plate was first sectioned, embedded into black epoxy resin, polished by a standard metallographic technique and then etched in a solution of 2% of Nital during 15 s. In this study, microstructure of the welded joints was characterized using VHX-1000 digital microscopy and scanning electron microscopy (SEM).

The measurements of micro hardness were carried out according to the FM-300e Tester in Vickers HV scale using a 1.0 kg load. To obtain the distribution of the hardness along the whole area of the cross section transverse to the welding direction, the measurements were realized for different depths. To get a minimum dispersion of the measurements, the spacing between each horizontal line is 1 mm (9 lines) and the spacing between each indent is 0.3 mm (50 points per line) as presented in Fig. 2.

The contour method was used to measure the longitudinal residual stress on the weld. It was developed by Prime (2001) in order to measure the residual stress field over a cross-section. This technique offers higher spatial resolution for thin plates as was verified by Richter-Trummer et al. (2008). The measuring protocol consists of

130 mm
200 mm
Welding direction
Z
Y
X
60°
2
1
10 mm
2 mm
2 mm
(a)
(b)

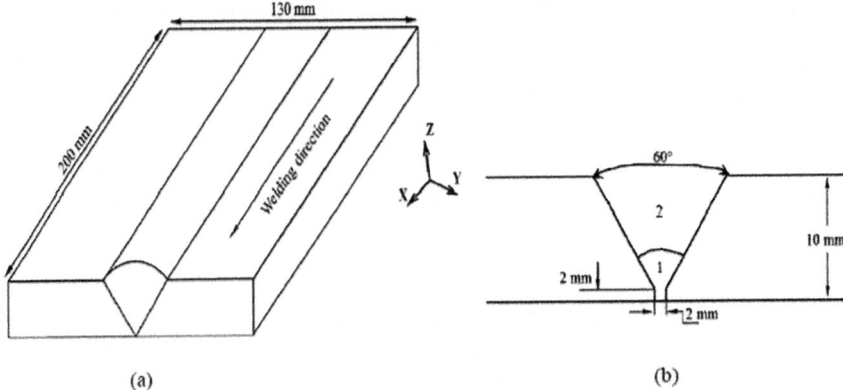

Fig. 1. Welding plate geometry (a) global view (b) front view

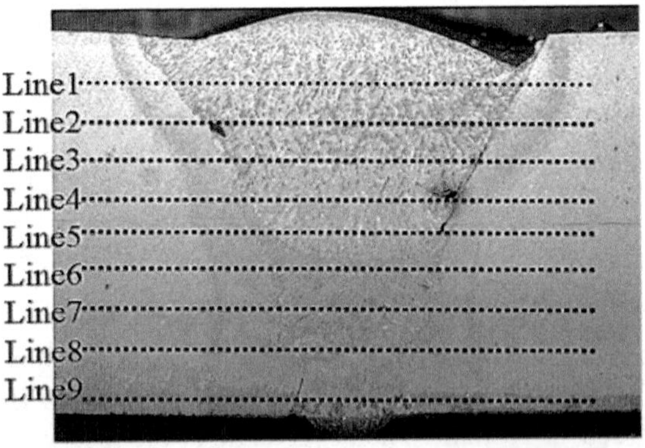

Line1
Line2
Line3
Line4
Line5
Line6
Line7
Line8
Line9

Fig. 2. Measurement lines of Vickers hardness

three main steps: (1) specimen cutting, (2) contour measurement and data processing, (3) numerical calculation. Firstly, Electrical Discharging Machining (EDM) method was used for the cutting process with a brass wire of 0.25 mm diameter and a cutting rate of 5 mm/min as shown in Fig. 3. Secondly; the coordinate measurement machine (CMM) with a touch probe of 1 mm diameter was used for scanning the profile of the cutting surface, as presented in Fig. 4

Numerical treatments were applied to the measured data to obtain a representative polynomial function of the measured values which will be our input in the finite element model. Finally, a three-dimensional elastic finite element analysis was performed to evaluate the residual stresses normal to the cut surface from the imposed displacements.

The three dimensional finite element model was presented in Fig. 5. Three additional displacement constraints were applied (Prime 2001) to prevent rigid body displacements in the y-z plane. The mesh was constructed using the three-dimensional 8 nodes C3D8R. The convergence was reached with 169,968 elements for a total of 183,855 nodes.

Fig. 3. Specimen cutting using EDM

Fig. 4. Contour measurement

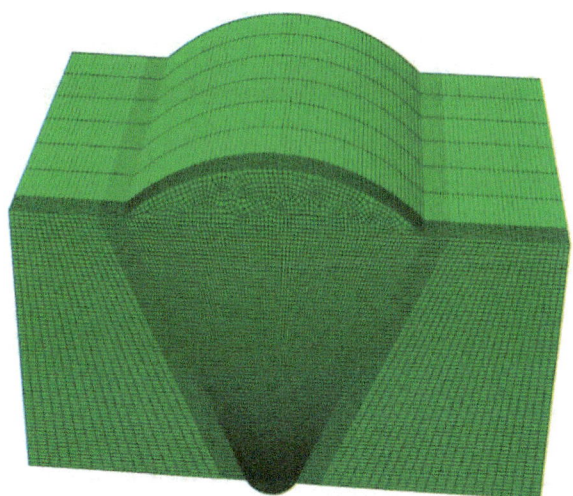

Fig. 5. Three dimensional finite element model

3   Results and Discussions

3.1   Macrostructure

Figure 6 shows the macrostructure of a cross section of the butt-welding plate. Three distinct areas can be clearly identifiable: (1) the base metal (BM), (2) the heat affected zone (HAZ) and (3) the weld metal (WM).

Defects can be visibly detectable in a digital macrograph. These defects are located at the boundary between the fusion zone and the HAZ. More precisely, they are located in the side of the WM as shown in Fig. 6. Two defects (porosity1, porosity2) were positioned at the right side of the WM and the third defect (porosity3) was located on the left side of WM. These elliptical defects were observed with a major axis of about 320, 300 and 80 μm respectively.

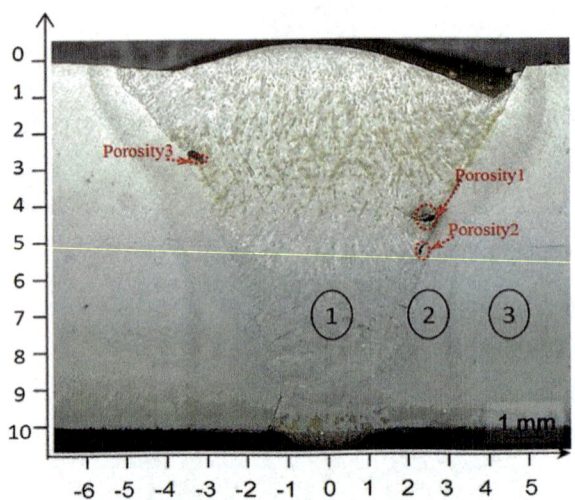

Fig. 6. Macroscopic view of the weld joint

3.2 Microstructure

Figure 7 shows the microstructure of the different regions of the welding joint. Significantly changes were observed while moving from the base HSLA metal to the fusion boundary especially the average grain size and the proportion of the phases. The grain size was determined with a planimetric method.

Figure 7a shows the microstructure of the base metal. A typical microstructure of ferrite can be observed with average grain size of 15 μm.

Figure 7b shows the microstructure of the heat-affected zone HAZ. The HAZ contain 82% of ferrite and 18% of coarse grained bainite the ferrite grain size is about 10 μm.

Figure 7c shows the microstructure of the weld metal. fine dendritic structure can be seen in the top of the weld centerline due to higher cooling rate with average grain size of about 5 μm. The WM reveals a coarser solidification microstructure composed of 70% coarse grained bainite and 30% very fine grained acicular ferrite. By moving to the bottom of the weld centerline, the grain size increased gradually.

Figure 7d shows the microstructure around the porosity1. A coarse grain region can be observed around the porosity with average grain size of about 22 μm. These coarse grains are surrounded by extremely fine grains resulting in an obvious structural inhomogeneity. The local microstructure was composed of about 68% acicular ferrite and 32% bainite

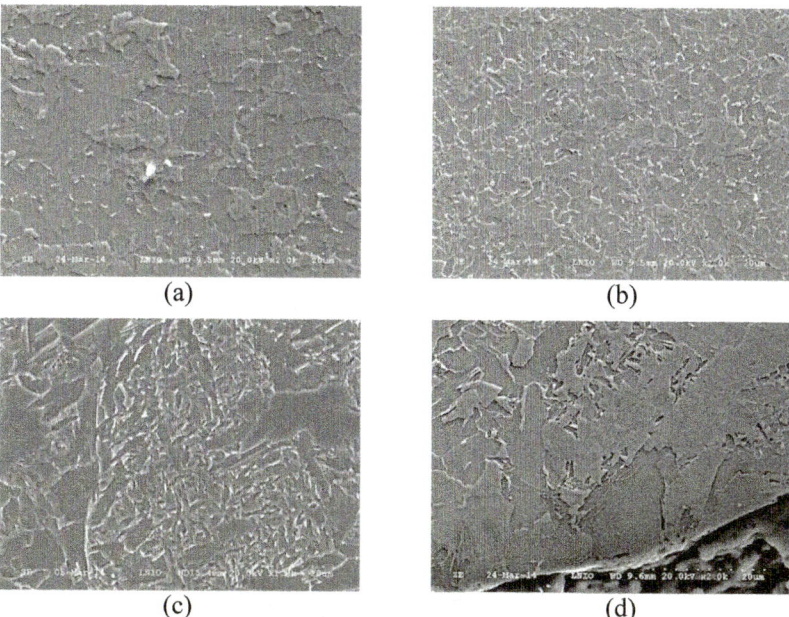

(a)   (b)

(c)   (d)

Fig. 7. Microstructures of the weld (a) The base metal (b) The HAZ area (c) The weld metal and (d) At the top of the porosity1 (SEM observations)

3.3   Micro-hardness Results

Figure 8 shows the distribution of Vickers micro-hardness in the whole area which is estimated according to the nine lines measuring strategy. In the base metal, the hardness remains at 215 ± 10 HV. Figure 9 presented the measurement results of line-1 and line-4 which are the lines located at 1 and 5 mm below the top of the weld respectively. As observed in Fig. 8, the hardness reaches the peak value of about

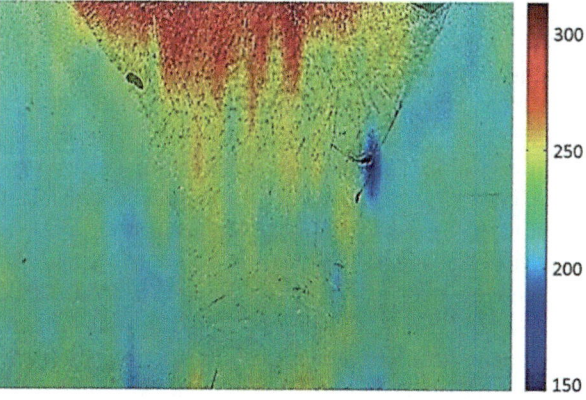

Fig. 8. Cartography of micro-hardness

312 HV at the top of weld metal. In the weld zone, high level of hardness might be correlated to the presence of bainite and acicular ferrite grain size. It is obvious that the hardness decreases rapidly near the fusion line which indicates the beginning of HAZ. According to line-4 of measurements, the hardness drops down dramatically to 147 HV at the boundary between the weld metal and the HAZ but in the side of the weld metal due to the presence of the porosity.

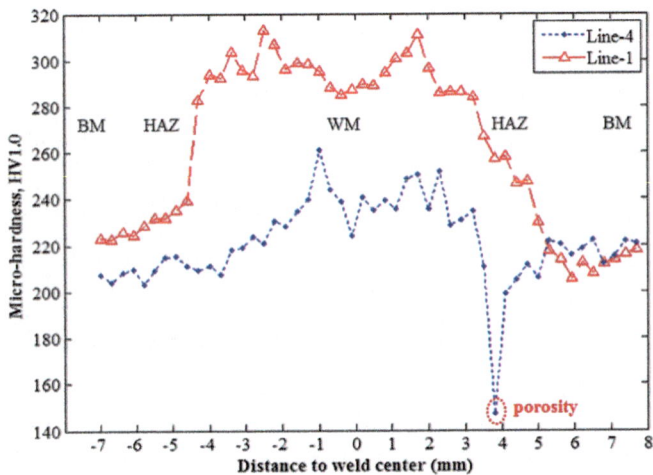

Fig. 9. Two lines of measurement of micro-hardness

3.4   Contour Method Results

After measuring the displacements of both cut surfaces, the average of the two measured data was calculated in order to remove the effect of shear stresses or any asymmetric effects in the cut. The numerical treatments (Frih et al. 2017), applied on the measured data using MATLAB, are the filter firstly and then the smoothing which will also allow us to obtain a representative polynomial function of the measured values as presented in Fig. 10.

The smoothing function represents our input in the finite element calculation. Displacements imposed are those calculated by smoothing: the approximate polynomial function estimations of the measured displacement value z in each node according to its coordinates x and y. Results given by numerical simulations were plotted in Fig. 11. The cartography of longitudinal stress, within the fusion zone, the heat affected zone and the base metal, were obtained. The longitudinal residual stress was not uniform throughout the thickness of the butt-welding joint.

The peak tensile residual stress was obtained at top surface line (517 MPa) and was larger than the maximum stress at mid-thickness surface (300 MPa). The peak measured stress was close to the yield strength of HSLA S500MC steel at room temperature. It can be seen that the maximum stress at the bottom surface was close to that measured at the top surface. Along the three lines, the highest tensile stress was observed at the weld centerline.

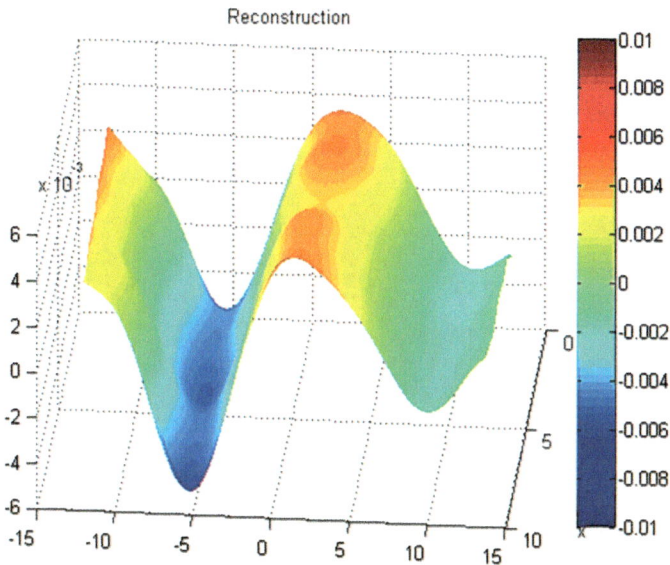

Fig. 10. The average of the smoothed and filtered measured data (mm)

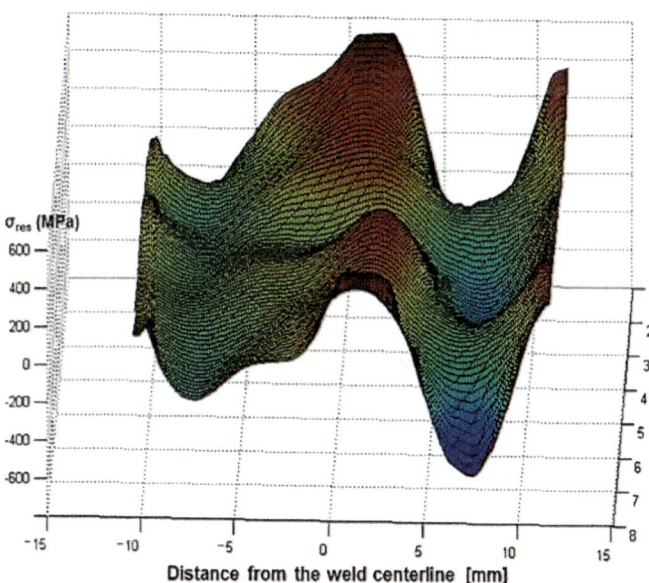

Fig. 11. Longitudinal residual stress distribution

As shown in Fig. 12, the correlation between results shows that the hardness and the residual stresses values were proportional to the percentage of bainite and inversely proportional to the ferrite grain size.

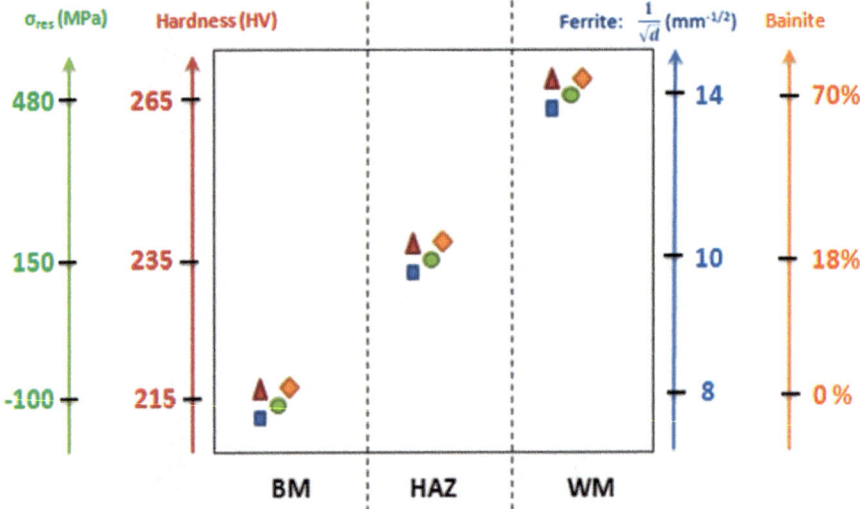

Fig. 12. Correlation between microstructure, hardness and residual stresses

4 Conclusion

The influence of a welding porosity on the microstructure, the hardness and the residual stress of a HSLA steel plate was studied. The main conclusions can be summarized as follows:

1. Very large grain size is observed around the lack of fusion defect which shows the lowest toughness and ductility in a welding structure.
2. Hardness tests show that: the maximum value of hardness was attained at the weld metal region, sharply decreasing in hardness from fusion boundary to fine grained HAZ, gradually increase hardness at the base metal zone and the lowest hardness was obtained around porosity.
3. The cross-sectional residual stress profile was measured using the contour method. Results show that the peak tensile stress is located near the weld centerline in the fusion zone and the highest compressive stress is situated in the heat affected zone.

Acknowledgements. This work is partially supported by Champagne-Ardenne region. The author gratefully acknowledge the helpful comments and suggestions of the reviewers, which have improved the presentation.

References

Manganello, M.: Microstructure and Properties of Microalloyed and other Modern HSLA Steels, pp. 331–343. ISS-AIME, Warrendale (1992)

Maddox, S.: Fatigue Strength of Welded Structures. Woodhead Publishing, Sawston (1991)

Microstructure, Hardness and Residual Stress Distributions in Butt-Welded Joint

Intissar Frih[✉]

Université de Technologie de Troyes, 12 Rue Marie Curie, 10004 Troyes Cedex,
France
frih.intissar@gmail.com

Abstract. This paper investigates the characterization of the microstructure changes and the distribution of hardness and residual stress of MIG-welded high-strength low-alloy steel. Residual stresses are experimentally measured by the contour method and the experimental values are numerically treated by MATLAB to find out a representative function which is used as an input of the finite element model. The microstructure of different regions of the weld joint is also investigated and shows the grains size change in the weld. Micro-hardness distribution shows a strong influence of the bainite and ferrite grain size. Residual stresses distribution shows that high tension ferrite grained weld metal is the most critical zone for cracking growth, mainly near the pre-existing porosity. The correlation between results shows that the hardness and the residual stresses values were proportional to the percentage of bainite and inversely proportional to the ferrite grain size. A particular attention is paid in this paper to the microstructure and the hardness around porosity defect. The results highlighted the interest of respecting proper welding procedures to avoid micro-porosities and to lower tension residual stress. Ultrasonic inspection should be obligatory performed in the weld zone to detect internal defect and identify the reliability of the piece.

Keywords: Welding · Porosity · Microstructure · Hardness · Residual stress

1 Introduction

High Strength Low-Alloy steel (HSLA) are among the nuances used for various applications (transport, civil engineering, offshore …) due to their good weldability, high yield strength, toughness and formability (Manganello 1992). However, the heterogeneity of temperature until welding, leads to the modification of the microstructure, the appearance of residual stresses in the structure and the formation of defects (porosities) which can deteriorate the performance of the welding structure. In fact, the modification of the grain size and transformation of phases during cooling have an impact on the levels of welding residual stresses and hardness that may cause fatigue strength diminution as demonstrated by Maddox (1991) and Zhang et al. (2011). Recently, several studies have treated the microstructure and mechanical properties of HSLA steel butt-welded joint (Thibault et al. 2009; Coelho et al. 2013). These authors have investigated the influence of the grain size and phases

© Springer Nature Switzerland AG 2019
T. Fakhfakh et al. (Eds.): ICAV 2018, ACM 13, pp. 310–319, 2019.
https://doi.org/10.1007/978-3-319-94616-0_32

Zhang, C., Vyver, S., Hu, X., Lu, P.: Fatigue crack growth behavior in weld-repaired high-strength low-alloy steel. Eng. Fract. Mech. **78**, 1862–1875 (2011)

Thibault, D., Bocher, P., Thomas, M.: Residual stress and microstructure in welds of 13%Cr–4% Ni martensitic stainless steel. J. Mater. Process. Technol. **209**, 2195–2202 (2009)

Coelho, R.S., Corpas, M., Moreto, J.A., Jahn, A., Standfuß, J., Kaysser-Pyzalla, A., Pinto, H.: Induction-assisted laser beam welding of a thermomechanically rolled HSLA S500MC steel: a microstructure and residual stress assessment. Mater. Sci. Eng. A **578**, 125–133 (2013)

Xue, Q., Benson, D., Meyers, M.A., Nestrenko, V.F., Olevesky, E.A.: Constitutive response of welded HSLA 100 steel. Mater. Sci. Eng. A **354**, 166–179 (2003)

Prime, M.: Cross-sectional mapping of residual stresses by measuring the surface contour after a cut. J. Eng. Mater. Technol. **123**, 162–168 (2001)

Richter-Trummer, V., Tavares, S.M.O., Moreira, P.M.G.P., De Figueiredo, M.A.V., De Castro, P.M.S.T.: Residual stress measurement using the contour and the sectioning methods in a MIG weld: effects on the stress intensity factor. Ciência e Tecnologia dos Materiais **20**, 114–119 (2008)

Frih, I., Montay, G., Adragna, P.-A.: Microstructure, hardness, and residual stress distributions in T-joint weld of HSLA S500MC steel. Metall. Mater. Trans. A **48**, 1103–1110 (2017)

Modeling of Viscoelastic Behavior of Flexible Polyurethane Foams Under Quasi-Static and Cyclic Regimes

Makram Elfarhani[1], Ali Mkaddem[2], Ahmed Al-Zahrani[2],
Abdessalem Jarraya[1,2(✉)], and Mohamed Haddar[1]

[1] LA2MP, National Engineering School of Sfax (ENIS),
Route Soukra, 3038 Sfax, Tunisia
ajarraya@uj.edu.sa
[2] Engineering College, FOE, University of Jeddah,
PO Box 80327, Jeddah 21589, Kingdom of Saudi Arabia

Abstract. This paper discusses the reliability of two approaches in modeling the Flexible Polyurethane Foam (FPF) behavior. FPFs are cellular polymers characterized by highly complex mechanical behavior including nonlinearity, viscoelasticity, hysteresis, and residual deformations. The review of this topic reveals that several studies have developed models based either on hereditary or on fractional derivation formulations. However, the viscoelastic behavior of the material integrates both short and long memory effects, which needs the combination of the two mathematical approaches to cover the full behavior of such a material. This work compares the two methodologies in identifying the parameters of foam behavior using the combined model. The approaches are based on experimental observations of the FPF behavior on compression (short memory effects) and cyclic (long memory effects) loadings. The relative inefficiency of the force difference method widely addressed in modeling processes was specially discussed.

Keywords: FPF · Viscoelasticity · Fractional derivative · Hereditary effect · Modeling

Nomenclature

a_i: Viscoelastic parameter
b_{cyc}:   Cyclic coefficients of the fractional derivative terms
c_j: Displacement residues
k: Viscoelastic damping coefficient
E: An elastic term
F_{L_i}: i^{th} Loading half-cycle
F_{U_i}: i^{th} Unloading half-cycle
P_j: Displacement eigenvalues
V_D: Viscoelastic damping force
V_R: Viscoelastic residual force

© Springer Nature Switzerland AG 2019
T. Fakhfakh et al. (Eds.): ICAV 2018, ACM 13, pp. 320–326, 2019.
https://doi.org/10.1007/978-3-319-94616-0_33

1   Introduction

Flexible polyurethane foam is a recent cellular material discovered in the middle of the previous century. Nowadays, most of the body comfort products are manufactured by using this material. In particular, soft foam is recommended in modern car seats because of its special mechanical behavior which integrates nonlinear elasticity with viscoelastic memory effects such as: the stress hysteretic damping, the dependence of loading rate, the residual deformations, and the dependence of the number of testing cycle.

Several investigations have been conducted to characterize this complex viscoelastic behavior by referring to two mathematical approaches: the hereditary models and the fractional derivative formulations.

The hereditary models are the most widely used constitutive representations used to characterize the relaxation of viscoelastic materials over short time. Mathematically, these representations assume that the material response is equal to the integral convolution of the displacement rate weighted with a relaxation kernel. Physically, the foam response at a given time during loading is influenced by accumulated effects of previous deformations states properly weighted by the relaxation function (White et al. 2000). Muravyov and Hutton (1997) supposed that the kernel is better expressed as a sum of exponentials. Ippili et al. (2003) and Joshi et al. (2010) used this representation to predict the viscoelastic behavior of polyurethane foams and identified its parameters from quasi-static compressive standards. Jmal et al. (2014) obtained good results using the hereditary model for three different types of soft polyurethane foam. The mean advantage of using the hereditary models is their simplicity and facility of calibration of its parameters with the experimental measurements.

The fractional derivative models are also popular and largely used to characterize mechanical behavior of viscoelastic materials and in particular to emphasize its dependence to past history of deformation. These formulations are obtained when considering derivatives on non-integer order in the stress–strain relationship (Bagley and Torvik 1983). Deng et al. (2006) established fractional derivative model with two orders to estimate the viscoelastic quasi-static compressive behavior of flexible foams and obtained accurate simulations of the experimental data.

In prior works (Elfarhani et al. 2016a, b), a fractional derivative and hereditary combined models was established to describe and quantify the memory effects of flexible foams under quasi-static uniaxial compression standards. We found that both approaches are complementary since reasonable results were drawn from the concordance between the physical meaning and the mathematical formulation structure of the hereditary and fractional derivative parameters. In first investigation, we considered the dependence of the material on compression rates, whereas in the second paper we focused our interest on the memory effects in the cyclic response of foam. Both studies are based on the same combined model but with two different parameters identification algorithms.

The main goal of this paper is to compare the two algorithms and to discuss the reasons of getting different estimation results of the viscoelastic residual force.

2 Combination of Fractional Derivative and Hereditary Formulations to Model FPF Viscoelastic Behavior Under Quasi-Static Compression Loading

2.1 Elastic-Viscoelastic Model

In quasi-static regime, flexible polyurethane foams manifest a highly nonlinear and viscoelastic behavior. If we adopt that the response of soft PU foam is an additive sum of a *viscoelastic* component and an *elastic* one (Deng et al. 2006), we can express the total foam responses during loading and unloading phases as:

$$F_L(x(t)) = E(x(t)) + V_D(x(t)) \tag{1}$$

$$F_U(x(t)) = E(x(t)) + V_R(x(t)) \tag{2}$$

Here E is the elastic component, V_D is the Damping component, and V_R is the Residual force. Indeed, in the elastic spring back part, the viscoelastic component manifests the residual force effects instead of pneumatic damping behavior.

2.2 Identification Methodology in the Quasi-Static Regime

The identification process includes four steps; in each one we identify the parameters of the elastic and viscoelastic components by reference to experimental data obtained from two compressions tests performed in the same foam block with two different loading rates 10 mm/min (Test1) and 25 mm/min (Test 2). In Fig. 1 we illustrate the identification methodology flowchart.

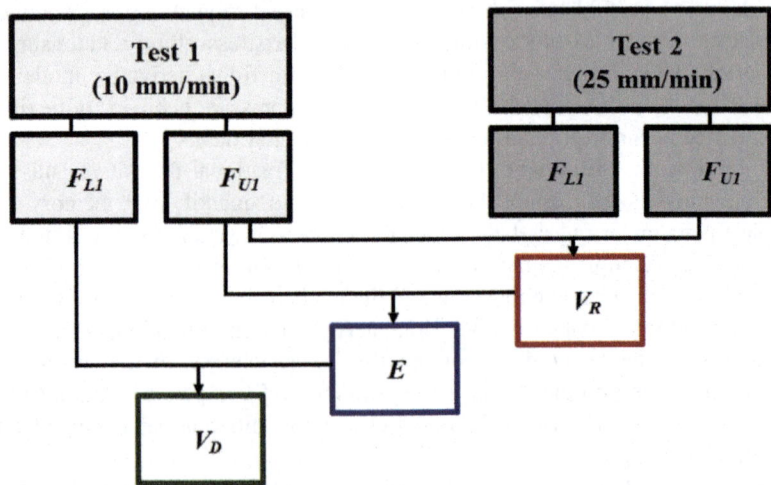

Fig. 1. Flowchart of parameters identification process in the quasi-static regime.

2.3   Quasi-Static Simulation Results

The optimization process described above allows to calibrate the parameters of each component of the elastic-viscoelastic model and then to obtain separately the simulation of each force as well as the reconstructed total response of the material.

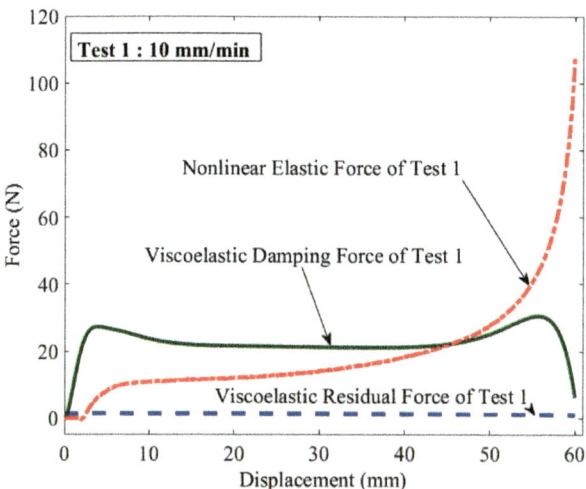

Fig. 2. Measured and predicted foam response components (Elfarhani et al. 2016a).

As shown in Fig. 2, it is clear that the identification process gives reasonable simulation results and the predictions of the foam response components are conform to the phenomenological hypothesis.

3   Combination of Fractional Derivative and Hereditary Models to Characterize FPF Memory Effects in Cyclic Loading

3.1   Combined Memory Model

Besides hysteretic loop and the stress-softening between loading and unloading phases, flexible polyurethane foams display memory effects of its loading history and it recovers when put to rest for enough time. Basing in the phenomenological curve identification and experimental observations, we can notice that the residual effects accumulate over cycles.

3.2   Optimization Methodology in the Cyclic Regime

The identification parameters method established for this cyclic model is primarily based on separating the experimental measurements of each component force apart. In Fig. 3, we illustrate the flowchart of the parameters optimization.

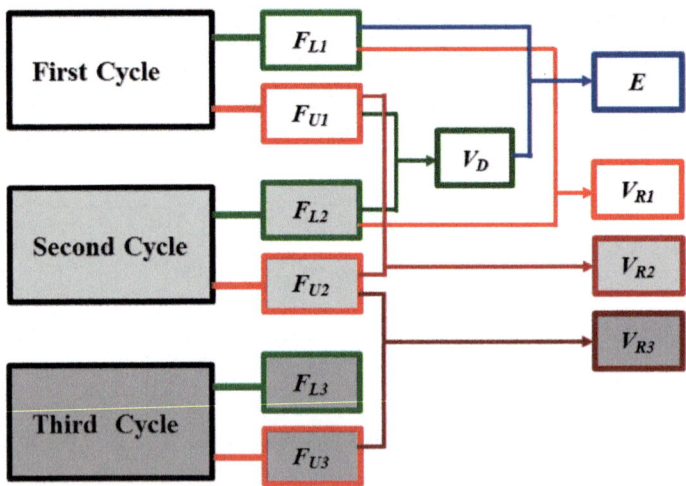

Fig. 3. Flowchart of the parameters optimization methodology using the cyclic model.

3.3   Cyclic Simulation Results

As illustrated in Fig. 4, it is noticeable that there is strong agreement between the components simulations and the phenomenological assumptions as it give clear physical significance.

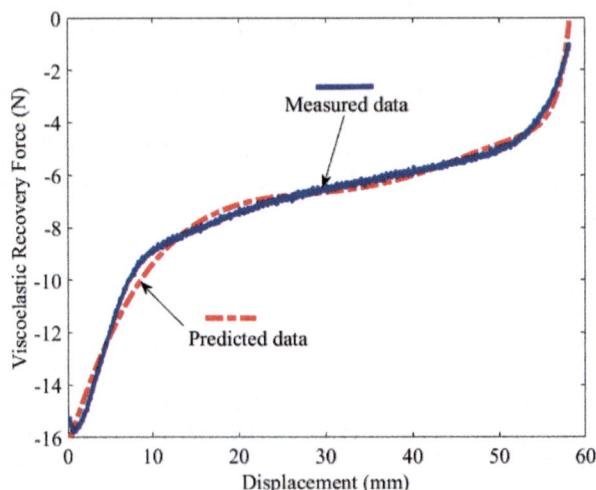

Fig. 4. Measurements and predictions of the viscoelastic residual force, (Elfarhani et al. 2016b).

4 Discussion: Comparison Between Proposed Models

Indeed, the two presented models have almost the same formulations, as they are both based on the same phenomenological assumptions; except that the multi-cycle model takes into account the accumulations of the residual effects overs cycles. Thus, we should rather compare the two optimizations process to distinguish the more reasonable and efficient one.

Mathematically, it was shown that the two presented models provide satisfactorily accurate simulations of the soft polyurethane foam responses. In fact, the fit statistics indicate that the first optimization algorithm explain 96% of the total variations in measurements about the average values and the second identification process consider at least 95% of the global variations in experimental data for the cyclic test.

Physically, we can notice that the curves of the viscoelastic damping forces obtained from the two identifications system have nearly the same shape. This remark also concerns the elastic component curves.

Concerning the viscoelastic residual force, the two algorithms allow to produce curves having nearly the same magnitudes with negative values (which indicate the resistance of the residual force to elastic effects). However the curves allures are significantly different. Actually, the cyclic model conduce to more guaranties values of residual parameters. The raison is primarily related to the use of the difference force method in the first algorithm. This method causes a mixing of the parameters values, since getting the optimum minimum is influenced mostly by the starting vector which is chosen randomly. Yet, in the first identification process the distinction of optimum values among local minima is based blindly on the fit quality and on the significance of curve allure. In the other hand, the cyclic algorithm has a major advantage, because it helps to extract the experimental measurements of each residual force apart, and to calibrate thereafter its parameters values separately. This important advantage serve to avert the admixture problem occurred often when using the difference force method between measurements of two different forces. From here, we propose to avoid the use of this method since it produces randomly distributed values of parameters.

To sum up, the second algorithm is more efficient and allows obtaining more guaranteed result, and the residual parameters are better identified by referring to the cyclic tests.

5 Conclusions

The memory combined model gives reasonably good results, and the cyclic identification algorithm help to characterize and quantify residual effects in soft foam. However, it would be important to check its ability to cover a wide range of loading rates. Moreover, it can be a good alternative to validate this model through cyclic compression tests with three different displacement rates. This standard allows for finding out the influence of compression rate in each viscoelastic component.

References

Bagley, R.L., Torvik, P.J.: A theoretical basis for the application of fractional calculus to viscoelasticity. J. Rheol. **27**(3), 201–210 (1983)

Deng, R., Davies, P., Bajaj, A.K.: A nonlinear fractional derivative model for large uni-axial deformation behavior of polyurethane foam. Signal Process. **86**, 2728–2743 (2006)

Elfarhani, M., Jarraya, A., Abid, S., Haddar, M.: Fractional derivative and hereditary combined model of flexible polyurethane foam viscoelastic response under quasi-static compressive tests. Cell. Polym. **35**(5), 235–269 (2016a)

Elfarhani, M., Jarraya, A., Abid, S., Haddar, M.: Fractional derivative and hereditary combined model for memory effects on flexible polyurethane foam. Mech. Time Depend. Mater. **20**(2), 197–217 (2016)

Ippili, R.K., Widdle, R.D., Davies, P., Bajaj, A.K.: Modeling and identification of polyurethane foam in uniaxial compression: combined elastic and viscoelastic response. In: Proceedings of the 2003 ASME Design Engineering Technical Conferences, DETC2003/VIB-48485, Chicago, Illinois, September 2–6 (2003)

Jmal, H., Dupuis, R., Aubry, E.: Quasi-static behavior identification of polyurethane foam using a memory integer model and the difference-forces method. J. Cell. Plast. **2011**, 447–465 (2014)

Joshi, G., Anil, K., Bajaj, A.K., Davies, P.: Whole-body vibratory response study using a nonlinear multi-body model of seat-occupant system with viscoelastic flexible polyurethane foam. Ind. Health **48**, 663–674 (2010)

Muravyov, A., Hutton, S.G.: Closed-form solutions and the eigenvalue problem for vibration of discrete viscoelastic systems. J. Appl. Mech. **64**, 684–691 (1997)

White, S.W., Kim, S.K., Bajaj, A.K., Davies, P., Showers, D.K., Liedtke, P.E.: Experimental techniques and identification of nonlinear and viscoelastic properties of flexible polyurethane foam. Nonlinear Dyn. **22**, 281–313 (2000)

Author Index

© Springer Nature Switzerland AG 2019
T. Fakhfakh et al. (Eds.): ICAV 2018, ACM 13, pp. 327–328, 2019.
https://doi.org/10.1007/978-3-319-94616-0

Printed by Printforce, the Netherlands